T0300985

# SWITCHING IN ELECTRICAL TRANSMISSION AND DISTRIBUTION SYSTEMS

# SWITCHING IN ELECTRICAL TRANSMISSION AND DISTRIBUTION SYSTEMS

**René Smeets**
*DNV GL – Energy, The Netherlands*

**Lou van der Sluis**
*Delft University of Technology, The Netherlands*

**Mirsad Kapetanović**
*University of Sarajevo, Bosnia and Herzegovina*

**David F. Peelo**
*Consultant, Canada*

**Anton Janssen**
*Liander Asset Management, The Netherlands*

WILEY

This edition first published 2015
© 2015 John Wiley & Sons, Ltd

*Registered office*

John Wiley & Sons Ltd, The Atrium, Southern Gate, Chichester, West Sussex, PO19 8SQ, United Kingdom

For details of our global editorial offices, for customer services and for information about how to apply for permission to reuse the copyright material in this book, please see our website at www.wiley.com.

*Library of Congress Cataloging-in-Publication Data*

Janssen, Anton, author.
  Switching in Electrical transmission and distribution systems / René Peter Paul Smeets, editor ; Anton Janssen, Mirsad Kapetanović, David F. Peelo, Lou van der Sluis.
    1 online resource.
  Includes bibliographical references and index.
  Description based on print version record and CIP data provided by publisher; resource not viewed.
  ISBN 978-1-118-70362-5 (ePub) – ISBN 978-1-118-70363-2 (Adobe PDF) – ISBN 978-1-118-38135-9 (cloth)  1. Electric switchgear.  2. Electric power distribution–High tension–Equipment and supplies.  I. Smeets, René Peter Paul, editor.  II. Kapetanović, Mirsad, author.  III. Peelo, David F., author.  IV. Van der Sluis, Lou, author.  V. Title.
  TK2831
  621.319′3–dc23
                                                                                                    2014023080

A catalogue record for this book is available from the British Library.

ISBN: 978-1-118-38135-9

Set in 10/12pt Times by Aptara Inc., New Delhi, India

Cover photo: Sam Rentmeester
Cover illustration: Phase to Phase
Cover design: Theo van Vliet

1  2015

To Geert Christiaan Damstra (1930–2012)

# Contents

# Preface

At the turn of the nineteenth century, a revolution took place in electrical engineering. In a rather short time, the transformer was invented, electric generators and motors were designed, and the step from DC to AC transmission was made. At the beginning of the twentieth century, the transmission voltages were steadily increased to reduce transmission losses. To improve operating efficiency, power systems began to be interconnected. Reserve power or spinning reserve could be then shared and capital expenditure could be reduced.

This is where "power" switching came in with its major task: isolating the faulted section of the system while keeping in service all healthy parts. Nowadays the power system can be regarded as one of the most complex systems ever designed, built and operated. Despite its complexity and robustness, the switching technology facilitates consumers to connect and disconnect electric loads in a rather simple and reliable way. Moreover, it protects the system from the effects of faults. However, this comes at a price since every change in the state of a system generates transients that may affect both the operating conditions of the system and its components.

With the first application of power switching in early electric systems, the development of standards for rating, testing and manufacturing high-voltage circuit-breakers began. In the United States, initiative was taken by a number of engineering and manufacturer trade organizations, such as the American Institute of Electrical Engineers (AIEE), dating back to 1884, later on merged into the Institute of Electrical and Electronics Engineers (IEEE) in 1963.

In Europe, the International Electrotechnical Commission (IEC) was founded in 1906, and the International Council on Large Electric Systems (CIGRE) started in 1921. CIGRE structured its organization by means of study committees in 1927. These study committees are responsible for the operation of working groups and task forces. Both collect field data and perform system studies, and their reports are used as input for creation and revision of IEC standards.

Over the years, many books and publications have been written on switching in electric transmission and distribution systems. A great deal of this knowledge results from the work of CIGRE and IEEE working groups, published as standards, technical brochures, reports and scientific papers.

For a utility engineer who wants to familiarize him- or herself with switching technology and the system aspects, the available literature is not easily accessible and sometimes difficult to comprehend. This book has been written to bridge the gap between the daily practice of utility engineers and the available literature.

The authors have served long periods in many working groups of CIGRE, IEC, and IEEE, and are familiar with switchgear manufacturing, power system aspects, and high power testing. In the relevant sections, contributions of the authors to CIGRE technical brochures, scientific literature, and standards are summarized.

The respective standards are referenced and run through the book like a continuous thread. This approach is intended to provide on the one hand guidance to the practical complexity of all essential switching operations and, on the other, a proper understanding of the standards and their background.

The book would not be complete without significant chapters on circuit-breakers and the switching media involved. Most of this material is gratefully taken from the book "High Voltage Circuit Breakers" by Professor Mirsad Kapetanović, issued by ETF – Faculty of Electrotechnical Engineering, Sarajevo (2011), commissioned by KEMA.

We acknowledge Zdeněk Matyáš M.Sc. who spent many hours harmonizing all our text, figures, and equations to be in line with the IEC standards and to achieve consistency in terminology.

Thanks are also due to Professor Viktor Kertész for the thorough checking of the mathematical sections, and we appreciate the contribution of Romain Thomas M.Sc. to the section on numerical simulation of transients.

We have dedicated our book to Professor Geert Christiaan Damstra (1930–2012), who has significantly contributed to the development of switching-, measurement-, and testing technology during his working life at the switchgear manufacturer Hazemeyer, KEMA, and the Eindhoven University of Technology. Being active and truly innovative in the high-voltage technique – literally to his last days – he has set an example to many.

René Smeets, Editor
*Arnhem, Spring 2014*

# 1

# Switching in Power Systems

## 1.1 Introduction

As electricity comes out of AC outlets every day, and has done so for more than 100 years, it is nowadays considered a commodity. It is a versatile and clean source of energy; it is rather cheap and 'always available'.

The purpose of a power system is to transport and distribute the electric energy generated in the power plants to the consumers in a safe and reliable way. Generators take care of the conversion of mechanical energy into electric energy, aluminium and copper conductors are used to carry the current, and transformers bring the electric energy to the appropriate voltage level. Society's dependence on this commodity has become extremely large and the social impact of a failing power system is unacceptable. The electrical power system is the backbone of modern society.

Switching operations in power systems are very common and must not jeopardize the system's reliability and safety. Switching in power systems is necessary for the following reasons and duties:

- Taking into or out of service some sections of the system, certain loads, or consumers. A typical example is the switching of shunt capacitor banks or shunt reactors, de-energization of overhead lines, transformers, and so on. In industrial systems, this type of switching is by far the most common of all the switching operations.
- Transferring the flow of energy from one circuit to another. Such operations occur when load current needs to be transferred without interruption from one busbar to another, for example, in a substation.
- Isolating certain network components because of maintenance or replacement.
- Isolating faulted sections of the network in order to avoid damage and/or system instability. The most well-known example of this is the interruption of a short-circuit current. Faults cannot be avoided, but adequate switching devices in combination with a protection system need to limit the consequences of faults.

Figure 1.1 provides an overview in orders of magnitude of the power switched in electrical-engineering applications.

*Switching in Electrical Transmission and Distribution Systems*, First Edition.
René Smeets, Lou van der Sluis, Mirsad Kapetanović, David F. Peelo, and Anton Janssen.
© 2015 John Wiley & Sons, Ltd. Published 2015 by John Wiley & Sons, Ltd.

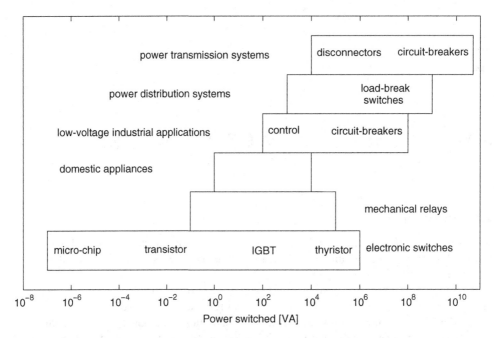

**Figure 1.1**   Overview of the power being switched in electrical-engineering applications.

Switching in electrical power systems re-configures the topology of an electrical network; it involves the making and breaking of circuits and causes a disturbance of steady energy flow. Therefore, transients have to be expected and are observed in the system during the change from the situation before to the situation after switching. Transients are abnormal patterns of current and voltage that have a limited duration. Attention should be paid to these phenomena because they very often exceed the values met during steady-state operation. Fundamentally in nature, any change of steady-state conditions generates transients.

The essential parameters in electrical systems are current and voltage. During switching operations, transients can be observed in both. Regarding operations related to switching-on (*making* or *energization*), the components of the system are mainly stressed by current-related transients. On the other hand, at switching-off operations (*breaking* or *de-energization*), voltage-related transients will especially stress the switching device performing the operation.

In a generalized concept, switching devices (dis)connect a source circuit to a load circuit (see Figure 1.2). Both circuits are a complicated combination of system components: lines, cables, busbars, transformers, generators, and so on. Through reduction of the complexity to relevant simple electrical elements, either lumped or distributed where necessary (refer to Section 1.3), the switching transients can be more easily understood.

## 1.2   Organization of this Book

The aim of this book is to describe and explain to technically interested and practically oriented readers the variety of switching processes and devices in electrical power systems, avoiding (as far as possible) deep physical details and formal mathematics – although both of these

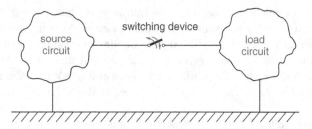

**Figure 1.2**   General concept of a switching device located between a source- and a load-circuit.

are in fact pillars in the resolution of problems encountered in the high-power switching technology. Numerous examples of measurements and observed effects have been selected from realistic tests of a wide variety of switchgear at the KEMA High-Power Laboratory of DNV GL – Energy, where real service conditions are simulated in powerful test-laboratories (see Section 14.2).

The book is divided roughly into two parts. The first part (Chapters 1 to 5) focuses on the switching phenomena and their description.

The second part (Chapters 6 to 14) describes the technology of the devices that must perform switching in all experienced varieties and their impact on the power system.

In Chapter 1, the necessary background on the practical aspects of switching is given. The origin and role of the two key phenomena governing the switching processes are described: the *switching arc* and the *transient recovery voltage* (TRV). Due to the nature of the transients that accompany switching, the general description must, in principle, be in terms of electromagnetic fields and travelling waves. However, at sufficiently low frequency, a simplification of the relevant circuits in terms of lumped elements can greatly facilitate mathematical formulation and calculation of the TRV characteristics in the majority of the practical cases.

Chapter 2 deals exclusively with faults in power systems. Essential transients of fault-current events are identified, together with their impact on network components. Data on fault statistics are summarized.

In Chapter 3, the switching of fault currents, correctly termed the *making* and *breaking operations*, resulting from various types of faults in power systems, is analysed. Since, in this case, the TRV plays a crucial role, the description of the TRV is given in terms of simplified $RLC$ circuits[1] – either with lumped elements or in terms of travelling waves where necessary. A systematic approach is taken to represent every case by the minimum possible number of elements, in order to facilitate understanding.

Relevant characteristics of the switching arc are considered, showing its influence on the interruption process.

Chapter 4 deals with switching of loads: overhead lines, capacitor banks and shunt reactors operated under normal condition. Although the currents to cope with are much lower than at faults, it is explained that due to the reactive nature of capacitive and inductive loads, this type of switching is sometimes an onerous challenge. The main issue here is the description of transients generated by the switching process.

---

[1] Oscillating circuits, each consisting of a single inductance $L$, capacitance $C$, and resistance $R$.

In Chapter 5, the calculation of the switching transients is treated. First, a formal analysis of the analytical solution of a few simplified circuits is given. Next, the background and the basics of some state-of-the-art numerical transient simulation programs are discussed.

Chapter 6 highlights the switching processes in gaseous media, such as air, the $H_2$ gas in oil circuit-breakers and $SF_6$ gas, the workhorse of present-day HV insulation and switching. Issues regarding health, safety and environment related to the 'electrical' use of $SF_6$ gas are treated in detail.

Chapter 7 focuses on the $SF_6$ circuit-breaker, the present-day technology of choice in HV systems. Technology and other relevant aspects of the circuit-breaker are described.

Chapter 8 describes the switching in vacuum, as presently applied on a large scale in distribution systems and slowly emerging in high-voltage systems as well.

Chapter 9 goes into the technology of the vacuum circuit-breaker. Recent information is added on the application of vacuum circuit-breakers for HV switchgear.

In Chapter 10, a variety of special switching situations and the technology of the appropriate devices are highlighted. A number of switching situations pose specific challenges to systems and switching devices. These are: generator circuit-breakers, switching with GIS- and air-break disconnectors, loop switching, switching in HV cable systems, the application of special by-pass switches in a series compensation capacitor bank, switching in the vicinity of shunt capacitor banks and switching in *ultra-high-voltage* (UHV) systems, high-speed earthing switches, *direct current* (DC) circuit-breakers and fuses.

Chapter 11 is devoted to switching overvoltages in systems. Practical values are given. Methods are discussed that enable reduction of the voltage stresses in particular situations. This is followed by an overview of controlled switching strategies that provide mitigation of unwanted transients.

Chapter 12 discusses the various investigations on reliability of circuit-breakers that have been conducted in the past 30 years. In addition, experiences regarding the endurance of switchgear against electrical and mechanical stresses are highlighted.

Chapter 13 deals with standardization and specification of switchgear. An explanation is given of the standardization framework for circuit-breakers that has been developed during the last half century in order to facilitate a system of quality assurance. In this chapter, as well as in other parts of the book, the worldwide accepted standardization system of the IEC (International Electrotechnical Commission)[2] will be followed mostly.

Chapter 14 describes the backgrounds of testing methods for circuit-breakers. A detailed analysis is given of the various possibilities and practices of testing of the switching and breaking capabilities of circuit-breakers.

Throughout the book, extensive reference is made to documents of the CIGRE ("Conférence International des Grands Réseaux Électriques" or "International Council on Large Electric Systems")[3], a non-profit association for promoting collaboration with experts from all around the world by sharing knowledge and joining forces to improve the electric power systems of today and tomorrow. CIGRE has over 12 000 members from the electrical energy industry. More than 3500 experts from all around the world work actively together in structured *Working Groups* (WGs) coordinated by the CIGRE *Study Committees* (SCs). Their main objectives are

[2] www.iec.ch
[3] www.cigre.org

to design and deploy the power system for the future, optimize existing equipment and power systems, respect the environment and facilitate access to information. For switching in power systems and its impact, the relevant study committees are A3 (High-Voltage Equipment), B4 (HVDC and Power Electronics), B5 (Protection and Automation) and C4 (System Technical Performance).

CIGRE documents can be accessed through www.e-cigre.org. A large volume of information is laid down in *Technical Brochures* (TB), the output documents of working groups, and in CIGRE meeting papers.

Another major source of reference is the IEEE ("Institute of Electrical and Electronics Engineers"). With more than 425 000 members IEEE is the world's largest professional association dedicated to advancing technological innovation.

IEEE has a number of societies. The one that is most closely related to the scope of this book is the PES ("Power and Energy Society") that covers planning, R&D, design, construction and operation of facilities systems for generation, transmission and distribution of electric energy. PES comprises 30 000 industry professionals, academics and students with a common interest in the electric power industry. It provides the world's largest forum for sharing the latest in technological developments in the electric power industry, for developing standards and for education.

Other societies related to switching technology are: the IAS ("Industry Applications Society"); the DEIS ("Dielectrics and Insulation Society") that deals with insulation materials and systems, dielectric phenomena and discharges in vacuum, gaseous, liquid and solid electrical materials; the IEEE PELS (Power Electronics Society), amongst others, on the development and practical application of power electronics technology; the NPSS (Nuclear and Plasma Sciences Society) that covers, amongst others, discharges in power switching devices.

IEEE documents are collected in the IEEE *Xplore*® Digital Library containing more than 3 million documents from IEEE and IEEE journals, transactions, magazines, letters, conference proceedings and active IEEE standards.

Within IEEE, the IEEE-SA (IEEE Standards Association) is developing and nurturing standards (see Chapter 13).

## 1.3  Power-System Analysis

Power-system analysis is a broad subject, too broad to be covered by a single textbook. Textbooks about the fundamentals of power-system analysis [1–4] give an overview of the structure of power systems (from the generation of electric energy, to the transmission and distribution to the customers) and take only the systems steady-state behaviour into account. This means that only the power-frequency phenomena are considered.

An interesting aspect of power systems is that the modelling of the system depends on the time scale that is being considered. In general, the time scales of interest are:

- Years, months, weeks, days, hours, minutes for steady-state analysis at power frequency (50 or 60) Hz. This is the time scale on which textbooks on the fundamentals of power-system analysis focus. The steady-state analysis covers a variety of topics, such as planning, design, economic optimisation, load flow / power flow computation, fault calculation, state estimation, protection, stability and steady-state control.
- Seconds for dynamic-behaviour analysis. The dynamic behaviour of electrical networks and their components is important in order to predict whether the system, or a major

part of it, remains in a stable state after a disturbance, for example, after a switching operation occurring at initiation or removal of a fault. The stability of power systems depends particularly on the ability of the installed control equipment to damp the electromechanical disturbances of the synchronous generators.

- Tens of microseconds to milliseconds for transients related to switching (kilohertz to tens of kilohertz). The insight into the transient behaviour of systems is important for understanding the effects of switching events (i.e. connection/disconnection of loads or fault clearing).
- Microseconds or less for disturbances having a disruptive origin (tens of kilohertz to several megahertz) like the effects of atmospheric disturbances (lightning strokes), breakdown phenomena causing excessive voltages and currents. Physically, the impact of these fast transients is mostly limited to the part of the system where the disturbance originates, that is, most often the affected part or component of the system itself and its immediate vicinity.

Although the power system itself remains unchanged when different time scales are considered, the components in the power system should be modelled in accordance with the appropriate time frame.

An illustrating example is the modelling of an overhead transmission line. For steady-state consideration at the power frequency of 50 Hz, the wavelength of the sinusoidal voltages and currents is 6000 km:

$$\lambda = \frac{c}{f} = \frac{3 \times 10^8}{50} = 6 \times 10^6 \text{ m} = 6\,000 \text{ km} \qquad (1.1)$$

$\lambda$ is the wavelength [m];
$c$ is the speed of light $\approx 3 \times 10^8$ m s$^{-1}$;
$f$ is the frequency [Hz = 1/s].

Thus, the transmission line is essentially of electrically small dimensions compared with the wavelength of the voltage or current. The generally valid Maxwell equations can therefore be approximated considering a quasi-static approach, and the transmission line can be rather accurately modelled by lumped elements. Kirchhoff's laws can be fruitfully used to compute the voltages and currents.

In contrast, when the effects of a lightning stroke have to be analysed, frequencies of 1 MHz and higher occur, and the typical wavelength of the voltage and current wave is 300 m or less. In this case, the transmission line is far from being electrically small, and it is no longer justified to use the lumped-element approximation. The distributed nature of the parameters of the transmission lines has to be taken into account, and has to be dealt with by travelling waves [5, 6].

Despite the fact that mainly lumped-element models are used in modelling, it is important to realize that the energy is mainly stored in the electromagnetic field surrounding the conductors with almost none in the conductors themselves. The Poynting vector, being the vector product of the electric field intensity vector $E$ and the magnetic field intensity vector $H$, indicates the direction and intensity of the electromagnetic power flow.

$$S = E \times H \qquad (1.2)$$

$S$ is the Poynting vector [W m$^{-2}$];
$E$ is the electric field intensity vector [V m$^{-1}$];
$H$ is the magnetic field intensity vector [A m$^{-1}$].

Due to the finite conductivity of the conductor material and the finite permeability of the transformer-core material, a small electric field component is present inside the conductor and a small magnetic field component results in the transformer core:

$$E = \frac{J}{\sigma} \qquad (1.3)$$

$J$ is the current density vector [A m$^{-2}$];
$\sigma$ is the conductivity [S m$^{-1}$].

$$H = \frac{B}{\mu} \qquad (1.4)$$

$B$ is the magnetic flux density vector [T = AH m$^{-2}$ = Vs m$^{-2}$];
$\mu$ is the permeability [H m$^{-1}$].

When one speaks of *electricity*, one thinks of current flowing through the conductors from generator to load. This approach is valid because the physical dimensions of the power systems are large compared with the wavelength of the currents and voltages. For steady-state analysis of the power flow at the power frequency 50 or 60 Hz, complex calculus with phasors representing voltages and currents can be used successfully. Switching transients, however, involve much higher frequencies, up to kilohertz and megahertz, and the complex calculus can no longer be applied. Now the differential equations describing the system phenomena have to be solved. In addition, lumped-element modelling of the system components has to be done with care if Kirchhoff's voltage and current laws are used.

In the case of a power transformer under normal power-frequency conditions, the trans-former ratio is given by the ratio of the number of turns of the primary and the secondary winding. However, for a lightning-induced voltage wave or a fast switching transient, the stray capacitance of the windings and the stray capacitance between the primary and secondary coil determine the transformer ratio. In these two situations, the power transformer has to be modelled differently. When one cannot get away with a lumped-element representation, wherein the inductance represents the magnetic field, the capacitance represents the electric field, and the resistance represents the losses, travelling wave analysis must be used. The correct 'translation' of the physical power system and its components into suitable models for the analysis and calculation of power-system transients requires insight into the basic physical phenomena. Therefore, it requires careful consideration and is not easy [7].

A transient occurs in the power system when the network changes from one steady state into another. This can be, for instance, the case when lightning hits the earth in the vicinity of a HV transmission line or when lightning hits a substation directly.

The majority of power-system transients are, however, the result of switching actions. Load-break switches and disconnectors switch off and on parts of the network under load and no-load conditions, respectively. Fuses and circuit-breakers interrupt higher currents and clear short-circuit currents in faulted parts of the system. The time period when transient voltage and current oscillations occur is in the range of microseconds to milliseconds. On this time scale, the presence of a short-circuit current during a system fault can be regarded as a steady-state situation, wherein the energy is mainly in the magnetic fields, and, after the fault current interruption, the system is transferred into another steady-state situation, wherein the energy is predominantly in the electric fields.

## 1.4   Purpose of Switching

### 1.4.1   Isolation and Earthing

Isolation of components from energized sections of the system is the simplest (no-load) switching operation. Isolation is usually necessary for safe maintenance, repair, and replacement of power-system components. Only after isolation and earthing, can personnel approach the equipment. In many countries, a visible break between live and workable parts is required.

To reduce the probability of breakdown to the absolute minimum, a very large contact distance, to be achieved with the switching device, is necessary. Such switching devices are commonly called *disconnectors* or *disconnecting switches*. These devices can operate in open air (such as in outdoor substations) or in an $SF_6$ (sulfur hexafluoride) environment, such as in *gas-insulated switchgear* (GIS) where the conductors and switchgear are insulated by pressurized $SF_6$ gas contained in metal tubes.

The no-load switching, that is, only isolation from energized sections, might seem to be an easy, straightforward operation. Nevertheless, due to the stray capacitance of the power system, a very small current always flows in energized systems. Because of this, disconnector switching is also associated with the extinction of an electric arc (see Section 1.5). Disconnector switching is discussed in detail in Section 10.3.2.

Earthing is the switching operation that connects a previously live part of the system to earth. In normal earthing operation, the section to be earthed is de-energized. In a faulty situation, when earthing is performed with energized sections or components of the power system, large currents can result, depending on earthing of the neutral of the power system. In any case, earthing switches must be capable of conducting the fault current, while special fast- and high-speed earthing switches have to perform the earthing operation under all (including faulty) conditions, see Section 10.4.2.

### 1.4.2   Busbar-Transfer Switching

For reliable operation of power systems many components and connections are installed in a redundant way. Busbars in substations are usually in a double arrangement. In cases when the flow of current has to be maintained but diverted (or commutated) from one busbar to another, switching devices, such as disconnectors are used to transfer the load current to the parallel busbar. Thus, the net load current will continue to flow uninterrupted. Because of the presence of the parallel busbar, current transfer up to a significant load current is relatively straightforward.

Bus transfer switching is treated in detail in Section 10.3.3.

### 1.4.3   Load Switching

Loads are regularly switched in power systems. For industrial systems, *contactors* are designed to switch normal loads, such as motors, pumps, furnaces, and so on, very frequently. In utility power systems, load-break switches and circuit switchers are the devices that can interrupt the load current – but not the (full) fault current. The frequency of normal-load switching in utility systems is usually very low. This is not the case for reactive-power installations, such as shunt capacitors and shunt reactors that are switched very frequently, often twice daily.

Unlike normal loads that mostly have a power factor close to one, shunt-reactor currents and capacitive currents have a phase angle of 90 electrical degrees between current and voltage. This has severe implication for the switching of these devices, as will be explained in Sections 4.2 and 4.3. Reactors can store energy in their magnetic field and capacitors store electric charge, the energy of which is released when the de-energization operation fails. The release of this energy in the system may have detrimental effects for the installed switchgear and other components.

## 1.4.4   Fault-Current Interruption

When a fault occurs in the power system, the associated short-circuit current is detected by protective relays which initiate circuit-breaker operation in order to interrupt the fault current (see also Section 2.1). The event is also known as *fault clearing*. The protective relays continuously monitor currents and voltages, collecting the information from the instrument transformers, that is, current and voltage transformers.

The time between the occurrence of a fault and its detection by the protection system, the *relay time*, is usually of the order of one to three half-cycles of the power-frequency of 50 or 60 Hz. The protection system issues a tripping command to the circuit-breaker(s) that should isolate the faulted section from the rest of the network. The tripping command activates the operating mechanism and through its kinematic chain makes the contacts in the circuit-breaker separate. After a certain *opening time*, the circuit-breaker arcing contacts will open in all three poles; this is usually referred to as *contact parting* or *contact separation*.

The pole of a circuit-breaker, or more generally of a switching device, is the part of the device that is located in one of the phases of the network, so there are three poles in a three-phase device.[4] A switching device is called single-pole if it has only one pole. If it has more than one pole, it may be called multi-pole (two-pole, three-pole, etc.) provided the poles are or can be coupled in such a manner as to operate together. The part of the pole in which the actual current is to be interrupted is generally called the *interrupter* or *interruption chamber*. It consists of contact system(s), a mechanical device supporting the arc-extinction process, and insulation. Depending on the rated voltage, a pole can consist of two or more interrupters placed in series in order to share the voltage. *Grading capacitors* across each interrupter have to take care of an equal voltage distribution across each interrupter.

So, a circuit-breaker may be designed as three single-pole switching devices or as a three-pole switching device and each pole will contain one or more interrupters. A three-pole device will be equipped with a single operating mechanism while single-pole devices will have one operating mechanism per pole or, at the highest system voltages, even several operating mechanisms per pole.

The electric arc in a circuit-breaker plays a key role in the interruption process and is therefore often called a *switching arc*. Upon contact separation, an arc is formed in the interrupter(s) of each pole. Actual interruption must wait for a zero crossing of the current. The arc is in essence resistive and therefore the arc voltage and the current reach the zero crossing at the

---

[4] Definition of *pole of a switching device*: *"The portion of a switching device associated exclusively with one electrically separated conducting path of its main circuit and excluding those portions which provide a means for mounting and operating all poles together"* (IEC 60050-441, "International Electrotechnical Vocabulary").

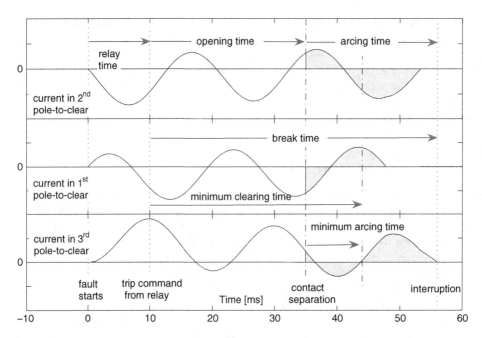

**Figure 1.3**   Circuit-breaker opening in a three-phase circuit and the IEC definitions of relevant times.

same instant. Around current zero (see Section 1.5), the energy input in the arc channel is rather low (at current zero there is even no energy input), and if the circuit-breaker design is such that the cooling by the extinction medium is adequate, the current can be interrupted. Depending on the type of circuit-breaker, the device may not be ready to interrupt at the first occurring current zero after contact separation. It takes a certain *minimum arcing time* before the circuit-breaker can actually interrupt the current, because sufficient cooling pressure of the extinction medium must be available and/or sufficient contact distance must be reached.

Then, after the minimum arcing time has elapsed, the current can be interrupted at the first following current zero. Current interruption will take place at the respective current zero of each of the three phases. When all the poles have interrupted, the fault has been cleared. The time between the instant of energizing the trip coil of the circuit-breaker and the current interruption in all phases is called the *break time*.

All relevant times are displayed in Figure 1.3 showing a three-phase-fault interruption sequence in an effectively earthed neutral system (see Section 3.3.2).

In the standard IEEE C37.04 [8], the rated interrupting time (i.e. the time between energization of the trip circuit and interruption in all phases) is expressed by the number of power-frequency cycles. A three-cycle breaker thus needs three power-frequency cycles to clear a fault.

## 1.5   The Switching Arc

When current flows through a circuit-breaker and the contacts of the breaker separate, the current continues to flow through the arc that starts at contact separation. Just before contact separation, the circuit-breaker contacts touch each other at a very small surface area, the contact bridge, and the resulting high current density makes the contact material melt. The melting contact material virtually explodes and this leads to a gas discharge in the surrounding medium, that is, air, oil, or $SF_6$.

The matter changes from a *solid state* to a *liquid state*. When more energy is added and the temperature increases, the matter changes from a liquid state to a *gaseous state*. A further increase in temperature gives the individual molecules so much energy that they dissociate into separate atoms, and if the thermal energy level is increased even further, the electrons in the outer shell(s) of the atoms acquire sufficient energy to become free electrons, leaving positive ions behind. The mixture of free electrons and ions is called the *plasma state*: a state of matter in which a certain portion of the particles is ionized. Because of the presence of free electrons and the heavier positive ions in the high-temperature plasma channel, the plasma channel is highly conductive and the current continues to flow through the arc plasma after contact separation. The electric arc is the plasma channel between the circuit-breaker contacts, a high-current electrical discharge in the extinction medium.

When considering current interruption, it is important to realize that an electric arc is always drawn at contact separation, and it appears immediately, automatically, and inevitably.

The electric arc is the only known element, apart from power semiconductors, that is able to change from a conducting to a non-conducting state in a short period. In HV circuit-breakers, the electric arc is a high-pressure arc burning in oil, air, or $SF_6$ (see Chapter 6). In medium-voltage (MV) circuit-breakers, the arc exists in vacuum or, more correctly, in the metal vapour released from the contacts (see Chapter 8).

Current interruption is performed by cooling the arc plasma so that it disappears at its most critical period of existence around the current zero.

Interruption of a short-circuit current is a very important function of a circuit-breaker. This function is verified in an extensive system of test-duties, set up by standardizing bodies, such as IEC and IEEE (see Section 13.1).

To understand the inevitability but also the advantage of an electric arc, consider a simple 50 Hz circuit with the r.m.s. inductive current $I = 100$ A at line-to-line voltage $U_r = 10$ kV (Figure 1.4).

Assume the hypothetical case that this current is interrupted immediately at contact separation without an arc at an instantaneous 'chopped' value of $i_{ch} = 100$ A as depicted in Figure 1.5.

The value of the inductance $L$ can be calculated straightforwardly from the r.m.s. values of the current and voltage:

$$L = \frac{U_r}{\sqrt{3} I \omega} = 0.184 \text{ H} \tag{1.5}$$

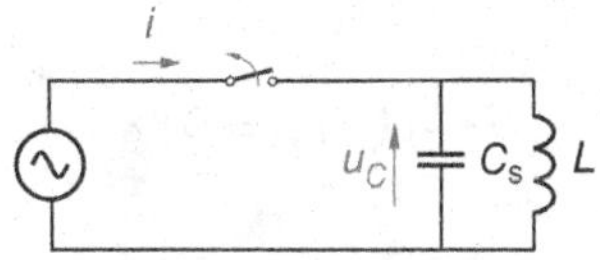

**Figure 1.4**  Equivalent circuit diagram for inductive-current interruption.

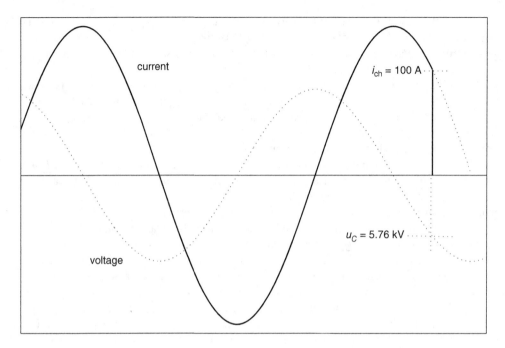

**Figure 1.5**  Current and voltage at current interruption without arc.

At the moment of interruption, the instantaneous voltage across the load $u_C$ is given by:

$$u_C = U_r \sqrt{\frac{2}{3}} \cos\left\{ \arcsin\left(\frac{i_{ch}}{I\sqrt{2}}\right) \right\} = 5.76 \text{ kV} \qquad (1.6)$$

In practice, the load reactance will always possess a certain stray capacitance $C_s$, which is taken in this example as a lumped stray capacitance $C_s = 5$ nF. The stray capacitance represents in fact the energy storage in the electric field caused by the voltage at the terminals of the load. At current interruption, this capacitance remains charged at the voltage $u_C$. Now the situation is that the reactor stores magnetic energy because of the current $i_{ch}$, and the capacitor stores electric energy because of its voltage $u_C$. In terms of energy, the magnetic energy stored in the disconnected load reactor is:

$$E_m = \frac{1}{2}L \cdot i_{ch}^2 = 920 \text{ J} \qquad (1.7)$$

whereas the electric energy stored in the capacitor is:

$$E_e = \frac{1}{2}C_s \cdot u_C^2 = 0.08 \text{ J} \qquad (1.8)$$

Since the circuit-breaker is assumed to have disconnected the load from the source, the stored energy cannot transfer back to the source and remains in the load. The energy in the reactor and in the capacitor varies widely, and there is an exchange of the energy between the two components tending to reach a state of balance. The exchange results in an oscillation. As a consequence, the total energy $E_m + E_e$ at a certain moment will be present only in the capacitor with zero current and energy in the reactor. At that moment, the voltage across the capacitor $u_{C,max}$ can be calculated as:

$$E_m + E_e = \frac{1}{2}C_s \cdot u_{C,max}^2 \qquad (1.9)$$

from which it results that $u_{C,max} = 607$ kV.

This is of course a rather unrealistic situation since 607 kV is a voltage unimaginable in a 10 kV system. The voltage cannot reach this value because a breakdown will occur at the dielectrically weakest location in the circuit. This location is the contact gap of the switching device.

The arc (*electric arc discharge*) will start because conducting plasma is created by the melting, evaporation and ionization of the metal originating from the last contact bridge through which all current is passing (see Section 1.4.4). Every time when the arc has a tendency to extinguish, for example when the current is too small to maintain the arc, there is still substantial energy trapped in the reactor and, consequently, the voltage appearing immediately across the contact gap will re-ignite the arc and re-create the conducting arc channel.

Therefore, in an inductive circuit, the arc is a continuously surviving discharge, lasting until the magnetic energy stored in the load reactor is released back to the source. Only at the instant of current zero ($i = 0$) is there no magnetic energy in the reactor and the arc disappears.

This shows the big advantage of the arc interruption over a sudden interruption: the arc allows a natural transfer of load energy back to the source, thus avoiding excessive overvoltages.

In AC systems, current zero crossings occur every half-cycle and all HV AC power switching devices interrupt the current at one of those current zeros.

Switching arcs are normally not visible in HV switching devices because they appear in a hermetically sealed interrupter. In simpler switching devices, however, the arc and its consequences can be observed. Figure 1.6 shows a switching arc in a load-break switch (see Section 1.4.3). In this example, a current of 700 A is interrupted in a 12 kV test circuit, which is a normal action at load-current switching. The impact of the arc can be seen: the pressure rise in the interruption chamber between open contacts in the atmospheric air causes a plasma jet to eject ionized gas to the outside, together with the debris from molten material of the contacts and the interruption-chamber walls. The interruption principle of confining the arc into a chamber with walls that release evaporation material that contributes to the cooling of the arc is sometimes referred to as the "deion principle" (see Section 6.2.3).

**Figure 1.6**   Switching arc of a load-break switch interrupting 700 A in a 12 kV test circuit.

From this example it is clear that arcs created in the largest single-interrupter circuit-breakers capable of breaking 63 kA in 550 kV systems, that is, having 4125 times greater apparent switching power than the load-break switch, are hugely violent phenomena.

Although current zero is the only opportunity for a switching device to interrupt a current, this does not imply that every current interruption is finally successful. The arc being present between the contacts may have disappeared, but the hot remnants, for example, ionized gases in $SF_6$ circuit-breakers and metal vapour in vacuum breakers, will reduce the dielectric strength, thus influencing the ability of the circuit-breaker to withstand the *transient recovery voltage* (TRV). The transient recovery voltage is the voltage that appears across the gap immediately following current interruption, as a reaction of the network to the new situation. A re-ignition may occur followed by another loop of power-frequency current. Eventually, after several unsuccessful attempts, the device may not be capable of interrupting the current and will explode, causing a short-circuit by itself.

## 1.6  Transient Recovery Voltage (TRV)

### 1.6.1  TRV Description

The TRV is the voltage across the open circuit-breaker contacts immediately after current interruption. The TRV, that is, $u_{ab}$, appears as the difference between the voltage-to-earth at the source side and the voltage $u_{bn}$ at the load side (see Figure 1.9):

$$u_{ab} = u_{an} - u_{bn} \qquad (1.10)$$

Thus, a TRV always consists of two components: a source-side component $u_{an}$ and a load-side component $u_{bn}$. In all cases, the TRV starts from zero at current zero, makes an excursion to the momentary power-frequency voltage, overshoots in a damped oscillatory manner, and continues to oscillate until a steady-state condition is reached. This steady-state situation is a power-frequency voltage, called the *recovery voltage* (RV).

The complete interruption process is outlined schematically in Figure 1.7. In this case the RV is equal to the source voltage.

The frequency of the TRV is determined by the relevant inductance and capacitance. The peak value in the undamped case is two times the peak source voltage; in practice, the peak value of the TRV is lower due to damping.

The TRV affects the interruption for two reasons:

- Determined by the oscillation frequency, the *rate-of-rise of recovery voltage* (RRRV) can be very high. This implies that very shortly after extinction of the arc, a high voltage appears across the contact gap. If there are still ionized, hot residues of the arc remaining to a certain degree, the arc will be re-established (will *re-ignite*) due to the impact of the TRV. In Section 13.1.2 it will be described that the standardized RRRV corresponds to the slope of the tangential line of the TRV wave shape, a value not necessarily equal to the highest value of the derivative of the TRV ($du/dt$), see Figure 1.8.
- The peak value of the TRV can be very high. In testing and standardization, the damping is expressed by the *amplitude factor* $k_{af}$, defined as the ratio between the transient peak value and the steady-state value; in Figure 1.8 the steady state voltage is the peak of the power-frequency voltage. The value of $k_{af}$ is in the range $1 < k_{af} \leq 2$.

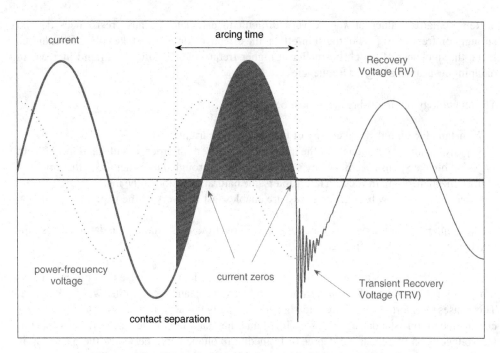

**Figure 1.7**   Current-interruption in a purely inductive AC circuit.

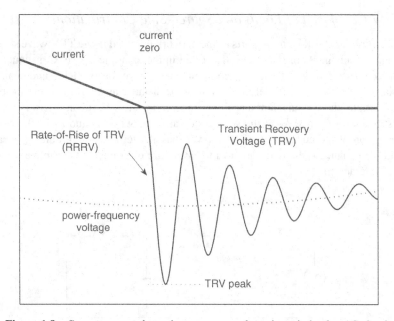

**Figure 1.8**   Current zero and transient recovery voltage in an inductive AC circuit.

Representative values of $k_{af}$ are very difficult to calculate, because resistances depend strongly on frequency. Due to the skin-effect, the effective conduction takes place in a surface layer, the thickness of which is smaller at higher frequencies, leading to a rapid increase in damping resistance at higher frequencies.

The technology of circuit-breakers must be able to:

- Withstand the high thermal energy of the arc before current zero.
- Rapidly remove the remnants of the arc after current zero in order to withstand the TRV. In gas/oil breakers, this is achieved by a forced gas flow through the former arc path, removing the ionized medium. In vacuum interrupters, the natural diffusion of the metal vapour plasma towards the very low background pressure enables fast recovery of the gap.

Background information on the awareness of TRV over the years and its standardized description can be found in Section 13.1.1.

Circuits subjected to a short-circuit are mainly inductive. This is because the current value is limited by the reactance $(X = \omega L)$, rather than by the resistance $R$. In other words, $R \ll \omega L$. This causes the short-circuit current to lag approximately 90° with respect to voltage. In 50 Hz circuits the ratio is standardized: $X/R = 14.14$ and the phase lag of the current with respect to the voltage is 85.9° (electrical degrees). In medium voltage cable networks the phase lag is smaller, which results in a less onerous interruption regarding TRV.

### 1.6.2   TRV Composed of Load- and Source-Side Contributions

In practical cases, there will be more parts of the network involved in the TRV wave shape than just the single $LC$ elements of the circuit described in the example above. Consequently, the TRV often comprises multiple-frequency components decisive for its RRRV and peak value.

As an example, the case is considered of a fault some distance away from the circuit-breaker. A simplified single-phase equivalent circuit is shown in Figure 1.9. It comprises two separate $LC$ circuits, one with $C_S$ and $L_S$ at the source side and one with $C_L$ and $L_L$ at the load side. After interruption of the current – which is lower than the terminal-fault current because of the load-side impedance – the two parts of the circuit are disconnected completely and have no electrical interaction.

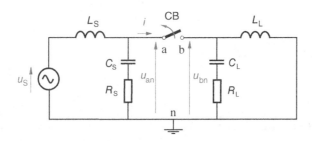

**Figure 1.9**   Equivalent circuit diagram for multi-frequency TRV.

The transients in both circuit parts behave independently and, in the construction of TRV, understanding of the initial, intermediate, and final conditions is very helpful:

- Initial condition (at current zero):
  Both transients start at the same voltage level:

$$u_{an}(0) = u_{bn}(0) = \frac{L_L}{L_S + L_L} \hat{U} \tag{1.11}$$

This is the voltage to earth at current zero; in the simple-circuit case it is the voltage across the capacitors $C_S$ and $C_L$ that are charged to equal voltage at current zero.
- Oscillation interval from current zero to the decay of the transients after which only power-frequency recovery voltage remains:
  The TRV component at the source side $u_{an}$ has an amplitude

$$u_{0,S} = \hat{U} - u_{an}(0) = \frac{L_S}{L_S + L_L} \hat{U} \text{ and frequency } f_S = \frac{1}{2\pi} \cdot \frac{1}{\sqrt{L_S C_S}} \tag{1.12}$$

The TRV component at the load side $u_{bn}$ has an amplitude

$$u_{0,L} = u_{bn}(0) = \frac{L_L}{L_S + L_L} \hat{U} \text{ and frequency } f_L = \frac{1}{2\pi} \cdot \frac{1}{\sqrt{L_L C_L}} \tag{1.13}$$

- Final condition after decay of transient components:
  The TRV component at the source side $u_{an}(t)$ will oscillate from the initial voltage $u_{an}(0)$ to the power-frequency voltage of the source $\hat{U} \cdot \cos(\omega t)$, ignoring the voltage drop across the capacitance, because $1/(\omega C_S) > \omega L_S$.
  The TRV component at the load side $u_{bn}(t)$ will oscillate from the initial voltage $u_{bn}(0)$ to zero (in the absence of an active source and/or a charge-storing element).
  Keeping these simple and clear rules in mind, understanding TRV in various situations is straightforward.

Considering the fact that in practical cases the frequency of the source- and load-side TRV component is much higher than the power frequency, the transient components can be treated independently from the power-frequency voltage.

In practical cases, in circuits with power factor close to zero, the equation of the TRV reads as follows:

$$u_{ab}(t) = \hat{U} \left[ \cos(\omega t) - \frac{L_S}{L_S + L_L} \exp(-\beta_S t) \cos(\omega_{0S} t) - \frac{L_L}{L_S + L_L} \exp(-\beta_L t) \cos(\omega_{0L} t) \right] \tag{1.14}$$

with $\beta_S$ and $\beta_L$ being the damping in the source- and load-side circuit:

$$\beta_S = \frac{R_S}{2L_S} \qquad \beta_L = \frac{R_L}{2L_L} \tag{1.15}$$

and $\omega_{0S} = 2\pi f_S$ and $\omega_{0L} = 2\pi f_L$ the respective angular frequencies.

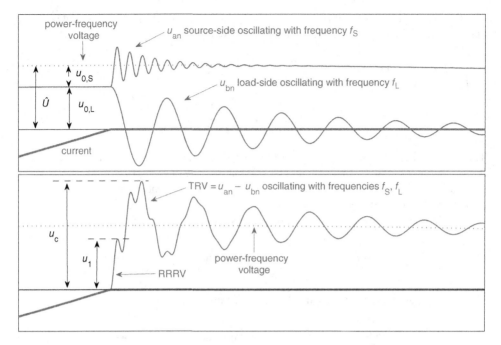

**Figure 1.10**   Double-frequency TRV with the source-side and load-side components.

The formal mathematical derivation of this generalized TRV equation can be found in Section 5.1.3.

Because of the presence of two frequencies, there is now a first local maximum value $u_1$ and a global maximum $u_c$.

The effect of the double-frequency nature of the TRV is twofold:

- The peak value $u_c$ is reduced (with respect to a single-frequency TRV) because the amplitude of each of the oscillations is smaller than $\hat{U}$;
- The rate-of-rise of TRV (RRRV) may increase because of a second component having a higher frequency.

In Figure 1.10, the transients at the load and source sides are given for different load and source impedances ($L_S = 0.5 \, L_L$). Both transients contribute to the TRV. As can be seen in this example, the initial rate-of-rise of TRV (RRRV) is determined by the TRV component of the highest frequency (here the source side), whereas its peak value will usually be determined by the oscillation having the largest amplitude (the load side in Figure 1.9). This is normally the circuit part experiencing the largest voltage drop.

A clear distinction can be made between a fault- and load-current interruption as detailed in Chapters 3 and 4 respectively:

1. In fault-current breaking, the current to be interrupted is mainly determined by the source impedance: $X_S \gg X_L$, as shown in Figure 1.11. Then, the TRV amplitude is mainly contributed by the source-side circuit ($u_{0,S} \gg u_{0,L}$).

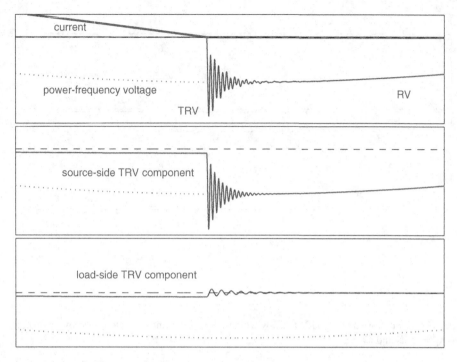

**Figure 1.11**  Double-frequency TRV fault switching with current mainly determined by the source reactance ($X_L = 0.1X_S$ and $C_L = 300$ nF).

2.  In load switching, the current is clearly determined by the load impedance, thus $X_S \ll X_L$ and $u_{0,S} \ll u_{0,L}$, leading to the situation in Figure 1.12 where the load oscillation is dominating the TRV wave shape.

Clearly, the TRV components originating from various parts of the circuit must be taken into account.

## 1.7  Switching Devices

Switching devices are devices designed to make and/or break the current in one or more electric circuits. Depending on their specific switching duty, a large number of devices can be identified, most of which will be treated in detail in this book.

When combined with their secondary equipment, switching devices are called *switchgear*, which is *"a general term covering switching devices and their combination with associated control, measuring, protective, and regulating equipment, also assemblies of such devices and equipment with associated interconnections, accessories, enclosures, and supporting struc- tures, intended in principle for use in connection with generation, transmission, distribution, and conversion of electric energy"* (definition by IEC [9]).

The most widely applied switching device is the *switch* that is defined as: *"A mechanical switching device capable of making, carrying and breaking currents under normal circuit*

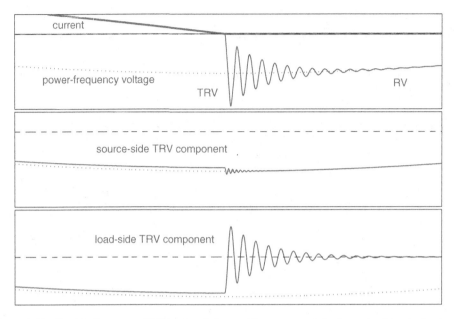

**Figure 1.12**   Double-frequency TRV load switching with current mainly determined by the load reactance ($X_L = 10X_S$ and $C_L = 30$ nF).

*conditions which may include specified operating overload conditions and also carrying for a specified time currents under specified abnormal circuit conditions such as those of short circuit. Note: A switch may be capable of making but not breaking short-circuit currents"* (definition by IEC [9]).

*Circuit-breakers* are indispensable switching devices in the power system. Their main task is to interrupt fault currents and to isolate faulted parts of the system. Besides short-circuit currents, a circuit-breaker must also be able to interrupt a wide variety of other currents at system voltage, such as capacitive currents, small inductive currents, and load currents. They are defined as *"mechanical switching devices, capable of making, carrying and breaking currents under normal circuit conditions and also making, carrying for a specified time and breaking currents under specified abnormal circuit conditions, such as those of short circuit"* (definition by IEC [9]).

Circuit-breakers are the only switching devices able to interrupt (break) fault currents in electrical power systems up to the highest short-circuit current levels [10–16, 21]. At present, in distribution networks operated at voltages below 72.5 kV, these short-circuit currents can go up to 63 kA in standard applications and up to 300 kA in generator applications; generator circuit-breakers are situated between the generator and the step-up transformer, in power stations (see Section 10.1). In transmission systems, generally operated at a voltage of 72.5 kV and above, a short-circuit current can arise up to 80 kA. In contrast to fuses and other fault current limiting devices, as discussed in Sections 10.12 and 10.12.1, after fault clearing, circuit-breakers are required to be ready for service again and even fulfil an operating sequence of several short-circuit current breaking and making operations.

Faults in power systems cannot be avoided. A wide-scale international survey on fault statistics shows that most faults occur on overhead lines, with several faults per 100 km overhead line per year [17] (see details in Section 2.4).

The general basic characteristics and requirements of circuit-breakers are the following:

- A good conductor when closed. In order to minimize energy losses during normal operation, the contact system must be designed to have very low resistance. In addition, during conduction of short-circuit current (when not called upon to interrupt), excessive generation of heat or other malfunction must not occur.
- A good insulator in open position. This requirement involves withstanding of high overvoltages due to switching and/or lightning overvoltages. Circuit-breakers in the open position do not have a safety insulating function, a feature achieved by disconnectors that are specified to withstand higher overvoltages across the open contacts.
- The transition from conductor to insulator (and vice versa) must occur sufficiently fast. The breaker must be able to act very quickly after being tripped, in order to minimize damage due to a short-circuit. In Section 1.5, the sequence of events is sketched during this transition.

  Interruption of any current (from very small transformer magnetizing currents up to massive terminal fault currents) must be assured. In addition, making (energizing a circuit) is a requirement. A making operation under short-circuit is not straightforward, it can cause damage to the contact system.
- The switching should not generate unacceptable transients in the network. The arc, initiated in the interruption chamber at contact separation, allows the release of the energy of the load back to the source (see Section 1.5).
- The breaker must be able to switch often without too much erosion. The arc is an intensive heat source and potentially damages the internal parts. This causes loss of material (erosion) from contacts and hot gas guiding elements (*nozzles* in gas circuit-breakers). Withstanding against repeated electrical arcing is called *electrical endurance*, refer to Section 12.2. Generally, the higher the current, the higher the erosion.
- The breaker must operate immediately, even after long idle time and in all possible climatic conditions.
- The breaker must be able to handle a large number of switching operations, all not necessarily with current. This capability is called *mechanical endurance* – refer to Section 12.2. It requires very high reliability of the mechanical components of the circuit-breaker. From international studies it is concluded that this requirement is very severe: most major and minor failures in switchgear have a mechanical rather than an electrical origin, refer to Section 12.1.3 [18].

Even though real HV circuit-breakers never realize the ideal transition from perfect conductor to perfect insulator instantaneously, it can be expected that the conductivity of their contact gap decreases by 13 to 15 orders of magnitude during a very short period of time, ($10^{-5}$ to $10^{-6}$) s, at the instant of the current interruption. At present, except for semiconductors, the only known medium that has the capability of such rapid change in conductivity is the arc plasma, after a change in temperature of only 1 to 2 orders of magnitude.

The highest technical requirements in terms of breaking current and rated voltage are set for HV transmission-line circuit-breakers. They must

- carry (without exceeding the temperature limits of their components) the normal rated current of up to 4 kA – at the highest system voltages even larger values; generator circuit-breakers in nuclear power stations even up to 40 kA;
- withstand (up to 3 s) high short-time currents $I_k \leq 100$ kA with the corresponding peak withstand currents $I_p \leq 250$ kA;
- switch all currents from several amperes to (80 to 100) kA short-circuit currents at rated voltages up to 1200 kV with a maximum break time of few cycles, which is necessary for reasons of system stability.

Circuit-breakers are often located in outdoor substations and exposed to all kinds of climates, from tropical heat with extreme humidity to arctic cold, down to –55 °C. They can also be exposed to extreme pollution. HV circuit-breakers may also be required to withstand earthquake stresses.

In the case of a short-circuit somewhere in the network, circuit-breakers are the last link in the chain of power-system protection and the only means of protecting the network. Therefore, they must comply with extremely high demands for operational reliability.

Unlike other components, circuit-breakers actively intervene in the flow of energy in electrical power systems and, therefore, not only cause but also have to withstand the electrical (and mechanical) stresses that arise due to their action.

An extended set of tests intended to verify all these requirements is defined and used in the industry. These tests, specifically designed to verify circuit-breaker performance, are collected in IEC 62271-100 [19] and IEEE C37-09 [20] (see Chapter 13).

## 1.8   Classification of Circuit-Breakers

Circuit-breakers can be classified by many different criteria, such as rated voltage class, type of installation, structural design, arc-extinction principle, extinction medium, and so on.

The basic classification of circuit-breakers relates to the voltage level for which they are designed. By this criterion, circuit-breakers are classified by IEC into two main groups:

- LV (low-voltage) circuit-breakers with rated voltages below 1000 V; and
- HV circuit-breakers with rated voltages 1000 V or more.

This classification of circuit-breakers is currently being used by both IEC and IEEE/ANSI standards.

In addition, the terms medium voltage (MV) for the range (1 to 52) kV, extra-high voltage (EHV) ranging (245 to 800) kV and ultra-high voltage (UHV) for rated voltages above 800 kV are in common use, though not defined in the standards.

HV circuit-breakers can be designed for either indoor or outdoor installation. Indoor circuit-breakers can be used only inside buildings or weather-resistant enclosures. MV indoor circuit-breakers are often designed for use inside a metal-clad switchgear enclosure. The essential differences between indoor and outdoor circuit-breakers are the external packaging and the

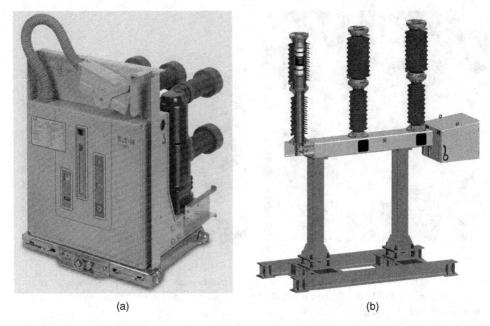

(a)                                                                (b)

**Figure 1.13**  Vacuum circuit-breakers: (a) 24 kV indoor circuit-breaker (reproduced with permission of Eaton Electric); (b) live-tank 72.5 kV vacuum circuit-breaker (reproduced with permission of Siemens AG).

enclosure materials that are used, as illustrated in Figure 1.13. In many cases, the interruption chambers and the operating mechanism are the same for both indoor and outdoor circuit-breakers. Therefore, they have identical or very similar switching capabilities.

There are two types of outdoor circuit-breakers, based on their structural design (Figure 1.14): dead-tank circuit-breakers and live-tank circuit-breakers.

Dead-tank type circuit-breakers are characterized by the following advantages over live-tank circuit-breakers:

- they have a lower centre of gravity with a high seismic withstand capability;
- multiple instrument transformers can be installed at both sides of the circuit-breaker at low potential; and
- they can be shipped completely assembled with factory-made adjustments.

Live-tank type circuit-breakers have the interrupter housing at a high potential above earth. They also have certain advantages over dead-tank circuit-breakers:

- lower cost, except for the current transformers;
- less technically complicated because the interrupter is far away from earth;
- smaller space requirements for the installation; and
- a smaller quantity of isolating medium (oil or gas) is required.

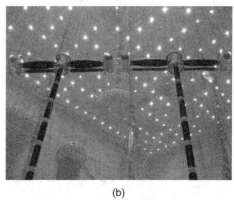

(a)                                                                                (b)

**Figure 1.14**   (a) 550 kV dead-tank single break type circuit-breaker in KEMA test-laboratory (reproduced with permission of Hitachi); (b) EHV-UHV four-break live-tank circuit-breaker (reproduced with permission of Alstom Grid).

Another important classification of circuit-breakers is by the medium for insulation and arc extinction. The evolution of circuit-breaker technology has been strongly related to the appearance of new media.

Oil and air were used during the early decades of electrification and they were prevalent throughout the first half of the twentieth century. Very reliable designs were developed, many of which are still in service today. They were manufactured until the 1980s when they were supplanted by circuit-breakers using vacuum and sulfur hexafluoride ($SF_6$) gas.

Vacuum and $SF_6$ made their appearance at about the same time – in the second half of the twentieth century. Today, they are absolutely dominant and leading technologies: vacuum for medium voltages and $SF_6$ for high voltages. For that reason, this book concentrates predominantly on these two technologies, on which all present-day products are based.

In Figure 1.15 an overview is provided of the various media and principles that were and are used in power switching devices.

HV circuit-breakers can also be classified by their operational arrangements. There are three-pole-operated circuit-breakers and single-pole-operated circuit-breakers, as illustrated in Figure 1.16.

A three-pole-operated circuit-breaker operates the interrupters for all three phases simultaneously with one operating mechanism. Such designs are normally used at medium voltages and are predominant up to ratings of 245 kV. The prevalence towards three-pole operation at lower rated voltages is primarily cost driven, as it requires only one operating mechanism for all three poles. Single-pole operated circuit-breakers at these voltages are used only if a single-phase auto-reclosure of transmission lines is required. Due to the mechanical coupling of all three poles, this type of circuit-breaker ensures better synchronism between poles during both closing and opening.

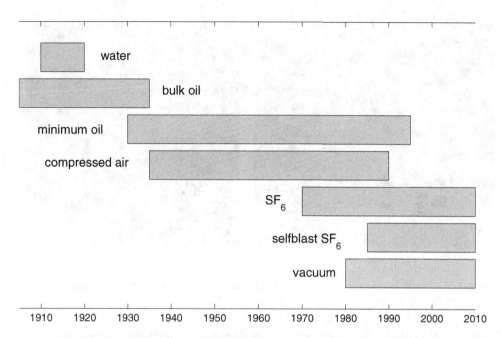

**Figure 1.15**  Use of arc-extinction media in switching devices over the years.

**Figure 1.16**  (a) Three-pole-operated single-break 170 kV live-tank $SF_6$ circuit-breaker; (b) single-pole-operated double-break 420 kV live-tank $SF_6$ circuit-breaker.

(a)                                                     (b)

**Figure 1.17**   (a) Three-phase-enclosed 123 kV GIS SF$_6$ circuit-breaker; (b) single-phase-enclosed 123 kV GIS SF$_6$ circuit-breaker.

Single-pole-operated circuit-breakers use three separate operating mechanisms to control each of the three phases separately. Such designs are used predominantly at voltages above 245 kV, mainly due to constraints of the physical size, the associated operating energies and forces involved in such large circuit-breakers.

Circuit-breakers for metal-enclosed SF$_6$ *gas-insulated switchgear* (GIS) represent a separate group of HV circuit-breakers. The distinctive advantages of GIS are:

• small size and therefore very small space requirements, which is particularly important in urban environments;
• extra safety for operating personnel protecting them from contact with live parts, since all live parts of the switchgear are contained in an earthed metal enclosure;
• full protection of HV parts against pollution, unaffected by weather, including higher altitudes;
• wide range of possible locations for installation and aesthetic compatibility with its surroundings;
• short on-site erection time, owing to extensive prefabrication and factory testing of complete bays;
• compact and modular design.

Two main enclosure-design principles are applied to the GIS circuit-breakers: three-phase enclosed and single-phase enclosed (Figure 1.17).

Single-phase enclosures are used, with few exceptions, for rated voltages above 170 kV, otherwise the diameter of the enclosures becomes too large. Up to ratings of 170 kV, the three-phase enclosures have certain advantages compared with single-phase enclosures:

• the total number of the enclosures is reduced to one-third;
• the bay width and floor area are considerably reduced;

- three-phase-enclosed GIS practically does not suffer from eddy currents and therefore electrical losses in the enclosure are negligibly small;
- larger gas zones ensure a slower rise of pressure in case of an internal fault;
- burn-through of the enclosure, in the case of an internal fault, is much less probable because a phase-to-earth fault usually turns into phase-to-phase arcing within several tens of milliseconds; and
- the probability of gas leakage is smaller due to the reduced number of gas zones and fewer compartments.

# References

[1] van der Sluis, L. and Schavemaker, P. (2001) *Electrical Power Systems Essentials*, John Wiley & Sons Ltd, Chichester, England, ISBN 978-0470-51027-8.

[2] Grainger, J.J. and Stevenson, W.D. Jr. (1994) *Power System Analysis*, Chapters 4, and 5, McGraw-Hill, New York, ISBN 0-07-061293-5.

[3] Duncan, J.D. (1993) *Power System Analysis and Design*, PWS Publishing Co., Boston, ISBN 0-53493-960-0.

[4] El-Hawary, M. (1995) *Electrical Power Systems*, IEEE Power System Engineering Series, IEEE, ISBN 0-7803-1140-X.

[5] Bewley, L.V. (1963) *Travelling Waves on Transmission Systems*, 2nd edn, Chapters 1–4, Dover Publications, Mineola, New York.

[6] Rüdenberg, R. (1962) *Elektrische Wanderwellen*, 4th edn, Parts I, II, IV, Chapters 1, 6, and 17, Springer-Verlag, Berlin.

[7] Rüdenberg, R. (1968) *Electrical Shock Waves in Power Systems*, Parts I, II, and IV, Chapters 1, 6, and 17, Harvard University Press, Cambridge, Massachusetts.

[8] IEEE Std C37.04-1999 (1999) IEEE Standard Rating Structure for AC High-Voltage Circuit Breakers.

[9] IEC 60050-441 (1984) International Electrotechnical Vocabulary – Chapter 441: Switchgear, controlgear and fuses.

[10] Browne, T.E. Jr. (1984) *Circuit Interruption*, Chapter 3, Marcel Dekker, New York, ISBN 0-824771-77-X.

[11] Garzon, R.D. (1997) *High-Voltage Circuit-Breakers*, Chapter 3, Marcel Dekker, New York, ISBN 0-8247442-76.

[12] Nakanishi, K. (ed.) (1991) *Switching Phenomena in High-Voltage Circuit-Breakers*, Marcel Dekker, New York, ISBN 0-8247-8543-6.

[13] Ragaller, K. (1977) *Current Interruption in High-Voltage Networks*, Plenum Press, ISBN 0-306-40007-3.

[14] Flurscheim, C.H. (ed.) (1975) *Power Circuit-Breakers*, Peter Peregrinus Ltd., ISBN 0-901233-62-X.

[15] Jones, G.R., Seeger, M. and Spencer, J.W. (2013) Chapter: Gas-filled interrupters - fundamentals, in *High-Voltage Engineering and Testing*, 3rd edn (ed. H.M. Ryan), IET, Stevenage, UK, ISBN 978-1-84919-263-7.

[16] Blower, E.A. (ed.) (1986) *Distribution Switchgear*, Collins, London, ISBN 0-00-383126-4.

[17] CIGRE Working Group 13.08 (2000) Life Management of Circuit-Breakers, Technical Brochure 165.

[18] CIGRE Working Group A3.06 (2012) Final Report of 2004-2007 International Enquiry on Reliability of High Voltage Equipment, Part 2 – Reliability of High Voltage SF$_6$ Circuit-Breakers, Technical Brochure 510.

[19] IEC 62271-100 (2012) High-voltage switchgear and controlgear – Part 100: Alternating-current circuit-breakers, Ed. 2.1.

[20] IEEE C37.09 (1999) IEEE Standard Test Procedure for AC High-Voltage Circuit-Breakers Rated on a Symmetrical Current Basis.

[21] Kapetanović, M. (2011) *High Voltage Circuit-Breakers*, ETF – Faculty of Electrotechnical Engineering, Sarajevo, ISBN 978-9958-629-39-6.

# 2

# Faults in Power Systems

## 2.1 Introduction

With the increasing dependence of our society on electricity supplies, the need to achieve an acceptable level of reliability, quality and safety at an economic price becomes important to customers. Mostly, the power system is well designed and adequately maintained to minimize the number of faults that can occur.

In normal operating conditions, a three-phase power system can be treated as a single-phase system when the loads, voltages, and currents are balanced. In practice, one can successfully use a single-line lumped-element representation of the three-phase power system for calculation.

A fault brings the system to an abnormal condition. Short-circuit faults are especially of concern because they result in a switching action, which often results in transient overvoltages. Both the high current and the switching overvoltages stress the equipment in the system.

Line-to-earth faults can result from direct contact of an overhead line to earthed objects, such as trees, falling limbs of trees, cranes and so on. Also, reduced isolation can lead to a short-circuit, for example in the case of a faulty component or initiated by ionized air above a bush-fire or pollution of insulators. Also, erroneous switching operations or human errors, for example, during maintenance or repair, can induce short-circuit.

The main failure mode, however, is the *lightning back-flashover* [1], in which a lightning stroke in an overhead shield wire or tower locally raises the potential of the system by its induced voltage, leading to flashover across line insulators. The majority of transmission-line faults are single line-to-earth faults (see Section 2.4.1).

Line-to-line faults can be the result of galloping lines because of high winds combined with ice loads or because of a conductor breaking and falling on a conductor below. Double line-to-earth faults result from causes similar to those of the single line-to-earth faults but are very rare. Three-phase faults, when all three phase conductors touch each other or fall to earth, occur in only a small percentage of the cases but are very severe faults for the system and its components.

*Switching in Electrical Transmission and Distribution Systems*, First Edition.
René Smeets, Lou van der Sluis, Mirsad Kapetanović, David F. Peelo, and Anton Janssen.
© 2015 John Wiley & Sons, Ltd. Published 2015 by John Wiley & Sons, Ltd.

In the case of a symmetrical three-phase fault in a symmetrical system, one can still use a single-phase representation for the short-circuit and transient analysis. However, for the majority of the fault situations, the power system has become unsymmetrical.

Symmetrical components and, especially, the sequence networks are an elegant way to analyse faults in unsymmetrical three-phase power systems because in many cases the unbalanced portion of the physical system can be isolated for study, the rest of the system being considered to be in balance [2]. This is, for instance, the case for an unbalanced load or fault. In such cases, the symmetrical components of the voltages and the currents at the point of unbalance are calculated and connected to the sequence networks. They are, in fact, copies of the balanced system at the point of unbalance (the fault point).

Protection systems are installed to clear faults, like short circuits, because short-circuit currents can damage the cables, lines, busbars and transformers.

The voltage and current transformers provide values of the actual voltage and current to the protective relay. The relay processes the data and determines, based on its settings, whether or not it needs to operate circuit-breakers in order to isolate faulted sections or components.

The classic protective relay is the electromagnetic relay which is constructed with electrical, magnetic and mechanical components. Nowadays digital relays are taking over as they have many advantages, such as: they can perform a self-diagnosis, they can record events and disturbances in a data base and they can be integrated in the communication, measurement and control environment of the modern substations.

A reliable protection is indispensable for a power system. When a fault or an abnormal system condition occurs (such as: over/under-voltage, over/under-frequency, over-current and so on) the related protective relay has to react in order to isolate the affected section while leaving the rest of the power system in service. The protection must be sensitive enough to operate when a fault occurs, but the protection should be stable enough not to operate when the system is operating at its maximum performance. There are also faults of a transient nature, a lightning stroke on or in the vicinity of a transmission line for instance, and it is undesirable that these faults lead to a loss of supply.

Therefore, the protective relays and circuit-breakers are usually equipped with *auto-reclosure* functionality. Auto-reclosure implies that the protective relay, directly after having detected an abnormal situation leading to the opening of the contacts of the circuit-breaker, commands the circuit-breaker to close again in order to check whether the abnormal situation is still there. In the case of a fault of a transient nature, the normal situation is likely to be restored again so that there is and was no loss of supply. When the abnormal situation is still there, the protective relay commands the circuit-breaker to open again so that either the fault is cleared or consecutive auto-reclosure sequences can follow. In most cases, so-called back-up protection is installed in order to improve the reliability of the protection system.

When protective relays and circuit-breakers are not economically justifiable in certain parts of the grid, in distribution systems fuses can be applied, see Section 10.12.2. A fuse combines the 'basic functionality' of the current transformer, relay and circuit-breaker in one very simple overcurrent protection device. The fuse element is directly heated by the current passing through and melts when the current exceeds a certain value, thus leading to an isolation of the faulted sections or components. After the fault is removed, the fuse needs to be replaced before the isolated grid section can be energized again.

## 2.2   Asymmetrical Current

### 2.2.1   General Terms

A fault current results from:

- a short-circuit in the system initiated by an external event or by an internal dielectric failure;
- closing a switching device into an already faulted system, such as an earthing connection that is erroneously not removed after maintenance.

Both situations are electrically equivalent, the only difference being the load current that is, in the latter case, present before the initiation of the fault current.

The transient that arises from making a circuit appears as a *DC component* of the current. The simplest but still relevant circuit is shown in Figure 2.1. Assuming a single-phase circuit with an AC voltage source of amplitude $\hat{U}$ and angular power frequency $\omega$, inductance $L$, and resistance $R$, the mathematical description is as follows (for the derivation, see Section 5.1.2).

If the zero-time reference ($t = 0$) is set equal to the instant of the short-circuit initiation, the voltage of the source can be described as:

$$u(t) = \hat{U} \cos(\omega t + \psi) \tag{2.1}$$

where $\psi$ is the *closing angle* of the voltage at the short-circuit initiation.

At $t = 0$, the short-circuit current starts to flow and is described by:

$$i(t) = \hat{I}\left[\cos(\omega t + \psi - \phi) - \exp\left(-\frac{t}{\tau}\right)\cos(\psi - \phi)\right] \tag{2.2}$$

where

$$\hat{I} = \frac{\hat{U}}{\sqrt{R^2 + (\omega L)^2}} \qquad \tau = \frac{L}{R} \qquad \phi = \arctan\frac{\omega L}{R} = \arctan(\omega\tau) = \arctan\left(\frac{X}{R}\right) \tag{2.3}$$

The fault current is offset by a decaying DC component, as expressed by the second right-hand term of Equation 2.2, having DC *time constant* $\tau$. In transient-behaviour studies, the ratio $X/R$ is often used (with $X = \omega L$) instead of the DC time constant. The advantage of this is that the power frequency $f = \omega/2\pi$ is included in $X/R$.

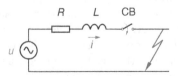

**Figure 2.1**   Basic circuit for making a short-circuit current.

Depending on the instant of short-circuit current initiation (with respect to the voltage waveform), two extremes are possible:

1.  Short-circuit initiation is at voltage zero ($\psi = \pi/2$).

    In this case, the DC component is at a maximum and causes a high *asymmetrical-current peak*. With the standardized DC time constant of $\tau = 45$ ms for HV applications, an asymmetrical peak of 2.55 $I_{sc}$ ($I_{sc}$ is the r.m.s. value of the short-circuit current) at 50 Hz will be reached. In 60 Hz circuits, the maximum value is 2.59 $I_{sc}$ because the first peak value occurs earlier than in the 50 Hz case, at a larger instantaneous value of the DC component. The *asymmetrical-current peak factor* $F_s$ is defined as:

    $$F_s = \frac{\max(|i(t)|)}{I_{sc}}$$

    (2.4)

    Note that the maximum possible DC component at $t = 0$ is 1.0 p.u. (1 p.u. $= I_{sc}\sqrt{2}$). This occurs when $\psi = \phi$ and only in pure inductive circuits and not in real circuits.

    The asymmetrical current flows through all components of the system. The most adverse effect is the action of electrodynamic forces due to the high current peak. Since these forces increase with the square of current, the asymmetrical-current peak factor has a major effect. Mechanical design of supports, spacers, bus-ducts, transformer windings, and so on should always take into account the worst case of full asymmetry, see Section 2.3.1. This case may be considered rare in practice, since most faults start with a dielectric failure, a flashover, most likely to happen at maximum voltage.

2.  Short-circuit initiation is approximately at voltage maximum ($\psi \approx 0$)

    In this case, the DC component is zero and the steady-state current is established immediately – the so-called *symmetrical current*. Perfect symmetrical conditions do not occur exactly at voltage maximum but at $\psi = \pi/2 - \phi$.

These two extreme situations are depicted in Figure 2.2.

As outlined above, the asymmetrical-current peak factor is a function of the DC time constant and the frequency. The DC time constant is directly related to the phase angle $\phi$ between the current and voltage through $\tau = (1/\omega)\tan \phi$.

In general, $F_s$ is a function of $\omega\tau$. A number of equations are in use to express this. One of these being accurate to within 0.5% [3] is:

$$F_s = \sqrt{2}\left[1 + \sin(\phi)\exp\left\{-\left(\frac{\pi}{2} + \phi\right)\frac{1}{\tan(\phi)}\right\}\right]$$

(2.5)

where $\phi$ is given in Equation 2.3.

The relationship between the asymmetrical-current peak factor $F_s$ and the DC time constant $\tau$ in 50 and 60 Hz circuits is represented graphically in Figure 2.3.

Figure 2.4 gives $F_s$ as a function of the ratio $X/R$. The value of $X/R = 14.13$ corresponds to $\tau = 45$ ms and $f = 50$ Hz.

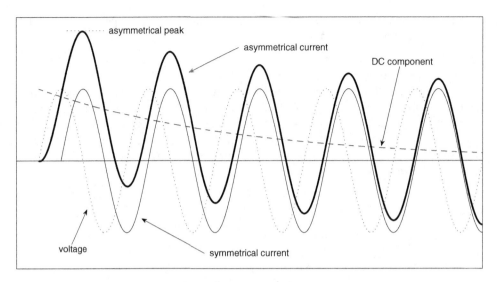

**Figure 2.2** The DC component in a single-phase case leading to a symmetrical current (when the short-circuit initiation is at maximum voltage) and asymmetrical current (when the initiation happens at voltage zero).

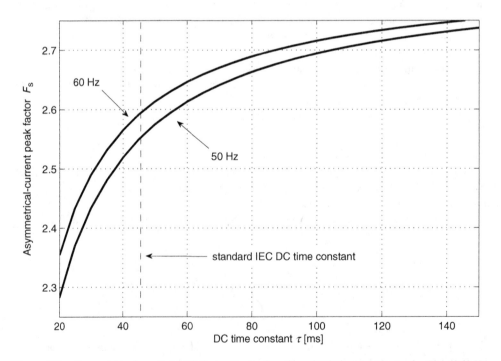

**Figure 2.3** Asymmetrical-current peak factor $F_s$ as a function of DC time constant $\tau$ for 50 and 60 Hz.

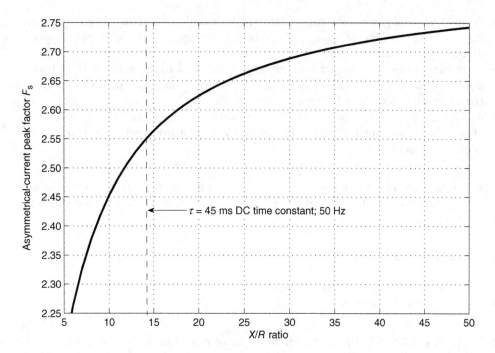

**Figure 2.4**  Asymmetrical-current peak factor $F_s$ as a function of $X/R$.

## 2.2.2   DC Time Constant

Although the value of the DC time constant is standardized in IEC [4] as $\tau = 45$ ms, there is a wide range of time constants in practical systems, mostly related to their various components. Guideline values can be found in Table 2.1 [5, 6].

Partly initiated by economical considerations, the DC time constants have increased [4] due to:

- The introduction of distributed energy generation within networks. Herewith, the large values of (sub)transient time constants of the generator can have significant impact;
- the tendency to use low-loss transformers with smaller copper losses leading to a decrease of resistance;

**Table 2.1**   Guideline values of DC time constant $\tau$ (in ms) for various circuit components.

| Component | Apparent power $S$ | | | |
|---|---|---|---|---|
| | 1 MVA | 10 MVA | 100 MVA | 1000 MVA |
| Overhead line at rated voltage $U_r$ (in kV) | $U_r < 72.5$ | $72.5 < U_r < 420$ | $420 < U_r < 525$ | $U_r > 525$ kV |
| | $\tau < 20$ | $15 < \tau < 45$ | $35 < \tau < 53$ | $58 < \tau < 120$ |
| Generator | $60 < \tau < 120$ | $200 < \tau < 600$ | $200 < \tau < 600$ | $300 < \tau < 500$ |
| Transformer | $20 < \tau < 40$ | $50 < \tau < 150$ | $80 < \tau < 300$ | $200 < \tau < 400$ |

- the installation of power transformers with a high short-circuit reactance in order to reduce the number of standard voltage levels in systems at the expense of larger tapping capability to reduce voltage drop;
- the tendency to use transmission lines with larger cross-sections and more conductor bundles in order to expand the transmission capability of existing lines. The transmission lines at UHV levels > 800 kV have DC time constants > 100 ms;
- the increasing use of reactive components for short-circuit limitation (e.g. series reactors), often used as a solution to postpone the investment in switchgear with higher short-circuit current breaking capacity.

The increased asymmetrical-current peak has led to the introduction of alternative time constants in addition to the conventional $\tau = 45$ ms in the IEC standard for circuit-breakers [6], depending on the rated voltage $U_r$:

- $\tau = 120$ ms for $U_r \leq 52$ kV, generally for transformer dominated networks;
- $\tau = 60$ ms for $(72.5 \leq U_r \leq 420)$ kV;
- $\tau = 75$ ms for $(550 \leq U_r \leq 800)$ kV; and
- $\tau = 120$ ms for $U_r > 800$ kV.

In the IEEE standard C37.09 [7], a preferred value of $\tau = 45$ ms is also recognized but any other value, suggested by the user of the equipment, is acceptable.

In the past, the DC time constant of 45 ms was assumed to be adequate in all cases. This is no longer true but almost all testing is carried out based on $\tau = 45$ ms. Various attempts have been made to deal with circuit-breaker stresses for higher time constants [8, 9].


## 2.2.3    Asymmetrical Current in Three-Phase Systems

In a three-phase situation, the asymmetrical-current peak has the same maximum value (of $2.55 \, I_{sc}$ at 50 Hz and $\tau = 45$ ms) as in single-phase when the making in all three phases occurs simultaneously. This maximum value can take place in one phase only (see the lower phase in Figure 1.3).

In the case of a delay in closing between the individual poles of a three-phase circuit-breaker, higher values of the asymmetrical-current peak are possible [10].

In systems with a non-effectively earthed neutral, a pole delay between closing can lead to values of the asymmetrical-current peak close to $F_s = 3$ at $\tau = 45$ ms. The theoretical maximum value is $F_s = 3.34$ in pure inductive circuits of any power frequency, or stated otherwise, at an infinite DC time constant.

The pole delay in non-effectively earthed systems is in fact the delay between the second pole-to-close and the third pole-to-close. At closing of the first pole, there is no power-frequency current in the unearthed situation. Figure 2.5 shows the effect of the pole delay at closing on the asymmetrical-current peak factor in a 50 Hz system.

Note that the IEC standard [4] allows a pole delay of up to a quarter of the power-frequency cycle in closing operations. This may imply that although a circuit-breaker has been tested with $F_s = 2.55$, its pole delay may give rise to a much higher value.

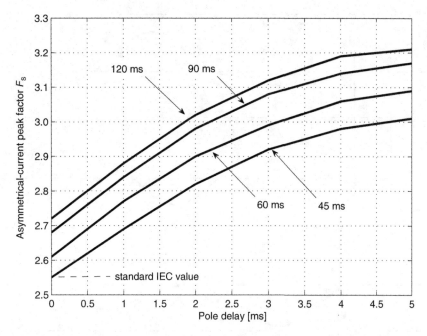

**Figure 2.5** Asymmetrical-current peak factor $F_s$ as a function of pole delay for a 50 Hz non-effectively earthed system.

The worst case mentioned above occurs when closing the second pole at a phase-to-phase voltage zero (between first- and second pole-to-close) and the third pole closing a quarter cycle later. The moment of closing of the first pole is irrelevant because no current will flow at closing. This situation is illustrated in Figure 2.6.

The actual value of $F_s$ depends on the earthing of the system.

Test laboratories make use of this in creating a "super-asymmetrical" current that can lead to *missing current zeros* in a certain span of time because the DC component can be made larger than the AC component. An example of how the combination of the maximum pole delay at making and high time constant (here $\tau = 133$ ms) can create missing zeros is shown in Figure 2.7. The probability of occurrence of missing zeros is higher at higher values of the DC time constant.

Missing or delayed current zero is a probable condition of generator-source fault current [11] (see Section 10.1.2) and a (remote) possibility with series compensated lines [12]. The problem with missing current zero is not in the first place the very high asymmetrical-current peak factor but the long arcing time because the circuit-breaker cannot interrupt until the current crosses zero.

## 2.3 Short-Circuit Current Impact on System and Components

The increase of short-circuit currents worldwide leads to a concern about the stress withstand capability of existing components [13]. In various expansion and upgrade projects of substations, full-scale laboratory tests were used to assess the impact of an increase in the short-circuit current beyond a given value, mostly an upgrade from 63 to 80 kA and in one

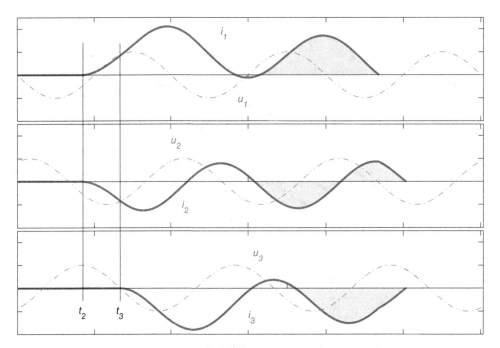

**Figure 2.6**    Three-phase short-circuit currents and maximum asymmetry. The times $t_2$, $t_3$ indicate the closing of the second and third pole. The first pole is closed before $t_2$.

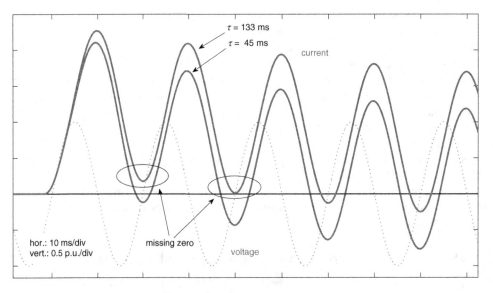

**Figure 2.7**    Missing current zeros.

case even to 100 kA. The demand for extremely high short-circuit current tests is notable in the last few years.

Very high levels of short-circuit currents result in severe stresses to various system components because of the electrodynamic forces and exposure to extreme thermal conditions of fault arcs.

**Switchgear:** Several EHV 80 kA circuit-breakers are on the market and tests up to 90 kA have been performed. A well-known way to increase the thermal interruption capability is to add capacitors across the breaking gap other than the grading capacitors, or at the line-side to earth of the circuit-breaker (see Section 3.6.1.6). The aim of this is to slow down the extremely rapid rise of the TRV that is associated with *short-line faults* (SLF) [14]. For SLFs, the initial rate-of-rise of TRV is proportional to the current, while terminal fault TRVs are not dependent on the fault current. If (sometimes large) capacitors are placed directly across the circuit-breaker gap, the most efficient solution, disconnectors have to interrupt the residual capacitive current.

In various situations, circuit-breaker contacts must remain closed under high-current conduction. Depending on the contact geometry, significant contact forces may be necessary to prevent contacts from "blow-off" during the current passage. This applies especially to butt contacts, as applied in vacuum interrupters. Contact forces over 10 kN are necessary for a peak current of 200 kA [15]. Vacuum interrupters have been reported capable of carrying 10 kA continuously and interrupting 100 kA [16].

The highest interruption currents are associated with generator circuit-breakers (see Section 10.1), which have been to be able to interrupt currents well above 200 kA [17].

Fault arcs inside switchgear are normally called *internal arcs*. Internal arcs lead to a rapid pressure rise, because 50 to 80% of the arc energy is converted into heat within the enclosure in which the arc burns [18].

In MV switchgear panels, the pressure relief is provided through ducts or burst discs that open very quickly after the occurrence of internal arcs. Internal arc tests [19,20] are designed to verify that the effects of internal arcs, mainly the exhaust of hot gases, do not lead directly to injury of personnel in the vicinity of the switchgear panel. The hot exhaust products of the fault arc must be guided in a controlled manner to the environment.

A typical test is shown in Figure 2.8. In such tests, an internal arc is initiated inside the switchgear panel by a thin wire [21]. Cotton cloth indicators, mimicking personnel clothing, facing the accessible parts are not permitted to be ignited by internal arc products.

In GIS, the main concern regarding the internal fault arcs is the burning of the arc through the enclosure. Due to the exothermic reaction of $SF_6$ (by-products) with aluminium, very high

**Figure 2.8** Internal-arc test.

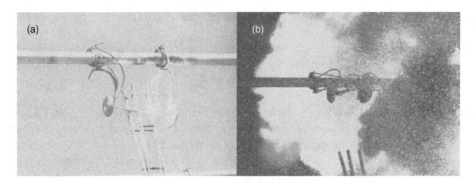

**Figure 2.9**  Forces on disconnectors due to the short-circuit current: (a) insignificant arcing at rated short-circuit current; (b) unintended contact separation, intense arcing at higher current asymmetry (other design than in (a)).

energy can be released at the arc foot-point, which may endanger locally the integrity of the GIS enclosure [21]. Relevant test procedures are prescribed in the standards [22].

**Disconnectors:** Disconnector contacts must remain closed during the conduction of the maximum short-circuit current. This is verified in a peak-current withstand test with a full asymmetrical-current peak, a test duration of typically 0.2 s, and short-time current test with a standard duration of 1 s, as specified in IEC 62271-102 [23].

Due to the very high electro-dynamic forces, contacts naturally tend to repel – unless a special design is applied using the inherent forces to achieve even more tight contact. Especially for pantograph disconnectors, the design must prevent any permanent deformation owing to the forces of contraction. Since the electrodynamic force is proportional to the square of the instantaneous current (in a single-phase case), the correct representation of the asymmetrical peak is essential in the peak-current tests.

Arcing must be avoided as this increases the contact resistance and may lead to welding of the contacts. Thermal withstand is verified in the short-time current test. Figure 2.9 shows a disconnector contact system under the passage of high short-circuit current. Note that most of the current flows through the left part of the contacts due to the additional stray inductance of the right-side current path.

Relevant tests may be performed on single-phase disconnectors, but the electrical environment should be three-phase, in order to create electrodynamic conditions equivalent to those in a substation.

**Arresters:** When surge arresters fail, the resultant fault arc must not lead to explosion of the arrester housing and body. Pressure relief tests have been defined in which a fault is initiated inside the arrester housing, either by a thin wire or by overvoltage. The design must ensure that the fault arc is expelled outwards in the very initial phase of the fault in order to avoid explosion of the body [24]. This process is visualized in Figure 2.10, showing the initial arc development during a peak current of 80 kA in a pressure relief test on a polymer surge arrester.

Even more severe stresses arise in case of a series capacitor-bank discharge current superimposed upon the power-frequency fault current. Arrester tests have been performed at the 245 kV level with 31.5 kA power-frequency fault current, superimposed by the peak discharge current of 447 kA at 3 kHz from a series compensation bank (refer to Section 10.5.1) [25].

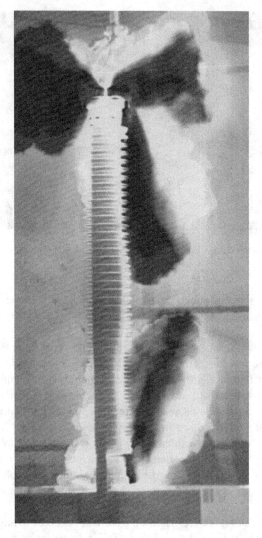

**Figure 2.10**   Pressure relief test on a polymer surge arrester.

**Transformers and reactors:** The fault current passing through a transformer is determined by the transformer's short-circuit impedance and is thus much lower than the examples given above. At the secondary terminal, however, large currents can result. Nevertheless, the effect of fault current on transformer windings is very severe because radial as well as axial electrodynamic forces act on the windings [26, 27]. From short-circuit test results on more than 200 large transformers of 25 MVA and above during 16 years it was found that approximately 25% fail to pass the short-circuit withstand test [28]. A frequently observed failure mode is the displacement of turns/windings, most commonly buckling of the winding due to radial forces. In addition, broken bushings and spraying of oil is observed regularly upon application of short-circuit current.

The test procedure is described in IEC 60075-5 [29].

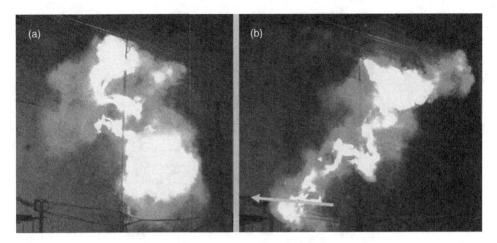

**Figure 2.11**   Power-arc test on a long insulator string.

Line traps, series reactors and shunt reactors also experience extreme forces during short-time current passage due to their circular and compact nature.

**Line insulators:** Insulator strings should not be damaged by the fault arcs burning between their arcing horns. These fault arcs are sometimes a result of back-flashover.

Power-arc tests verify that insulator strings remain structurally intact under thermal stresses of the nearby arc. The mechanical stresses due to the weight of the conductors and the electrodynamic stresses, dependent on an adequate choice of the fault-current path, have to be simulated in tests. Figure 2.11 gives an impression of effects observed in tests with a 50 kA power arc across a very long string insulator. Power-arc tests with 100 kA [30] have shown a significant increase in broken insulator sheds with increasing current. Proper design of arcing horn contacts will keep the erosion caused by arcing within acceptable limits. It has been demonstrated that the material losses due to arc jets increase in proportion to the integrated current.

In a number of cases (see Figure 2.11 right), the arc was observed to migrate from an arcing horn and to commutate to the bundle conductor. This resulted in damage to the conductor. Movement of a free burning fault arc away from its power supplying source is a natural phenomenon, well known from internal arcs in GIS [31].

It arises from the *Lorentz force* described by vector $\boldsymbol{F}$ that acts upon a given current-carrying filament where the current $I$ is flowing in the direction of vector $\boldsymbol{l}$ along the filament, and the *magnetic induction* described by vector $\boldsymbol{B}$, generated by the current loop. If the conductor is straight, the equation for the Lorentz force simplifies to:

$$\boldsymbol{F} = I(\boldsymbol{l} \times \boldsymbol{B}) \tag{2.6}$$

The motion is illustrated in the test example shown in Figure 2.12. The arc is initiated by a power source exploding a thin wire between the conductors at the left side and is travelling to the right. This behaviour is observed in the figure, visualizing a general phenomenon that the

**Figure 2.12**   Travelling arc between parallel conductors.

fault-current loop tries to increase its surface. Therefore, this causes any arc always to travel in the direction of power flow, away from the source.

Note also that the forces on the fault-current carrying conductors repel these because the currents are in opposite directions. Such repulsion does not appear in the conductor sections not yet touched by the arc because there is not yet any current in this section.

If currents through the neighbouring conductors have the same direction, these conductors tend to contract. Note the contraction of the four-bundle conductor supplying the current in the test depicted in Figure 2.13.

**Overhead line components:** Conductors move violently under short-circuit current passage. Conductor bundles tend to contract under the action of fault currents. This causes significant mechanical stresses to bundle spacers. In the case of inadequate design, or when upgrading the rated fault current, there is a risk of loss of spacers, as can be seen in Figure 2.13. In addition, due to the collision of the conductors with each other, the parts facing the centre of, for example, quadruple conductors are deformed. It was observed that significant deformation of aluminium conductor steel reinforced (ACSR) conductors resulted from 100 kA tests, a fault-current peak of 270 kA, duration 0.34 s on a 190 m long line sample [30].

Because of the touching of the conductors, fusion, and subsequent commutation of unbalanced current, arcing can occur between the conductors. This is shown in Figure 2.14 for a

**Figure 2.13**   Spacer removal due to the contraction of a four-conductor bundle.

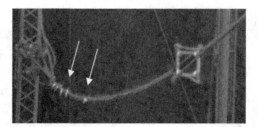

**Figure 2.14**   Contraction of a four-conductor bundle with arcing.

peak current of 130 kA. Together with nicking and non-elastic deformation, this may lead to fatigue of the material and reduction of its tensile properties. Also, an increase in the sag of 2 m, in a 500 to 800 m span length, was observed after short-circuit tests with 100 kA [30].

Deformation of jumpers and bundle spacers has also to be considered. Jumpers (in a 500 kV line) can move violently and they have been observed to collide with an insulator string.

**Busbars and conductors:** Busbars and their supports can behave unexpectedly when short-circuit forces increase beyond the design values [32]. Tests have been performed on substation mock-ups, where the essential components (busbars, supports, disconnectors) have been subjected to a peak current of 80 kA. In one case, the busbars collided because of lack of mechanical support strength (see Figure 2.15).

Because of the very high d$i$/d$t$ at extreme fault currents, eddy currents may be induced in unexpected parts of the station. As an example, Figure 2.16 shows a generator busduct, the connecting element between a power-plant generator and a step-up transformer, under test with three-phase r.m.s. currents of (275 and 850) kA asymmetrical peak. All evidence of arcing present in this high-speed recording image originates from induced currents outside the test-object, mainly in and between segments of the metal laboratory floor.

Also flexible loops, intended to allow movement, can be short-circuited due to mechanical forces and/or induced voltage flashover. Both events lead to severe arcing. An example of this can be seen in Figure 2.17.

In tests with 100 kA current in overhead lines it was observed that line sections, lying 6 m below and in parallel to the energized lines on a partly metallic floor, became damaged beyond repair due to induced current arcing.

**Figure 2.15**   Substation-busbar collapse due to short-circuit forces (reproduced with permission of FGH Engineering & Test GmbH).

**Figure 2.16**   Arcing by induced currents in the metal laboratory floor during generator busduct testing.

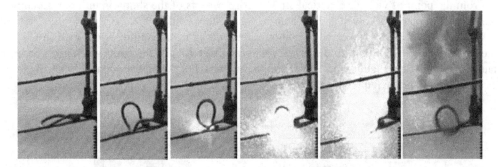

**Figure 2.17**   Video stills taken during a 80 kA test showing arcing and damage to a flexible loop connection.

## 2.4   Fault Statistics

### 2.4.1   *Occurrence and Nature of Short-Circuits*

The most reliable and extensive source of information on the electrical arcing stresses that circuit-breakers experience in service is the worldwide survey that WG 13.08 of CIGRE conducted in the late 1990s [33]. Information was collected on the number of faults in HV networks and especially on overhead lines (OH-lines). The whole population covered 900 000 circuit-breaker-years,[1] 70 000 overhead-line-years in the voltage class of 63 kV and above.

Thirteen countries in four continents were involved. Statistical information on the number of phases involved in faults and the auto-reclosure effectiveness of removing faults was collected.

Most short-circuits in a transmission network occur on the overhead-lines, that is, more than 90% of all faults. Table 2.2 gives an overview by voltage level.

As a very general rule of thumb, based on the statistical distributions of the number of faults per 100 km of an overhead line and the line lengths per voltage class, the average number of

---

[1] Circuit-breaker years: the number of circuit-breakers considered times the number of years in service.

**Table 2.2**   Number of short-circuit faults per 100 km × year of overhead lines [33].

| Number of faults | Voltage class [kV] | | | | | |
|---|---|---|---|---|---|---|
| | <100 | ≥100 to 200 | ≥200 to 300 | ≥300 to 500 | ≥500 to 700 | ≥700 |
| Average | 11.5 | 5.1 | 3.1 | 2.0 | 2.0 | 1.7 |
| 90 percentile | 17.3 | 8.3 | 4.8 | 3.3 | 4.2 | not available |

short-circuits is 1.7 per year (i.e. for all voltage classes). The 50 and 90% percentiles are 1.2 and 3.3 faults per year, respectively.

The reported number of faults per transformer is 3 to 4 per 100 transformer-years. The number of faults per substation busbar(s) is 2 to 3 per 100 substation-years. The number of faults per cable circuit is also very low in comparison with the number of faults on overhead lines.

From the point of view of electrical stresses, two aspects of the short-circuit are relevant: the number of phases involved and the auto-reclosure duty used.

The number of phases involved in a short-circuit on an overhead line is depicted in Figure 2.18. It can be seen that for the lower voltages, about 70% of the faults are single-phase faults, 20% two-phase faults, and 10% three-phase faults. For the higher voltages, 90% of the cases are single-phase faults.

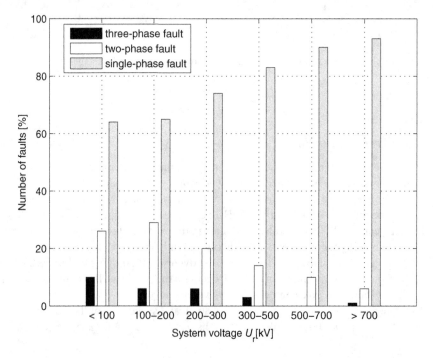

**Figure 2.18**   Number of phases involved in overhead-line short-circuit faults.

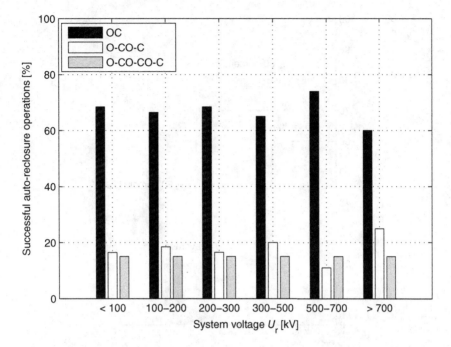

**Figure 2.19**   Percentage of successful auto-reclosure operations.

The percentage of successful auto-reclosure operations is high, as can be seen in Figure 2.19. Approximately 80% of the faults disappear within one OC (open—close) operation. Another 5% are cleared within an O-CO-C operation, and about 15% of all faults are permanent.

## 2.4.2   Magnitude of Short-Circuit Current

Three short-circuit current parameters are important in this respect:

$I_{sc}$  the rated short-circuit breaking current $I_{sc}$ of the circuit-breaker – that is, the maximum current that the circuit-breaker has to interrupt in a system of the circuit-breaker rated voltage, as demonstrated by type tests;

$I_{tf}$  the prospective (expected) maximum possible short-circuit current $I_{tf}$ at the specific location of the circuit-breaker, that is, the current mostly obtained by short-circuit calculation or analysis, that can arise at the worst-case condition for a terminal fault at the specific location;

$I_{act}$  the actual short-circuit current. This is not a single value, but it depends on the nature and location of the fault, the impedance of the overhead line, impedance of the substation, and so on.

In normal practice, $I_{act} \leq I_{tf} \leq I_{sc}$.

It can be concluded from the international survey [19] that the average value of all reported maximum possible short-circuit currents in a terminal-fault situation, expected from

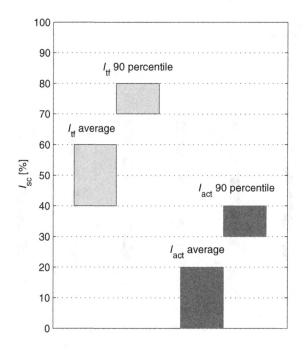

**Figure 2.20**   Average values and 90-percentiles of expected and actual short-circuit current magnitude.

short-circuit current calculations, is 40 to 60% of the rated short-circuit breaking current of the circuit-breaker. The 90-percentile is 70 to 80% of the rated short-circuit breaking current of the circuit-breaker. From examples, it was concluded that the average actual short-circuit current is around 20% of the rated short-circuit breaking current of the circuit-breaker. The 90-percentile is 30 to 40% of the rated short-circuit current of the circuit-breaker. These values are visualized in Figure 2.20.

# References

[1]   CIGRE Working Group C4.407 (2013) Lightning parameters for engineering applications, CIGRE Technical Brochure, 549.
[2]   van der Sluis, L. (2001) *Transients in Power Systems*, John Wiley and Sons, Ltd, ISBN 0-471-48639-6.
[3]   Slamecka, E. and Waterscheck, W. (1972) *Schaltvorgänge in Hoch- und Niederspannungsnetzen*, Siemens Aktiengesellschaft Erlangen, ISBN 3 8009 1106 X.
[4]   IEC 62271-100 (2012) High-voltage switchgear and controlgear – Part 100: Alternating-current circuit-breakers, Ed. 2.1.
[5]   CIGRE Working Group 13.04 (1997) Specified time constants for testing asymmetric current capability of switchgear. *Electra*, **173**, 19–31.
[6]   van den Heuvel, W.M.C., Janssen, A.L.J. and Damstra, G.C. (1989) Interruption of short-circuit currents in MV networks with extremely long time constant. *IEE Proc.-C*, **136** (2), 115–119.
[7]   IEEE C37.09 (1999) IEEE Standard Test Procedure for AC High-Voltage Circuit-Breakers Rated on a Symmetrical Current Basis.
[8]   Shimato, T., Chiyajo, K., Nakanishi, K. *et al.* (2002) Evaluation of interruption capability of gas circuit-breakers on large time constants of DC component of fault currrent. CIGRE Conference, Paper 13–304.

[9]  Fairey, T. and Waldron, M. (2005) Short-circuit currents with high DC time constants: Calculation methodology and impact on switchgear specification and rating. CIGRE SC A3 & B3 Joint Colloquium, Paper 104, Tokyo.

[10] Kersten, W.F.J. and van den Heuvel, W.M.C. (1991) Worst case studies of short-circuit making-currents. *IEE Proc.-C*, **138**(2), 129–134.

[11] Kulicke, B. and Schramm, H.-H. (1980) Clearance of short-circuits with delayed current zeros in the Itaipu 500 kV-substation. *IEEE Trans. Power Ap. Syst.*, **PAS-99**(4), 1406–1412.

[12] Bui-Van, Q., Khodabakhchian, B., Landry, M. *et al.* (1997) Performance of series-compensated line circuit-breakers under delayed current zero conditions. *IEEE Trans. Power Deliver.*, **12**(1), 227–233.

[13] Janssen, A.L.J., van Riet, M. and Smeets, R.P.P. (2012) Prospective Single and Multi-Phase Short-Circuit Current Levels in the Dutch Transmission, Sub-Transmission and Distribution Grids. CIGRE Conference, paper A3-103.

[14] Urai, H., Ooshita, Y., Koizumi, K. *et al.* (2008) Estimation of 80 kA Short-Line Fault Interrupting Capability in an SF6 Circuit-Breaker Based on Arc Model Calculations. Int. Conf. on Gas Disch. and their Appl.

[15] Barkan, P. (1985) A New Formulation of the Electromagnetic Repulsion Phenomenon in Electrical Contacts at very high Current. Proc. 11th Int. Conf. on Elec. Cont., pp. 185–188.

[16] Yanabu, S., Tsutsumi, T., Yokokura, K. and Kaneko, E. (1989) Recent developments in high-power vacuum circuit-breakers. *IEEE Trans. Plasma Sci.*, **17**(5), 717–723.

[17] Smeets, R.P.P., Barts, H.D. and Zehnder, L. (2006) Extreme Stresses of Generator Circuit-Breakers. CIGRE Conference, Paper A3-306.

[18] CIGRE Working Group A3.24 (2014) Tools for the Simulation of Pressure Rise due to Internal arcs in MV and HV Switchgear. Technical Brochure.

[19] IEC 62271-200 (2003) AC metal-enclosed switchgear and controlgear for rated voltages above 1 kV and up to and including 52 kV.

[20] IEEE Std. C37.20.7 (2001) IEEE Guide for Testing Medium Voltage Metal-Enclosed Switchgear for Internal Arcing Faults.

[21] Smeets, R.P.P., Hooijmans, J., Bannink, H. *et al.* (2008) Internal Arcing: Issues Related to Testing and Standardization. CIGRE Conference, Paper A3-207.

[22] IEC 62271-203 (2003) Gas-insulated metal-enclosed switchgear for rated voltages above 52 kV.

[23] IEC 62271-102 (2001) High-voltage switchgear and controlgear – Part 102: Alternating current disconnectors and earthing switches.

[24] Smeets, R.P.P., Barts, H., van der Linden, W., and Stenström, L. (2004) Modern ZnO surge arresters under short-circuit current stresses: test experiences and critical review of the IEC standard. CIGRE Conference, Paper A3–105.

[25] Dubé, J.-F., Goehler, R., Hanninen, T. *et al.* (2012) New achievements in pressure-relief tests for polymeric-housed varistors used on series compensated capacitor banks. IEEE PES General Meeting, pp. 1–8.

[26] CIGRE Working Group 12.19 (2002) The Short-Circuit Performance of Power Transformers. CIGRE Technical Brochure 209.

[27] Bertagnolli, G. (2006) *Short-circuit Duty of Power Transformers*, ABB, Milano, Italy, 3rd ed.

[28] Smeets, R.P.P. and te Paske, L.H. (2012) Sixteen Years of Test Experiences with Short-Circuit Withstand Capability of Large Power Transformers. Conf. of the Electr. Pow. Supply Ind. (CEPSI), paper 229, Bali.

[29] IEC Standard 60076-5 (2006) Power Transformers - Part 5, Ability to withstand short-circuit.

[30] Yamada, T., Saito, K., Ito, S. *et al.* (1998) Short-circuit Testing of Overhead Line Components up to 100 kA. 12th Conf. of the Elec. Pow. Supply Ind. (CEPSI), Paper 42–11, Pattaya.

[31] Boeck, W. and Krüger, K. (1992) Arc Motion and Burn-through in GIS. *IEEE Trans. Power Deliver.*, **7**(1), 254–261.

[32] CIGRE Working Group 23.11 (2002) The mechanical effects of short-circuit currents in open air substations. CIGRE Technical Brochure 105 (1996) and part II, CIGRE Technical Brochure 214.

[33] CIGRE Working Group 13.08 (2000) Life management of Circuit-Breakers. CIGRE Technical Brochure 165.

# 3

# Fault-Current Breaking and Making

## 3.1 Introduction

This chapter deals with breaking and making of current in faulted networks. Commonly, the term *fault-current breaking* or *fault-current interruption* is used for switching operations where faults are involved. The circuit-breaker is the only switching device to perform this duty. For switching in a faulted network, the term *making* is in common use. The making in faulted networks occurs, for example, when, after maintenance work, earth connections are erroneously left installed.

Fault-current making by a switching device (a circuit-breaker or an earthing switch) leads to similar system transients as occur after the initiation of a fault in a healthy system: the system cannot distinguish whether the circuit-breaker closes a system already faulted or whether an external effect, such as lightning-induced flashover, caused a fault.

## 3.2 Fault-Current Interruption

In general, a fault-current breaking can be classified as:

- *Short-circuit fault interruption* (see items 1 and 2 below); or
- *Out-of-phase fault interruption* (see item 3).

*Short-circuit faults* are related to the current flowing in an erroneous path, mostly initiated by climatic, environmental, or mechanical interaction with power systems, such as lightning strokes, line galloping due to storms and wind, falling trees or branches, animals, fire, dredging accidents, and so on.

Different locations of faults call for the following fault characteristics:

1. Short-circuit current, single-, double-, or three-phase, directly at the circuit-breaker terminals. This type of fault is called a *terminal fault* (TF).

*Switching in Electrical Transmission and Distribution Systems*, First Edition.
René Smeets, Lou van der Sluis, Mirsad Kapetanović, David F. Peelo, and Anton Janssen.
© 2015 John Wiley & Sons, Ltd. Published 2015 by John Wiley & Sons, Ltd.

Transformer- and reactor-limited faults are also faults in substations. In both cases, the series element (transformer, reactor), though limiting the short-circuit current has considerable impact on the TRV.

2. Single-phase-to-earth fault on an overhead line (*line fault*). A special case of this is the *short-line fault* (SLF), a fault hundreds of metres up to a few kilometres from the circuit-breaker location.

*Out-of-phase faults* are caused by asynchronous operation of different parts of the power system:

3. Fault current, due to an out-of-phase condition of two networks, to be interrupted, for example, by a circuit-breaker connecting the two networks. The AC recovery voltages are present at both sides of the circuit-breaker. The difference in phase angle can be up to 180° in theory.

In the cases above, the current lags the voltage by almost 90°, causing the TRV to reach a maximum value. This is because the transient component of the TRV is superimposed on the peak value of the power-frequency component, as shown in Figure 1.7.

## 3.3 Terminal Faults

### 3.3.1 Introduction

Terminal faults are caused by short-circuits occurring directly at the terminals of a circuit-breaker. In practice, these faults are extremely rare compared with faults some distance away from the circuit-breaker on an overhead line or cable.

In this case, there is virtually no impedance between the circuit-breaker and the short-circuit point, and it represents the highest fault current the circuit-breaker may face. The *rated short-circuit breaking current* of the circuit-breaker, a name-plate rating, shall be specified to be higher than the terminal-fault current.

The terminal fault is the fault resulting in maximum permissible thermal and electrodynamic stresses of the interruption chamber due to the arc and the high current. Tests are designed to verify the interruption capability under full short-circuit conditions with a variety of arcing times and degrees of fault-current asymmetry.

The TRV is very simple in this case. The equivalent circuit can be simplified as shown in Figure 3.1. Herein, the load circuit is a direct short to earth, whereas the source-circuit impedance reduces to the short-circuit reactance and a (stray) capacitance.

Taking this approach, $u_{bn} = 0$ and the full transient- and power-frequency recovery voltage is supplied by the load side: $u_{ab} = u_{an}$. This is envisaged in Figure 3.2.

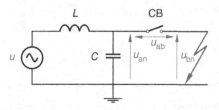

**Figure 3.1**  Equivalent circuit for a circuit-breaker terminal fault.

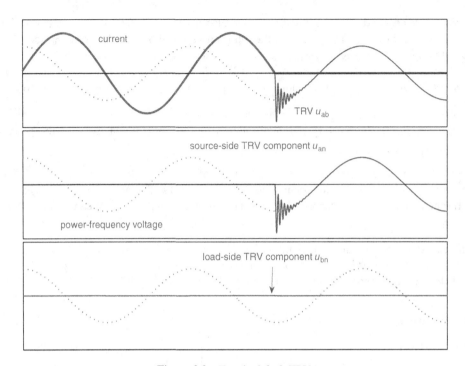

**Figure 3.2**   Terminal-fault TRV.

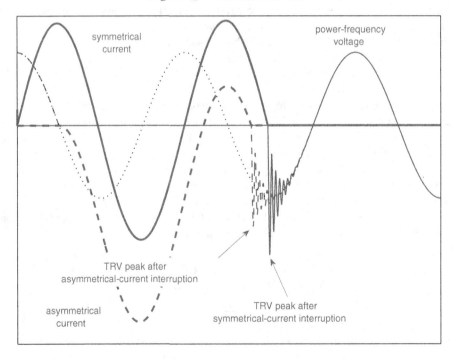

**Figure 3.3**   TRV at interruption of a symmetrical and asymmetrical current.

The maximum TRV occurs at 90° phase difference between the current and voltage. In inductive circuits this can only occur with symmetrical currents. In asymmetrical conditions current zero does not coincide with the voltage maximum, thus creating lower TRV peak values. An example of this is given in Figure 3.3, where the relation between the TRV peak value and the instantaneous voltage at the moment of interruption is demonstrated.

## 3.3.2  Three-Phase Current Interruption

In current interruption of a three-phase fault current, the earthing of the power system plays an essential role, both in the arc duration and in the TRV.

### 3.3.2.1  Three-Phase Interruption in Non-Earthed Systems

A very simplified but essentially adequate example of a three-phase system is shown schematically in Figure 3.4. In the case where there is no neutral connection (the reactance $X_n$ is infinite), the power system is described in the standards as a *non-effectively-earthed system*. In such a system, the floating potential of the neutral N can take a value – notably, after the first pole clears – and its voltage contributes to the load-side AC component of the TRV.

One of the circuit-breaker poles will be the first to encounter a current zero and if the interruption is successful, this pole is called *first-pole-to-clear*.

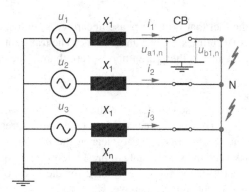

**Figure 3.4**  Equivalent circuit of a three-phase fault interruption.

After interruption of this pole, the following changes occur in the system, having repercussions on the overall interruption:

- The current in the first-pole-to-clear is interrupted ($i_1 = 0$), causing the current in the two other phases to become equal in magnitude and opposite in polarity. The circuit has changed from a three-phase to a two-phase system. The two-phase currents $i_2$ and $i_3$ can be calculated easily as:

$$i_2 = \frac{u_2 - u_3}{2X_1} \qquad\qquad i_3 = -i_2 \qquad\qquad (3.1)$$

where $X_1$ is the short-circuit reactance of each of the phases, and $u_2$, $u_3$ are the phase-to-earth voltages of the source. From this it follows that currents in the second- and third-pole-to-clear are reduced by a factor of $0.5\sqrt{3} = 0.87$ with respect to the first-pole-to-clear current. Furthermore, current zeros of $i_2$ and $i_3$ coincide, implying that current zero is advanced in one phase ($i_2$) and delayed in the other ($i_3$) by 30 electric degrees in pure inductive circuits.

- The source voltages in the phases are balanced and equal to zero, but a net voltage-to-earth will appear at the fault-side neutral. The voltage across the circuit-breaker can be calculated in the following way:

$$u_2 + i_2 X_1 = u_{b1,n}$$
$$u_3 - i_2 X_1 = u_{b1,n}$$
$$u_{b1} = 0.5(u_2 + u_3) \tag{3.2}$$
$$u_{a1} = u_1$$

The source is balanced and this gives $u_1 + u_2 + u_3 = 0$, or $u_2 + u_3 = -u_1$. By simple substitution, it follows:

$$u_{a1,b1} = 1.5\,u_1 \tag{3.3}$$

This result expresses the fact that in three-phase unearthed systems, the first-pole-to-clear has a recovery voltage of 1.5 times the per-unit voltage. The transient recovery voltage will oscillate around this level and is, therefore, more severe than in the case of a single-phase interruption.

The parameter that is commonly used to describe the influence of earthing on TRVs is the *first-pole-to-clear factor* $k_{pp}$. This is a dimensionless constant, expressing the ratio of the power-frequency voltage across the first-pole-to-clear at the instant of interruption, the so-called *recovery voltage* (RV) and the steady-state or pre-fault power-frequency voltage[1]. Thus, $k_{pp}$ is the multiplication factor of the recovery voltage for the first-pole-to-clear, compared with the single-phase case. For the situation above, $k_{pp} = 1.5$.

The second- and third-pole-to-clear interrupt at the same instant and will share a recovery voltage of $u\sqrt{3}$, equal to the phase-to-phase or line voltage imposed across the series combination of both circuit-breakers. Thus, assuming equal voltage distribution across each of the circuit-breakers, each pole recovers against a momentary recovery voltage of $0.5\,u\sqrt{3}$, having an amplitude of 0.87 p.u. at last-poles current zero in a pure inductive case. After interruption of the last two poles, the RV across all poles is restored to 1 p.u. The current-interruption process in this case is illustrated in Figure 3.5.

---

[1] Definition in IEC 62271-100, Ed. 2.0, clause 3.7.152 reads: *"first-pole-to-clear factor (in a three-phase system), when interrupting any symmetrical three-phase current, the first-pole-to-clear factor $k_{pp}$ is the ratio of the power-frequency voltage across the first interrupting pole before current interruption in the other poles, to the power-frequency voltage occurring across the pole or the poles after interruption in all three poles".*

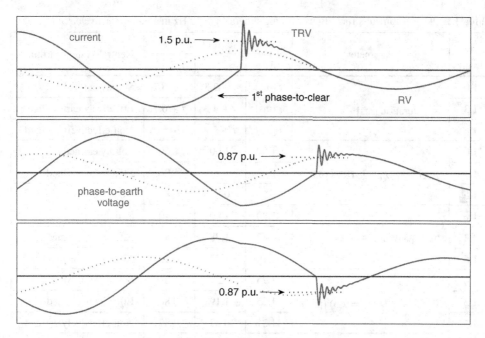

current          1.5 p.u. ──▶          TRV

◀── 1$^{st}$ phase-to-clear          RV

0.87 p.u. ──▶

phase-to-earth
voltage

0.87 p.u. ──▶

**Figure 3.5**   Three-phase fault interruption.

Thus, the TRV related to the first-pole-to-clear is more severe than in a single-phase case, namely by a factor of 1.5 regarding the RRRV and its peak value, whereas the last poles-to-clear have a reduction equal to $0.5\sqrt{3} = 0.87$ p.u. regarding the TRV severity.

### 3.3.2.2   Three-Phase Current Interruption in other than Non-Earthed Systems

In reality, non-effectively-earthed systems cover only a portion of the complete range of power systems. Generally, only MV systems are non-effectively earthed. The IEC 62271-100 standard assumes that systems with rated voltages below 100 kV have non-effectively-earthed neutral. Systems having rated voltage of 245 kV and above are generally operated with effectively-earthed neutral, implying that the neutral is earthed through an impedance causing the system to have *zero-sequence impedance*. In the intermediate range, covering rated voltages of 100, 123, 145, and 170 kV, both non-effectively-earthed neutral and effectively-earthed neutral systems are standardized.

In the present section, a general approach is chosen for the derivation of interruption characteristics towards system earthing. In the IEC standard, however, only three values have been standardized for $k_{pp}$:

- for non-effectively-earthed neutral systems, it has been shown in Section. 1.3.2.1 that $k_{pp} = 1.5$;
- for effectively-earthed neutral systems, the first-pole-to-clear factor is standardized as $k_{pp} = 1.3$;
- for the ultra-high-voltage (UHV) ratings of 1100 and 1200 kV, $k_{pp} = 1.2$.

**Table 3.1** TRV parameters for various earthing situations at 50 Hz in a pure inductive circuit.

| $k_{pp}$ [p.u.] | Parameter | Pole-to-clear | | | Neutral-system earthing |
|---|---|---|---|---|---|
| | | 1st | 2nd | 3rd | |
| 1.0 | Arc duration [ms] | $t$ [ms] | $t + 3.3$ | $t + 6.7$ | Solidly earthed |
| 1.3 | | $t$ [ms] | $t + 4.3$ | $t + 6.7$ | Effectively earthed |
| 1.5 | | $t$ [ms] | $t + 5.0$ | $t + 5.0$ | Non-effectively earthed |
| 1.0 | Recovery Voltage (RV) [p.u.] | 1.00 | 1.00 | 1.00 | Solidly earthed |
| 1.3 | | 1.30 | 1.27 | 1.00 | Effectively earthed |
| 1.5 | | 1.50 | 0.87 | 0.87 | Non-effectively earthed |
| 1.0 | TRV peak ($k_{af}^* = 1.4$) [p.u.] | 1.40 | 1.40 | 1.40 | Solidly earthed |
| 1.3 | | 1.82 | 1.78 | 1.40 | Effectively earthed |
| 1.5 | | 2.10 | 1.22 | 1.22 | Non-effectively earthed |
| 1.0 | RRRV ($Z_{0s}/Z_{1s}^{**} = 2.5$) [p.u.] | 1.25 | 1.33 | 1.5 | Solidly earthed |
| 1.3 | | 1.25 | 1.19 | 0.86 | Effectively earthed |
| 1.5 | | 1.25 | 0.87 | 0.87 | Non-effectively earthed |
| 1.0 | $di/dt$ at current zero [p.u.] | 1.00 | 1.00 | 1.00 | Solidly earthed |
| 1.3 | | 1.00 | 0.89 | 0.57 | Effectively earthed |
| 1.5 | | 1.00 | 0.87 | 0.87 | Non-effectively earthed |
| 1.0 | Peak current in last loop [p.u.] | 1.00 | 1.00 | 1.00 | Solidly earthed |
| 1.3 | | 1.00 | 1.00 | 0.89 | Effectively earthed |
| 1.5 | | 1.00 | 0.87 | 0.87 | Non-effectively earthed |
| 1.0 | Duration of last loop [ms] | 10.00 | 10.00 | 10.00 | Solidly earthed |
| 1.3 | | 10.00 | 10.95 | 9.98 | Effectively earthed |
| 1.5 | | 10.00 | 11.66 | 8.33 | Non-effectively earthed |

*$k_{af}$ is the overshoot factor of the oscillating TRV (see Chapter 13), called *amplitude factor* - in IEC 62271-100, 3.7.153, taking into account damping. In the theoretical undamped case, $k_{af} = 2$.
**$Z_{0s}/Z_{1s}$ is the ratio between zero- and positive-sequence surge impedance. IEC 62271-100, 4.102.3, chooses this factor as "approximately 2".
Refer to Table 3.2 for definitions of all 1 [p.u.] values.

The hypothetical case of a solidly earthed neutral system (with zero earthing impedance) implies $k_{pp} = 1.0$. In this case, there is no interaction between the phases, and the three-phase system can be regarded as three independent single-phase systems.

In Table 3.1, the relevant parameters during current interruption are listed for $k_{pp} = 1.0$, 1.3, and 1.5 p.u., as standardized by IEC [1] and IEEE [2].

In Figure 3.6, three cases of the neutral-system earthing are shown in a combined display. Note that these values apply to a pure inductive circuit and symmetrical current. Values will

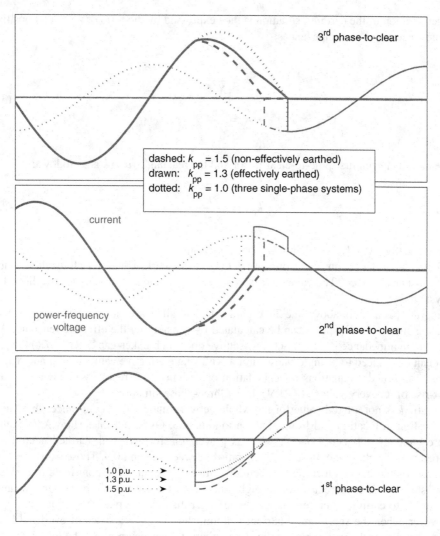

**Figure 3.6** Current interruption in systems with different earthing conditions characterized by the first-pole-to-clear factor $k_{pp}$ = 1.0, 1.3, and 1.5 p.u.

change in a practical circuit with a non-zero power factor and under asymmetrical fault-current conditions.

Regarding TRV behaviour, in general terms, three-phase systems are characterized by their positive-sequence reactance $X_1$ and surge impedance $Z_{1s}$. The reactances are given schematically by the physically present reactances ($X_n$ and $X_1$, see Figure 3.4), or more frequently by the zero-sequence reactances and surge impedances [3]:

$$X_0 = X_1 + 3X_n$$

$$Z_{0s} = Z_{1s} + 3Z_n$$

(3.4)

Characterization of the earthing situation is then expressed by the ratio of zero- and positive-sequence reactances and impedances:

$$k = \frac{X_0}{X_1}$$

$$\kappa = \frac{Z_{0s}}{Z_{1s}}$$

(3.5)

The relation between first-pole-to-clear factor and the reactance ratio $k$ is as follows:

$$k_{pp} = \frac{3k}{2k + 1}$$

(3.6)

yielding usually $1.0 \leq k_{pp} < 1.5$ in simplified situations.

Effectively-earthed neutral networks are defined as networks where for all locations under all circumstances $k < 3$.[2] The value of $\kappa$ is usually in the range $2 \leq \kappa \leq 2.5$ for overhead-line HV systems.

Using the parameters above, the TRV parameters of all phases after interruption of their respective circuit-breaker poles can be calculated by considering the effective reactance $X_{eff}$ and the surge impedance $Z_{eff}$ that each consecutive opening breaker-gap "sees". With this, the circuit can be reduced to a single-phase circuit with single effective values of $Z_{eff}$ and $X_{eff}$.

Table 3.2 provides equations for calculation of the current $I$, recovery voltage $U_{RV}$, and rate-of-rise of recovery voltage (RRRV) in any three-phase situation.

The ratio $k$ is not a fixed value for the whole network under all circumstances, but varies from location to location and from situation to situation. Overhead lines show $X_0/X_1$ ratios that are typically between 2 and 3, whereas power transformers with earthed neutral and equipped with a delta-winding show $X_0/X_1$ ratios between 0.5 and 0.9. Those overhead lines and transformers that conduct the short-circuit current, or a part of it, contribute to the ratio $k$. If most of the fault current flows through transformers, the ratio will be rather low and a single-phase-to-earth fault current may become larger than a three-phase fault current.

However, when the short-circuit current is dominated by the reactance of overhead lines, the ratio $k$ will be relatively high and the single-phase-to-earth current will be lower than the three-phase fault current.

In Figure 3.7 the dependence of interrupted fault current and recovery voltage on $k$ is visualized for each pole.

The ratio $k$ will change when certain power plants or overhead lines or transformers are out of service. The ratio $k$ may even be different per substation bay, as the fault-current contribution of that particular bay depends on the situation and exact location of the fault. Moreover, the ratio $k$ changes during the process of fault clearing because of successive tripping of the circuit-breakers. Some results of studies on actual and future single- and three-phase fault currents are reported [4].

---

[2] The value of $k_{pp} = 1.3$ is a rounded value. Since IEC defines $k$ to be smaller than 3, the maximum value that IEC considers in reality is $k_{pp} = 1.27$.

**Table 3.2** Interruption parameters (in [p.u.]) in three-phase systems for first, second, and third pole-to-clear.

| Pole-to-clear | Interruption parameter | | | | |
|---|---|---|---|---|---|
| | Reactance [p.u.] $X_{\text{eff}} = X_1$ | Surge impedance [p.u.] $Z_{\text{eff}} = Z_{1s}$ | Current [p.u.] $I = \dfrac{U}{X_1}$ | Recovery voltage (RV) [p.u.] $U_{\text{RV}} = I\,X_{\text{eff}}$ | Rate-of-rise of recovery voltage (RRRV) [p.u.] $RRRV = Z_{\text{eff}}\dfrac{\mathrm{d}i}{\mathrm{d}t} = Z_{\text{eff}}I\omega\sqrt{2} = Z_{1s}\dfrac{U}{X_1}\omega\sqrt{2}$ |
| | | | $1 \le k_{pp} < 1.5$ [p.u.] | | |
| 1st | $\dfrac{3k}{2k+1}$ | $\dfrac{3\kappa}{2\kappa+1}$ | 1 | $\dfrac{3k}{2k+1}$ | $\dfrac{3\kappa}{2\kappa+1}$ |
| 2nd | $\dfrac{2k+1}{k+2}$ | $\dfrac{2\kappa+1}{\kappa+2}$ | $\dfrac{\sqrt{3}\sqrt{k^2+k+1}}{2k+1}$ | $\dfrac{\sqrt{3}\sqrt{k^2+k+1}}{k+2}$ | $\dfrac{2\kappa+1}{\kappa+2}\dfrac{\sqrt{3}\sqrt{k^2+k+1}}{2k+1}$ |
| 3rd | $\dfrac{k+2}{3}$ | $\dfrac{\kappa+2}{3}$ | $\dfrac{3}{k+2}$ | 1 | $\dfrac{\kappa+2}{3}\dfrac{3}{k+2}$ |
| | | | $k_{pp} = 1.5$ [p.u.] (non-effectively-earthed neutral) | | |
| 1st | 1.5 | $\dfrac{3\kappa}{2\kappa+1}$ | 1 | 1.5 | $\dfrac{3\kappa}{2\kappa+1}$ |
| 2nd | 1 | 1 | $\dfrac{1}{2}\sqrt{3}$ | $\dfrac{1}{2}\sqrt{3}$ | $\dfrac{1}{2}\sqrt{3}$ |
| 3rd | 1 | 1 | $\dfrac{1}{2}\sqrt{3}$ | $\dfrac{1}{2}\sqrt{3}$ | $\dfrac{1}{2}\sqrt{3}$ |

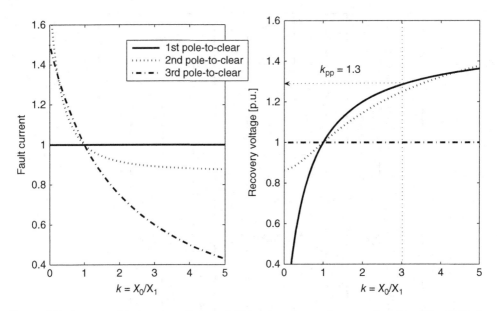

**Figure 3.7**   Interrupted fault current (1 p.u. = $U/X_1$) and recovery voltage as a function of $k = X_0/X_1$ for first, second and third pole-to-clear.

## 3.4   Transformer-Limited Faults

Transformer-limited faults (TLF) are short-circuit conditions predominantly determined by a transformer in a given circuit. In that case, the transformer impedance is the dominant factor for the value of the short-circuit current [5]. The transformer's $X/R$ ratio determines the peak factor of the asymmetrical short-circuit current and the $X_0/X_1$ ratio determines the magnitude of the single-phase earth-fault versus multi-phase fault currents. At fault-current interruption, the high-frequency behaviour of the transformer, characterized by stray capacitances and leakage inductances, is essential for the TRV waveshape.

Several conditions of TLF can be distinguished regarding a circuit-breaker at a certain voltage level. The possibilities are explained in Figure 3.8.

Depending on the fault location, two situations can arise:

- The fault is at the terminals or at the busbars of the transformer at the other voltage level than the circuit-breaker in question. Such a fault is called a *transformer-secondary Fault* (TSF).
- The fault is at the busbars on the same voltage level as the circuit-breaker. This fault is called a *transformer-fed fault* (TFF).

Since the transformer impedance is the dominant parameter determining the symmetrical short-circuit current, the voltage drop across the transformer in the case of a TLF is close to 100%, while the fault current will be in the range 10 to 30% of the rated short-circuit breaking current of the circuit-breaker. This short-circuit current is thus relatively low, and is in contrast to the terminal-fault situation, in which the circuit-breaker has to cope with a much larger

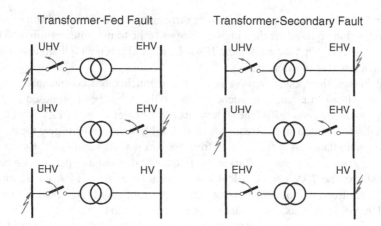

**Figure 3.8**    Transformer-secondary faults and transformer-fed faults.

fault current fed by multiple sources that are connected to the busbar. Therefore, in the case of TLFs, the short-circuit current is not a significant stress factor, although the *X/R* ratio is very high. On the other hand, the TRV, especially the RRRV, may become more severe than specified in the IEC standard for the test-duties covering the lower fault currents of 10% (T10) and 30% (T30) of the rated short-circuit breaking current. Verification of the capability of a circuit-breaker to interrupt TLFs is covered by the test-duty T30 for circuit-breakers applied at 52 kV and below on transformer secondary terminals (see IEC 62271-100, Annex M) and by T10 for other applications.

## 3.4.1    Transformer Modelling for TRV Calculation

The simplest representation of the TRV when clearing a transformer-limited fault is outlined in Figure 3.9, with $L_{tr}$ the leakage inductance of the transformer and $C_{tr}$ its stray capacitance.

The transformer and the connections between transformer and circuit-breaker are represented by a single-frequency circuit, characterized by a single capacitance and a single inductance. The circuit essentially consists of the transformer's short-circuit inductance, the transformer's dominating stray capacitance and the capacitance of the connected equipment [6]. The stray capacitance is neither well-defined nor documented in the frequency range of switching voltages. Transformer specialists are more interested in the faster transient phenomena that occur at lightning strokes or during switching operations of disconnectors in GIS. For modelling purposes, they will therefore accommodate measurements of fast and very fast

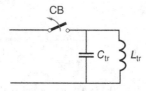

**Figure 3.9**    Simple equivalent circuit for a transformer-limited fault.

transients up to the megahertz range. Moreover, their main interest is in overvoltages between windings and within coils, rather than in overvoltages at the transformer terminals. The bandwidth relevant to TRV phenomena is some 100 kHz, at maximum, with dominant frequencies around ten to a few tens of kilohertz.

It is well known that power transformers show a much more complicated response to impulses than related to a single-frequency model only. It has been proposed [7, 8] to use *frequency response analysis* (FRA) measurements for the calculation of TRV waveforms that include deviations from the single-frequency model. FRA measurements give the transformer impedance or admittance as a function of frequency in a wide bandwidth. By multiplying the impedance $Z(\omega)$ by the current $I(\omega)$ in the frequency domain and performing an inverse Fourier transform, the TRV can be calculated in the time domain. FRA measurements are usually available from the manufacturer or utility, but care should be taken for proper and adequate FRA measurements, for instance representing the first-pole-to-clear. The effect on the frequency response of an unloaded transformer versus a short-circuited transformer, as registered by FRA measurements, can be seen in Figure 3.10.

The upper curve shows the FRA response (impedance) of an unloaded transformer, that is, with the secondary (LV) terminals open, while the lower curve gives the FRA response of a transformer with LV terminals short-circuited. The equivalent inductance in the open situation is 64 H, against 1.09 H in the shorted situation. In the upper frequency range, the curves coincide and an equivalent capacitance of 217 pF can be derived.

From the figure it is clear that the lowest frequency range of the FRA measurement shows the predominant characteristic of an inductance, providing that the increase of the impedance with

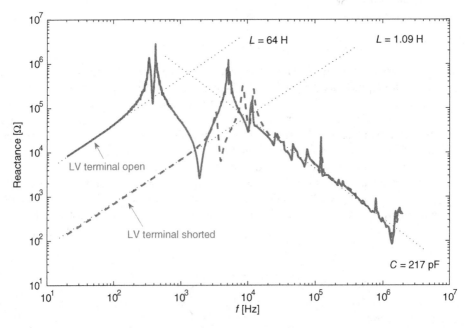

**Figure 3.10**   FRA pattern of a 420 kV / 80 MVA transformer with LV terminals open versus short-circuited.

frequency is linear. The upper straight line represents the (core) inductance of an unloaded transformer, whereas the lower line represents the leakage inductance associated with the shorted transformer. In the higher-frequency range, representation of a pure capacitance will give a characteristic that decreases linearly with the frequency, as can be recognized in both curves, but it is not as clearly illustrated as in the inductive part of the FRA measurement. The deviations from the straight line represent resonances of local $LC$ circuits.

The inductance and capacitance deduced in this way from the FRA measurement form the main elements for the single-frequency model. The damping, that may be represented by a resistance $R$ parallel to $L_{tr}$ and $C_{tr}$ in Figure 3.9 can be read off from the peak value of the FRA measurement around the resonance frequency due to the main inductance and the main capacitance: $R = 200$ k$\Omega$.

The characteristic impedance is:

$Z = \sqrt{L_{tr}/C_{tr}} = 40$ k$\Omega$. The ratio $R/Z = 5$ gives the *amplitude factor* for a single-frequency response:

$k_{af} = 1.73$, as can be learned from reference [9].

When comparing TRVs calculated from a single $LC$ model with TRVs calculated from FRA measurements, the TRV from the single $LC$ model turns out to be more severe, with both higher RRRV and higher amplitude factor. The main reason for the lower amplitude factor in the case of the multi-frequency model is the interference between multiple resonance frequencies around 10 kHz.

Above approximately 500 kHz the FRA measurement may be replaced by just a straight line representing a single stray capacitance. At the lower-frequency range of the FRA plot, the graph may be approximated by a straight line representing the leakage inductance.

More details can be found in references [10–12].

## 3.4.2 External Capacitances

The values of capacitances of transformers range from several hundreds of picofarad to 1 nF per phase for power transformers with power up to a few hundreds of MVA, and from several nF to some tens of nF for transformers with power exceeding 500 MVA.

The connections between the circuit-breaker and transformer may consist of cables or gas-insulated busbars, but mostly they are air insulated conductors. The distance between the circuit-breaker and the transformer may vary from some tens up to a hundred metres. The connected equipment, such as current transformers, surge arresters, disconnectors and earthing switches, is not numerous and has limited capacitance to earth. GIS substations tend to have larger capacitances because of the bushings and cable connections. The capacitance to earth of current and voltage transformers may be very small, of the order of 150 pF, but much higher values are also possible. Capacitive voltage transformers, when applied, show higher capacitances in the range of several nanofarads (nF). The minimum values of the external capacitances range from a few hundred picofarads to some nanofarads with the higher values for the higher rated voltages.

When an external $C_{ext}$ is incorporated in the FRA plot of $Z(\omega)$, this can be identified by a shift of the high-frequency part to a lower frequency. The resonance frequencies will be reduced by a factor $\sqrt{C_r/(C_r + C_{ext})}$ with $C_r$ being the dominant capacitance at a specific resonance frequency. The resonance peak value of $Z(\omega)$ is determined by the equivalent resistance and will not change, but the ratio $R/Z$ will increase as $Z = \sqrt{L/(C_r + C_{ext})}$ is smaller than $\sqrt{L/C_r}$. A smaller $R/Z$ in the parallel damped circuit means a higher quality factor and thus higher amplitude.

If the external capacitance is large enough, it will force the system to a more or less single-frequency response, dominated by the transformer short-circuit inductance and the transformer capacitance increased by the external capacitance [13]. This results in a rather simple TRV waveshape having a single frequency. It may have higher amplitude, unlike the case of the superimposed oscillations of various frequencies in the multi-frequency case.

For voltages between 100 and 800 kV, a minimum additional capacitance of 0.5 to 1 nF may be assumed. In many cases, distances will be longer and the capacitances of the HV equipment involved will be larger. A more representative value is of the order of several nanofarads. Such capacitances have a noticeable effect on the TRV frequency.

## 3.5   Reactor-Limited Faults

Series reactors may be applied for load-flow management. For instance, they can limit fault currents when applied in neutral earthing (see Section 10.12.1), or transient currents, such as those met in motor starting or in arc-furnace control [14]. In general, reactors provide limitation of the current, hence the term *reactor-limited fault* (RLF). However, the application of the series current-limiting reactors naturally increases the RRRV [15].

The basic equivalent circuit for these situations is given in Figure 3.11.

As usual, the load-side component of the TRV is determined by the load-side network, in this case the oscillating circuit providing the natural frequency of the reactor $f_R$:

$$f_R = \frac{1}{2\pi}\sqrt{\frac{1}{L_L C_L}} \tag{3.7}$$

with $L_L$ the inductance and $C_L$ the stray capacitance of the series-reactor.

Due to the very small stray capacitance of certain current-limiting reactors, the natural frequency of transients produced by these reactors can be very high. A circuit-breaker installed immediately in series with such a reactor will face a high-frequency TRV when clearing a terminal fault (reactor at the source side of the circuit-breaker) or clearing a fault behind the reactor (reactor at the load side of the circuit-breaker). The resulting TRV frequency generally exceeds by far the standardized TRV values regarding RRRV.

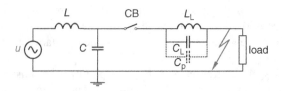

**Figure 3.11**   Basic equivalent circuit for reactor-limited fault interruption.

In these cases, it is necessary to take mitigation measures, such as the application of capacitors in parallel to the reactors or connected to earth. In medium voltage systems, the available mitigation measures are very effective and cost efficient [16].

The general effect of additional parallel capacitance $C_p$ on the natural frequency, and thus on the TRV load-side component, is a reduction of its rate-of-rise because the frequency of the load is reduced in the simple approach from $f_R$ to $f_{Rp}$:

$$f_{Rp} = \frac{1}{2\pi}\sqrt{\frac{1}{L_L(C_L + C_p)}} \tag{3.8}$$

Application of parallel capacitance is sometimes used to reduce the RRRV, for example, if the RRRV in service exceeds the tested values of the TRV rate-of-rise [17].

An example is shown in Figure 3.12, where a capacitor bank ($C_p$ = 50 nF) in parallel with the reactor ($C_L$ = 0.2 nF) can reduce the TRV in a 380 kV system to an acceptable limit, that is, below the standardized IEC TRV envelope T30 (see Section 13.1.2), at 30% of the rated short-circuit breaking current of the circuit-breaker rated at 420 kV. The reactor is designed to limit the fault current to 0.3 $I_{sc}$.

Note that though the RRRV is reduced significantly, application of capacitance leads to an increase in the TRV peak due to the much lower damping at the lower oscillation frequency.

It is strongly recommended to apply such measures, unless it can be demonstrated by tests, or by application of a definite-purpose circuit-breaker that the circuit-breaker can successfully clear faults with the actually occuring high-frequency TRV [18].

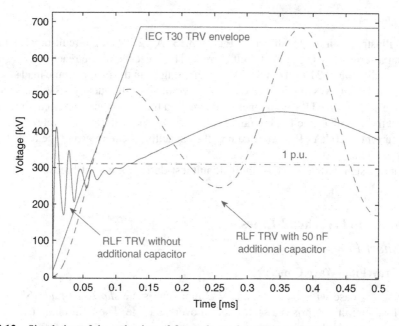

**Figure 3.12**  Simulation of the reduction of first pole-to-clear TRV oscillation frequency of a 420 kV circuit-breaker by 50 nF of capacitance parallel to a current limiting reactor when interrupting the limited fault current of 0.3 $I_{sc}$.

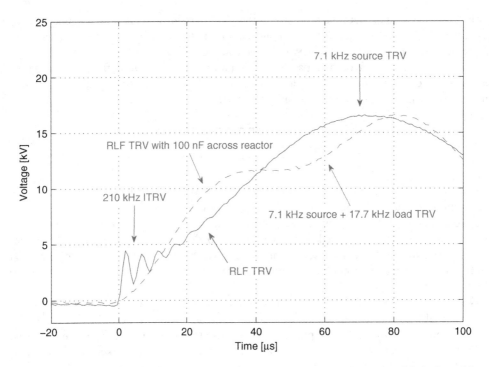

**Figure 3.13** Measured initial TRV from series reactor (drawn trace) and TRV modified after adding 100 nF across the reactor (dashed trace).

This is illustrated in a test, shown in Figure 3.13. A 12 kV SF$_6$ circuit-breaker is tested to interrupt a series- reactor-limited fault current. The presence of the reactor causes a high-frequency oscillation of 210 kHz in the TRV originating from the reactor at the load side. After a capacitor of 100 nF was applied across the reactor, the high-frequency TRV was eliminated and only the source-side TRV component remained. In this case, the breaker can interrupt 8.1 kA without the 100 nF capacitor, but fails to interrupt 10.9 kA, although its rated short-circuit breaking current is 20 kA [19]. After adding the capacitor, it can interrupt the higher current without a problem.

There are no standardized reactor-limited fault test-duties for circuit-breakers.

## 3.6    Faults on Overhead Lines

### 3.6.1    Short-Line Faults

#### 3.6.1.1    Travelling-Wave Concept

The most severe case with a very fast TRV component is the *short-line fault* (SLF). In this case, the load circuit comprises a section of an overhead line. The fault on an overhead line takes place a short but already significant distance of a few kilometres away from the terminals of a circuit-breaker in a substation. Figure 3.14 shows the situation. Obviously, both the source

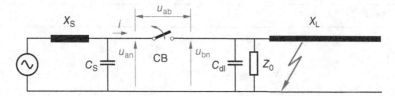

**Figure 3.14**   Short-line fault (SLF) equivalent single-phase circuit.

side (represented by the usual $LC$ circuit) and the load side (the short-circuited line section) will contribute to the TRV wave shape.

A section of a transmission line cannot be represented by a combination of lumped parameters $L$ and $C$ as in Figure 1.9 because the length of the line section is comparable to the wavelength of the voltage transients. In this frequency range it must be treated as a distributed combination of inductance per metre $L'$, capacitance per metre $C'$, and frequency-dependent resistance $R'$. With the distributed character of parameters, travelling waves will dominate the transient response of the line [20].

In order to calculate the line-side contribution to the TRV, the voltage drop across the line must be considered.

At current zero, assuming an inductive current, the source voltage is close to its peak value, and the initial voltage $u_{01}$ at the line-side terminal of the circuit-breaker is given as:

$$u_{bn}(0) = u_{01} = \hat{U}\frac{X_L}{X_S + X_L} = X_L \,\hat{I} = \omega L_L I\sqrt{2} = L_L \frac{di}{dt} \tag{3.9}$$

The voltage at the location of the fault is zero. The initial voltage profile across the faulted line section decreases linearly along the section from $u_{01}$ at the open circuit-breaker terminal to zero at the fault location, considering a single line-to-earth fault. The line is therefore left charged at the instant of current interruption. The distribution of the voltage across the line causes a movement of charges, appearing as a travelling voltage wave (see Figure 3.15).

The time variation of the voltage $u_{bn}(t)$ can be constructed considering two identical triangular voltage profiles – each one having the length of the faulty line and amplitude equal to $u_{01}/2$ at the line-side terminals of the circuit-breaker, and zero at the location of the fault [21]. These two fictitious profiles overlap each other at the instant of current interruption. From that moment on, both profiles shift out from the initial position as travelling waves with the same propagation speed $v$, but in opposite directions.

These travelling waves will be reflected at each end of the faulted line section in the following way: voltage will be totally reflected and doubles in magnitude at the open end, the circuit-breaker side, and will reverse in sign at the short-circuited fault location. The resulting voltage at any moment and location along the line section is then the sum of the two instantaneous voltages.

This yields a triangular-shaped TRV component with an initial magnitude $u_{01}$, gradually decaying to zero.

The propagation speed $v$ of the travelling wave is [20]:

$$v = \sqrt{\frac{1}{L'C'}} \tag{3.10}$$

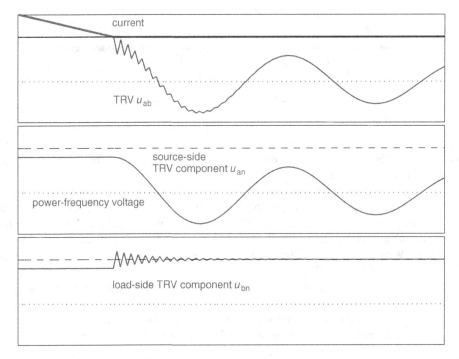

**Figure 3.15**  Short-line fault TRV and its components. The dashed line indicates voltage zero. The power-frequency voltage is almost constant on this time scale and appears as the straight dotted line $u(t)$.

The period $(T_L)$ of the triangular oscillation on the short-circuited line section with $l$ the length of line to the fault location is:

$$T_L = 4\frac{l}{v} \tag{3.11}$$

Combining Equations 3.9, 3.10, and 3.11, and using $L_L = l\,L'$, the initial rate-of-rise of voltage immediately after interruption $(i = +0)$ at the line side can be calculated:

$$\left.\frac{du}{dt}\right|_{i=+0} = \frac{u_{0l}}{\frac{1}{4}T_L} = \frac{\omega L_L I \sqrt{2}}{\frac{l}{v}} = \sqrt{\frac{L'}{C'}}\,\omega I \sqrt{2} = Z_0 \left.\frac{di}{dt}\right|_{i=-0} \tag{3.12}$$

hereby introducing the line *surge* (or *characteristic*) *impedance* $Z_0$, defined as:

$$Z_0 = \sqrt{\frac{L'}{C'}} \tag{3.13}$$

the value of which is standardized as $Z_0 = 450\ \Omega$ for overhead lines in the standards of IEC and IEEE.

In terms of the initial response to a fast transient, the surge impedance represents the "instantaneous" resistance. This approach is valid as long as the transient phenomenon has a shorter duration than twice the transit time of a wave travelling on the line from its beginning to the end.

Thus, the initial rate-of-rise of the line-side TRV component $(du/dt)_{i=+0}$ just after current zero is related to the rate-of-decline of the current $(di/dt)_{i=-0}$ just before current zero through the surge impedance:

$$\left.\frac{du}{dt}\right|_{i=+0} = Z_0 \left.\frac{di}{dt}\right|_{i=-0} \tag{3.14}$$

The first peak of the line-side TRV component will occur at a time $t_p$:

$$t_p = \frac{2l}{v} = \frac{2l}{\sqrt{\dfrac{1}{L'C'}}} = \frac{2l}{\dfrac{1}{L'}\sqrt{\dfrac{L'}{C'}}} = \frac{2lL'}{Z_0} = \frac{2L_L}{Z_0} \tag{3.15}$$

It will reach the first local maximum value $(\hat{U}_{\text{line}})$ of:

$$\hat{U}_{\text{line}} = u_{\text{bn}(t=t_p)} = k_{\text{af}}u_{0l} = k_{\text{af}}L_L \left.\frac{di}{dt}\right|_{i=-0} \tag{3.16}$$

with $k_{\text{af}}$ the amplitude factor, representing the damping of the line. This is standardized in IEC for overhead line oscillations as $k_{\text{af}} = 1.6$.

Commonly, the length of the faulted overhead line section from the circuit-breaker to the fault is not expressed in (kilo)metres, but rather in terms of the percentage reduction $\Delta I_\%$ of the terminal fault current $I_{\text{sc}}$, which in the standards is specified as equal to the rated short-circuit breaking current. If the current that actually flows in the faulted line is denoted by $I$, then:

$$\Delta I_\% = 100\,\frac{I}{I_{\text{sc}}} = 100\,\frac{L_S}{L_L + L_S} \tag{3.17}$$

Testing of the short-line fault interruption capability of circuit-breakers is standardized with $\Delta I_\% = 90\%$ and $\Delta I_\% = 75\%$, hence the name *short-line fault* or *kilometric fault*.

The actual length of "short" lines can be calculated as follows:

$$L_L = lL' = L_S\,\frac{100 - \Delta I_\%}{\Delta I_\%} \quad \rightarrow \quad l = \frac{L_S}{L'}\,\frac{100 - \Delta I_\%}{\Delta I_\%} = \frac{1}{L'}\,\frac{U_r}{\omega\sqrt{3}I_{\text{sc}}}\,\frac{100 - \Delta I_\%}{\Delta I_\%} \tag{3.18}$$

**Example** Taking $L' = 1.5$ mH km$^{-1}$, Equation 3.18 yields a line length of $l = 908$ m for a 50 Hz, 420 kV system with the rated short-circuit current 63 kA when the short-circuit current reduction is taken as $0.9 \times 63$ kA or $\Delta I_\% = 90$.

On the source side, the TRV component $u_{\text{an}}$ can be assumed to be a single- or multiple-frequency oscillation and the TRV across the circuit-breaker gap consists of a steep-rising triangular line-side component and a multi-sinusoidal source-side component. Due to the

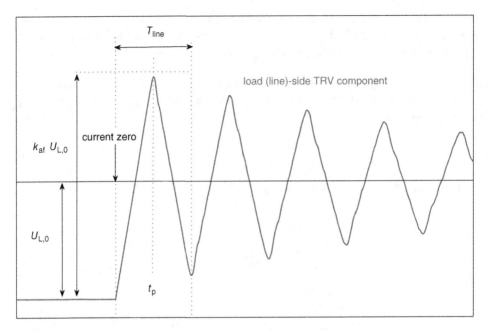

**Figure 3.16**  Time-extended initial TRV component of the line side.

summation of the triangular and multi-sinusoidal TRV component, the actual first TRV peak is somewhat higher than the peak line-side value $\hat{U}_{line}$ given by Equation 3.16 [22].

A typical example can be seen in Figure 3.16.

### 3.6.1.2   Initial TRV

Apart from overhead lines, all other conductors beyond a certain length and surge impedance also produce travelling waves as a response to a sudden change in voltage pattern. These also have an impact on the circuit-breaker's capability to interrupt.

Apart from sections of overhead line, another type of conductor creating these triangular wave-pattern TRVs is the busbar, normally connecting the source side of the circuit-breaker to other components of the substation. Busbars have surge impedance values of several hundreds of ohms. This implies that a triangular TRV wave component also exists on the source side. Because the length of busbar systems is much shorter than overhead lines, the frequency of this TRV component is extremely high (> 1 MHz), giving rise to an *initial transient recovery voltage* (called ITRV in the literature). The ITRV must be taken into account in testing, at least in cases where there is a direct connection between the interruption chamber and the busbar.

When the interruption chamber is placed inside an earthed enclosure, in dead-tank-type or GIS-type circuit-breakers, the stray capacitances of the bushing reduce the ITRV to insignificant proportions.

In other cases, notably for live-tank circuit-breakers, the ITRV must be considered.

Cables are another example of transmission lines that develop travelling waves. However, due to the small value of the surge impedance of several tens of ohms as compared with several hundreds of ohms in overhead lines and busbars, the rate-of-rise of the cable-side TRV component is of little concern and no test procedure is defined.

### 3.6.1.3 Time Delay

The influence of local stray capacitance(s) $C_{dl}$ at the line entrance (see Figure 3.14), such as capacitance of current and voltage transformers, bushings, support insulators, and so on, produces a slower rate-of-rise of the voltage during the first few microseconds of the TRV. This effect is called *time delay*, and is quantified by the delay time $t_d$. Thus, the presence of capacitance to earth between fast TRV-producing circuit elements (such as overhead lines) and circuit-breakers reduces the RRRV.

The value of the delay time in test circuits, as adopted in the standards, is extremely small. As an example, the delay time $t_d$ is defined as $t_d = 0.1$ μs for circuit-breakers with rated voltage up to and including 170 kV, and $t_d = 0.2$ μs above this rating. Using the simple relationship $t_d = C_{dl}Z_0$, it can be understood that a capacitance with a value in excess of 0.44 nF already causes an unacceptably large delay time in a test-circuit.

In Figure 3.17, the effect of the capacitance $C_{dl} = 1$ nF is shown. Graphically, the time delay is quantified by the intersection of the TRV tangent with the zero line.

Apart from the delaying effect, the addition of capacitance leads to an increase in the TRV peak value.

In the practice of testing, in order to minimize the time delay, the TRV components have to be brought either very close to the circuit-breaker under test or complicated compensating circuits must be applied.

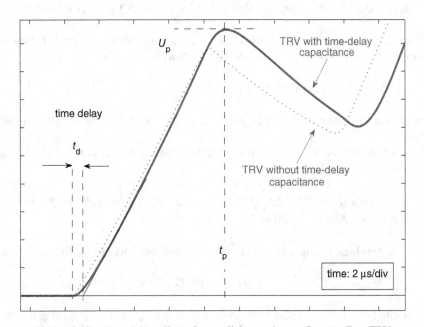

**Figure 3.17**  Time delay effect of a parallel capacitance $C_{dl} = 1$ nF on TRV.

It must be noted that the time delay is a fictitious quantity having a meaning for an ideal circuit-breaker without arc voltage. As such, the time delay is defined for test-circuits that have to produce an inherent sufficiently steep prospective TRV rise. This is demonstrated by injecting the test-circuit with a voltage source. This voltage source must first charge the time-delay capacitor, before TRV can start to rise. Procedures are described in IEC 62271-100, Annex F.

In reality, the TRV does not have to charge the time-delay capacitance since the arc voltage takes care of this already, long before current zero. During real interruptions, the TRV never starts with charging a capacitance, it immediately rises from the instant of current zero without any delay.

The actual effect of the time-delay capacitance is in the interaction between the arc and the circuit before current zero (see Section 1.6.1.5). Higher capacitance drains more current out of the arc immediately before current zero.

In substations more parasitic capacitance will usually be present than the minimum values that can be deduced from the time delay specified in the standards.

### 3.6.1.4   Impact of Short-Line Faults on Circuit-Breakers

The real challenge of a short-line-fault interruption is the very fast TRV that rises immediately after interruption. This implies the instantaneous appearance of a significant voltage across the circuit-breaker gap that is still recovering from severe arcing before current zero.

Regarding the example of Section 1.6.1.1, a circuit-breaker with rated voltage $U_r = 420$ kV and $I_{sc} = 63$ kA, the derivative of the current in a 90% short-line fault at 50 Hz is:

$$di/dt = (\Delta I_\%/100)2\pi f \sqrt{2} I_{sc} = (90/100)2\pi 50 \sqrt{2}\,63 \times 10^3 = 25.2 \times 10^6 \text{ A s}^{-1} = 25.2 \text{ A } \mu\text{s}^{-1}$$

Through the surge impedance $Z_0 = 450\ \Omega$, the value specified in the standards for short-line faults, the initial rate-of-rise of the line-side TRV, following Equation 3.12, results in:

$$du/dt = Z_0(di/dt) = 450(25.2 \times 10^6) = 11.3 \times 10^9 \text{ V s}^{-1} = 11.3 \text{ kV } \mu\text{s}^{-1}$$

This rate-of- rise will continue linearly until the first peak value of the line-side TRV component is reached.

With the line-to-line voltage $U_r = 420$ kV, the equivalent inductance of the line $L_L$ can be calculated from Equation 3.18:

$$L_L = [U_r/(\sqrt{3}\omega I_{sc})](100 - \Delta I_\%)/\Delta I_\% = [(420 \times 10^3)/(\sqrt{3} \times 2\pi 50 \times 63 \times 10^3)]$$
$$\times (100 - 90)/90 = 1.36 \text{ mH}$$

Therefore, the first local maximum value $\hat{U}_{line}$ of the line-side TRV is (see Equation 3.16):

$$\hat{U}_{line} = k_{af} L_L(di/dt) = 1.6(1.36 \times 10^{-3})(25.2 \times 10^6) = 54.8 \text{ kV}$$

This implies that 4 μs before current zero, the instantaneous value of the current in the arc was 100 A, whereas 4 μs after current zero, the voltage across the gap is already 45 kV. During the 8 μs in between the conditions in the gap have to change drastically from a very high to a very low conductivity.

Coping with these rapid changes is the main challenge for the circuit-breaker designers in order to guarantee the short-line-fault current-interruption capability. Recovery of the gap is required in an extremely short time.

The residual plasma, still present in the gap, has a potentially inhibiting effect on the recovery. Forced flow of $SF_6$ in an appropriate pattern and flow mode, well defined by the nozzle and pressures at both sides of it inside the interruption chamber, should ensure very fast removal of the arc residue.

Electrically, the recovery conditions in the gap can be characterized by the *post-arc current* $i_{pa}$. This current is a very small current driven by the TRV through the residual plasma after disappearance of the arc. The product of the instantaneous values of the post-arc current $i_{pa}(t)$ and the TRV across the contacts $u_{ab}(t)$ is the actual power input into the post-arc plasma. When more power is supplied than can be removed via cooling $P_{cool}$ by the external gas flow, ionization will dominate over de-ionization, causing the arc to *re-ignite*:

$$u_{ab}(t)\, i_{pa}(t) > P_{cool}(t) \tag{3.19}$$

This event is called *thermal re-ignition*. Figure 3.18 shows two examples of measurements around current zero: (a) successful interruption and (b) thermal re-ignition.

Measurement and evaluation of current-zero parameters, such as post-arc current and arc conductivity, can provide a substantial aid for an adequate design of circuit-breakers regarding the thermal interruption capability [23].

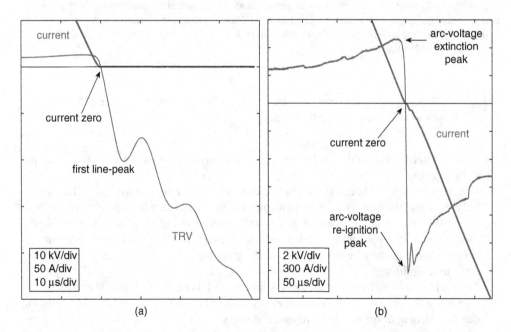

**Figure 3.18**  Current-zero measurements showing: (a) successful interruption; (b) thermal re-ignition.

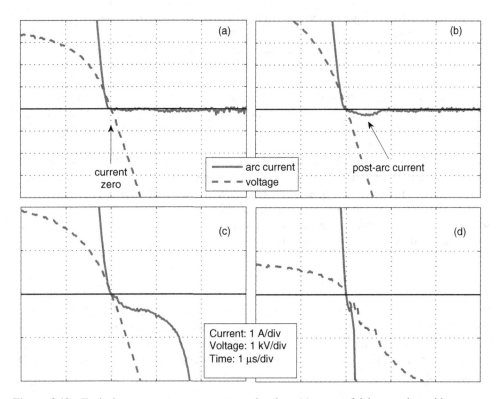

**Figure 3.19** Typical current-zero measurements showing: (a) successful interruption with post-arc current below the measurement threshold; (b) successful interruption with maximum post-arc current of 200 mA; (c) thermal re-ignition after approximately 1.5 μs (post-arc current develops into a failure); (d) immediate re-ignition (failure).

Figure 3.19 presents current-zero measurements in a HV $SF_6$ circuit-breaker [24]. It shows four typical examples of current-zero behaviour:

(a) Normal interruption, in the example the post-arc current is smaller than a few hundred milliamperes.
(b) Normal interruption, but post-arc current can now clearly be distinguished. This is an indication that interruption is close to the limit of the circuit-breaker's thermal capability.
(c) Thermal re-ignition as a result of a post-arc current too high for the circuit-breaker's thermal interruption capability. A tendency to interrupt can be identified clearly, but the energy supply in the post-zero period by the combination of post-arc current and TRV leads to re-ignition.
(d) Thermal re-ignition, far below the limit of the circuit-breaker's capability. This is often the case when the pressure of the cooling effect of the $SF_6$ gas flow is too low, for example, due to widening of the nozzle by repeated arcing stresses – a process known as *ablation*.

A very good indicator of the thermal re-ignition margin is the arc conductivity $g(t)$, given by the ratio of the arc current $i_a(t)$ and the arc voltage $u_{ab}(t)$ across the contacts:

$$g(t) = \frac{i_a(t)}{u_{ab}(t)}$$

Note that before current zero the physical variables are the arc current and arc voltage, while after current zero these are the post-arc current and the TRV, until re-ignition.

From numerous measurements a clear relationship has been established [25] between the probability of successful interruption and the value of the arc conductivity $g(t = -200 \text{ ns}) = G_{-200}$ at the moment of 200 ns before current zero. Arc conductivity traces taken from the measurements in Figure 3.19 are shown in Figure 3.20.

In this case, as in a large number of $SF_6$ circuit-breakers, interruptions are characterized by a value of $G_{-200}$ below a certain threshold value $G_{lim}$, whereas cases with thermal re-ignitions show $G_{-200}$ values above this threshold.

On a microscopic scale, thermal re-ignition is caused by the directed motion of charge carrying particles under the influence of the TRV. Charge carriers are present in the hot arc residue and they possess a significant amount of thermal energy. By collisions and ionization, a new, fairly conductive path can arise.

Immediately after current zero, the particle density is high, causing a limited mean free path between collisions. The energy that each particle can gain in the electric field from the TRV is therefore limited. Over time, the particle density decreases, the mean free path increases, and the thermal energy decreases due to expansion of the plasma and to the interaction with cold $SF_6$ gas.

The race between the increase in electric energy on the one hand, and the decrease in thermal energy, enhanced by the injection of cold gas in the gap, and the number of particles on the other hand, will ultimately determine the success or failure of interruption.

The process described above is relevant during the first few microseconds after current zero. This is why the thermal stresses associated with the very fast rising TRVs represent a very severe challenge to any gas circuit-breaker, even when the first peak of the TRV is no more than several tens of kilovolts.

The given description is relevant for current interruption in high-pressure gaseous media ($SF_6$, air, hydrogen released by oil, etc.). For vacuum circuit-breakers, dealing with very different recovery processes, much less is known about the stresses that very steep TRVs impose on the interrupter gap. Experience shows that very steep TRV stresses are generally more easily dealt with by vacuum circuit-breakers than by those based on $SF_6$.

The TRVs associated with interruptions of line faults are characterized by three parameters:

- RRRV – this parameter increases when a short-line fault occurs closer to the circuit-breaker (at shorter lines) because the current increases, following Equation 3.12.
- TRV frequency – the frequency of the triangular TRV pattern increases at shorter line-length, since the travelling time of the waves on the line becomes shorter, following Equation 3.11.
- Value of the first peak $\hat{U}_{line}$ – this parameter decreases when a fault occurs closer to the substation, since the initial voltage drop across the line is smaller, as follows from Equation 3.16.

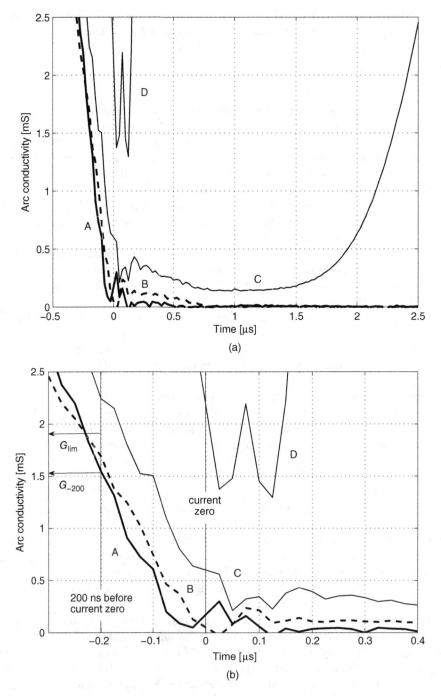

**Figure 3.20**   Arc conductivity $g(t)$ around current zero evaluated from the post-arc current measurements from Figure 3.19a–d.

From studies of the recovery of gas-filled circuit-breakers, it was found that the most onerous combination of current, RRRV, and $\hat{U}_{line}$ (under the physical processes in gases) occurs at short-line fault, with "short" being defined in the standards as $\Delta I_\% = 90$ and 70% [26]. Especially in the range $89 \leq \Delta I_\% \leq 93$, the line length appears critical and the maximum of cooling power is needed in order to interrupt the current under these conditions.

As discussed earlier, an excess of post-arc current is the prime cause of a thermal re-ignition. For $SF_6$ circuit-breakers, the post-arc current is in the range of a few tens to a few hundreds of milliamperes, whereas in vacuum interruption it can reach several amperes.

### 3.6.1.5 Interaction between Circuit-Breaker and Circuit

In the current zero region, there exists a strong interaction between the circuit-breaker arc and the circuit in which it is embedded. This interaction is of particular importance for the short-line-fault interruption, since this switching duty is critical owing to its very fast rate-of-rise of the TRV in the first microseconds. There is no other switching duty in which the fast removal of the arc residue is of such crucial significance.

For gas circuit-breakers, the arc voltage plays a major role, because it is the main driver of the interaction.

The *arc voltage* is the voltage across the arc, and is a result of the physical interaction between the charge-carriers in the arc discharge. It basically consists of a small cathode and anode voltage drop localized in the immediate vicinity of the contacts and a much larger voltage drop across the arc's conducting body.

Because of the principal differences between the physics of the high-pressure arc discharge [27–29] that exist in gas circuit-breakers and the metal-vapour arc [30–32] in vacuum circuit-breakers [33], the arc-circuit interaction is essentially different.

### 3.6.1.6 Arc-Circuit Interaction in Gas Circuit-Breakers

Whether interruption succeeds or not is a matter of (sub-)microsecond processes, and parasitic circuit elements with their high-frequency transients play a decisive role here.

In gas circuit-breakers, the arc voltage is usually in the range of a few hundred volts to several kilovolts, depending on the following factors:

- *The length of the arc.* Generally, the voltage drop across the arc is proportional to the length of the arc. The length of an arc can be much longer than the contact distance (gap length). Especially at lower currents, the arc in gases makes random-like excursions by curling and bulging (see sketch Figure 3.21). During the existence of the arc, these bulges can be shortened again, which is visible on arc-voltage measurements as a sudden collapse of the arc voltage. This mechanism is outlined schematically in Figure 3.22, in which an example of the measured arc voltage collapse can be seen.
- *The gas in which the arc occurs.* The arc voltage depends on a number of physical parameters of the surrounding medium [27, 34]. Arcs in compressed air have higher arc voltage than arcs in $SF_6$.
- *The contacts on which the arc foot-points reside.* The material of the arcing contacts has a minor influence on arc voltage, since it influences only the voltage in the anode and cathode

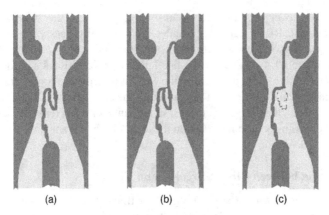

(a)                    (b)                    (c)

**Figure 3.21** Arc motion between arcing contacts inside an insulating nozzle of a HV circuit-breaker.

voltage-drop region. The main voltage drop of gaseous arcs is across the arc body, keeping the effect of the contact material marginal.

- *The cooling of the arc.* The internal power of the arc is the product of the current and the arc voltage. In the case of higher thermal losses, the arc increases its power by increasing the arc voltage. Especially at the lower level of the current near current zero, the arc has in itself an unfavourable ratio of surface area to volume, is very thin, and cooling rapidly increases the arc voltage.

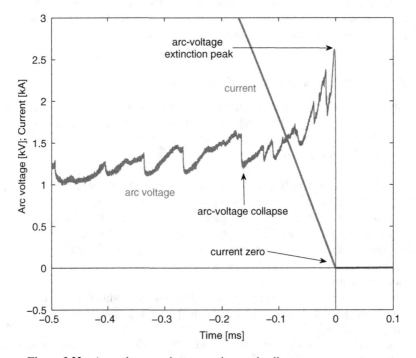

**Figure 3.22** Arc-voltage random excursions and collapses near current zero.

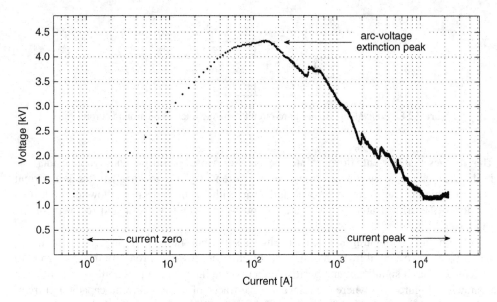

**Figure 3.23**   An example of arc voltage versus current characteristic.

- *The current through the arc.* Arcs in gases have a negative volt-ampere characteristic. This implies that arc voltage increases with decreasing current and vice versa. As with cooling, for the sake of maintaining its energy balance, the arc reacts to a lower supplied current by increasing its arc voltage. An example is shown in Figure 3.23.

The most prominent feature of the AC voltage is the arc-voltage extinction peak, the last reaction of the arc at a very low current, a few microseconds before current zero. The arc-voltage extinction peak appears as a steep spike just before current zero. The peak voltage, reaching up to several kilovolts, is much higher than the arc voltage of the stable burning high-current arc. For arc-circuit interaction, this arc-voltage extinction peak is of crucial importance since it charges the parallel capacitance $C_p$ across the circuit-breaker. This capacitance appears as a parasitic element from the bushings of dead-tank and GIS circuit-breakers, resides in nearby instrument transformers, and so on. The capacitance may also be that of capacitors applied intentionally between terminals and earth or across each interrupter; in the latter case as grading capacitors to distribute the voltage equally across each interrupter.

The current $i_C$ through the capacitor $C_p$ under the influence of changing arc voltage $u_a$ can be described as:

$$i_C = C_p \frac{\mathrm{d}u_a}{\mathrm{d}t} \tag{3.20}$$

Noting that the fault current $i_{\text{circuit}}$ is not perceptibly influenced by the arc voltage, the balance of currents in Figure 3.24 is:

$$i_{\text{circuit}} = i_a + i_C \tag{3.21}$$

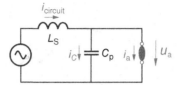

**Figure 3.24**   Equivalent circuit of arc-circuit dynamic interaction around current zero.

This implies that current, flowing into the parallel capacitor during the dynamic increase in the arc voltage, is subtracted from the arc current. By this mechanism, the arc current is approaching current zero at a lower $di/dt$ than without the capacitor, allowing more time to cool in the critical area. As a result, the post-arc current is lower and recovery of the gap faster.

The current commutating into the capacitor is small but, in the current-zero region, it still shows up as a substantial fraction of the arc current. The slower approach of the arc current to current zero has a significant and positive effect on the interruption probability. The situation is shown in Figure 3.25 where the relevant wave-traces of the arc current, capacitor current, and arc voltage are calculated for the case of $C_p = 1$ and $0.1$ nF.

The effect on the near-zero current is clear in Figure 3.26 for the same capacitance values as in Figure 3.25.

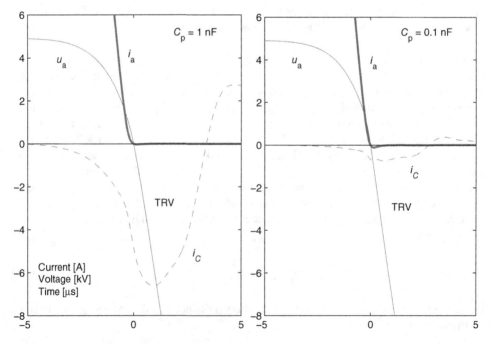

**Figure 3.25**   The effect of parallel capacitance $C_p = 0.1$ and $1$ nF across a circuit-breaker on $di/dt$ around current zero.

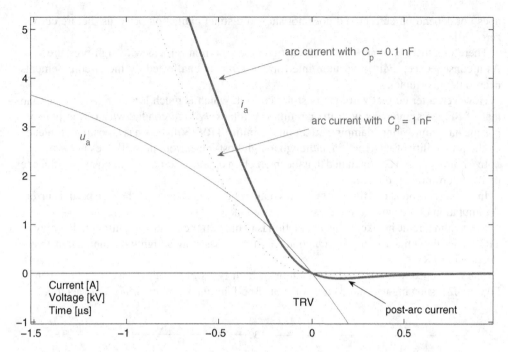

**Figure 3.26** Parallel capacitance $C_p = 0.1$ and 1 nF across circuit-breaker and the effect upon $di/dt$ around current zero (extension).

Capacitors are sometimes applied deliberately to exploit the arc-circuit interaction and increase the thermal interruption capability of circuit-breakers by the mechanism described above.

With arc models [34, 35], the effect of capacitance, and in fact of many other parameters, on the interruption capability of circuit-breakers can be simulated.

In addition, capacitors are also beneficial after current zero in terms of reduction in the rate-of-rise of TRV.

The drawback of applying capacitors directly across the circuit-breaker is that the disconnection of the circuit-breaker is not complete. After interruption, a small power-frequency current is still flowing through the capacitor, which has to be interrupted by a disconnector switch. In order to avoid this, capacitors can be placed across the line with one terminal connected to earth. There is no residual current in this case, but the method is less effective in increasing the thermal interruption capability.

### 3.6.1.7 Arc-Circuit Interaction in Vacuum Circuit-Breakers

For vacuum arcs, the arc voltage is only a few tens of volts. Basically, at low currents, near current zero, the arc voltage in vacuum only consists of a *cathode drop* with a negligible voltage drop across the arc column. Since near to current zero the vacuum arc is merely a collection of a number of cathode spots (see Section 8.3.3) in parallel operation, the arc voltage does not depend on current and does not show a distinct extinction peak just before current zero [30].

Therefore, there is no arc-circuit interaction before current zero, as with high-pressure arcs. As a consequence, $di/dt$ in vacuum interruption remains unaffected by the circuit elements near to the interrupter.

However, after current zero, the post-arc current, which is much larger in magnitude than that of $SF_6$ arcs, interacts with the circuit. The relatively high conductivity of the post-arc plasma has a mitigating, damping effect on the initial TRV [36]. This has a positive influence on the probability of interruption, although a high post-arc current in itself does not.

In summary, it can be concluded that the interaction of arc and circuit has a positive influence on the interruption probability.

In $SF_6$ circuit-breakers, this interaction is mainly before current zero, through partial current commutation of arc current into a parallel capacitance.

In vacuum circuit-breakers, this interaction is mainly after current zero, through damping of TRV under the influence of post-arc conductivity and assisting a high withstand against very steep rising TRV.

The difference in current-zero behaviour of both types of circuit-breakers can be seen in Figure 3.27, showing measured cases of short-line-fault interruption in $SF_6$ and vacuum.

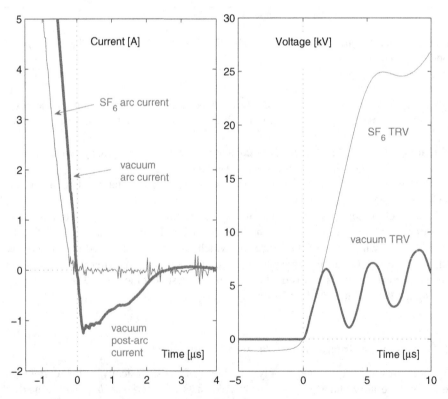

**Figure 3.27**   Current-zero behaviour of vacuum and $SF_6$ circuit-breakers. Tests are performed in circuits with a different source voltage.

## 3.6.2 Long-Line Faults

When clearing a fault current at some significant distance from the circuit-breaker, so-called *long-line fault* (LLF) phenomena, comparable to those occurring at a SLF are involved. Travelling waves between a fault location and the circuit-breaker largely determine the TRV waveform. The line-side component of the TRV shows the well-known triangular shape, coming from the original voltage distribution along the line, the negative reflection at the fault location, and the positive reflection at the circuit-breaker terminal. As with SLF, the last pole that clears a three-phase-to-earth fault faces the largest equivalent surge impedance. The SLF test duty is limited to the last pole phenomena or a single-fault-to-earth clearing and the test current is fixed to 90% or 75% of the rated short-circuit breaking current.

But there are also differences between LLF and SLF. The LLF current is mainly determined by the reactance of the overhead line and, as the $X_0/X_1$ ratio of overhead lines is between 1.5 and 3.0, the three-phase short-circuit current is always considerably larger than the single-phase-to-earth fault current.

Consequentially, the RRRV for the first-pole-to-clear is higher than that of the last pole-to-clear a three-phase fault at a certain location. The peak value of the TRV for the first pole is also higher than that of the last clearing pole. Suppose that the faults in individual phases are located so that the last pole interrupts a fault current of the magnitude as the first pole. This implies that the fault location of the first-pole-to-clear is much farther away and the longer travel time gives again a higher peak value for the first-pole TRV than for the last clearing pole.

In addition, because of the distance to the fault, the first pole encounters at the line side a considerable power-frequency voltage induced by the other phase fault currents, leading to a line-side amplitude factor $k_{af}$ much larger than 1.6, actually $k_{af}$ can be up to 2.4 [37]. Therefore, the first pole-to-clear of a LLF is considered to give the most severe TRV conditions. The high peak values of the TRV when clearing LLF were the reason for recent adaptation of the test-duty T10 parameters in IEC 62271-100 [10].

A particular phenomenon is the effect of a circuit parallel to the faulty circuit, as in the case of a double-circuit overhead line or multiple single-circuit overhead lines in parallel. When clearing a LLF, the busbar part of the TRV sends a switching surge into the system. This switching surge will pass the parallel circuit and arrive at the busbar terminal of the fault clearing circuit-breaker at the other end. Therefore, the total TRV of that circuit-breaker consists of three parts:

- the triangular waveform from the travelling waves on the faulty line;
- the response of the system at the busbar side of the circuit-breaker; and
- the switching surge associated with the other-side fault clearing.

In Figure 3.28 it is depicted how these three parts interact depending on the fault location along the line.

## 3.7 Out-of-Phase Switching

### 3.7.1 Introduction

*Out-of-phase* (OOP) switching conditions occur at a coupling of two network parts of equal operating voltages, the equivalent sources of which have different phase angles, partly or

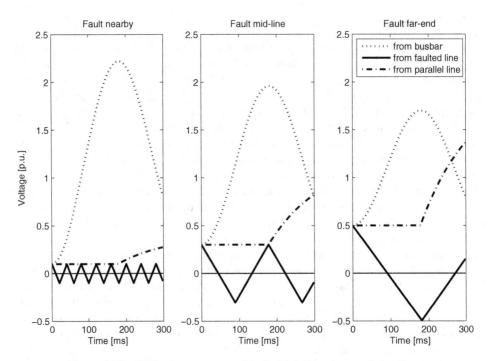

**Figure 3.28**  TRV components associated with faults on transmission lines.

entirely 180° out of phase. A difference in the phase-angle of the rotating vectors representing the source voltages causes out-of-phase currents across the connection, which must be interrupted by a circuit-breaker at either side of the connection. Somewhere along the connection the voltage vector will almost vanish and this location is called the *equilibrium point*.

In practice, the following situations call for the need to separate the networks or network parts:

1. The equilibrium point is in the connection of a single generator/transformer unit, as the generator operating point became unstable, for example due to clearing of a nearby fault taking longer than a critical clearing time;
2. The equilibrium point is on an overhead line, as may occur with system stability problems due to reactive power unbalance, overloading, load rejection or other major disturbances.

Note that, unlike in the faults discussed in Sections 1.3 and 1.4, where usually an external arc is the consequence of the fault, this is not the case here. The out-of-phase current is far smaller than the rated short-circuit breaking current $I_{sc}$ of the circuit-breaker, but may be larger than the transient or even the subtransient short-circuit current of a generator (see Sections 3.7.2 and 10.1.2).

As regards the TRV, the specialty of this switching duty is the presence of active sources on both sides of the circuit-breaker. This is explained in Figure 3.29 with sources $S_1$ and $S_2$.

Considering the fault-switching duties discussed before, in all cases the load-side TRV component decays to zero. In the out-of-phase situation, however, the $S_2$-side TRV component

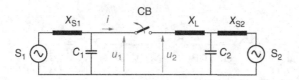

**Figure 3.29**  Equivalent single-phase circuit of an out-of-phase fault.

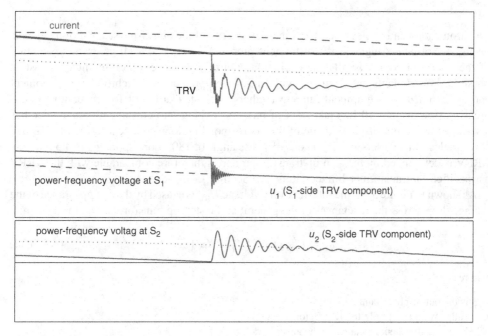

**Figure 3.30**  Out-of-phase TRV and its components. The dashed and dotted traces show the power-frequency voltages of source $S_1$ and $S_2$ respectively.

will decay to the power-frequency RV of the $S_2$-side source. This is outlined in Figure 3.30 where the voltage phase difference between both sources is assumed to be 90° and the short-circuit reactances are considered equal.

As a result, the out-of-phase switching duty is characterized by a very high TRV peak with a moderate RRRV and a moderate current. Because the TRV of the out-of-phase test duty shows the highest peak value of all switching duties, it is often used as a reference for other special switching conditions, such as clearing long-line faults or faults on series-compensated lines.

### 3.7.2    Switching between Generator and System

Switching between the generator and power system may take place at the HV side or at the MV side of the transformer when a step-up transformer is applied. Switching may be faced during system disturbances or during tripping of the power plant, but can also happen during

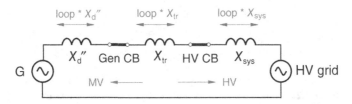

**Figure 3.31**   Principal lay-out for calculation of generator circuit-breaker recovery voltage.

synchronization and mis-synchronization. The severity of the out-of-phase condition depends on the out-of-phase angle between the generator and network as well as on the excitation of the generator rotor. Usually the excitation control will reduce the rotor field as fast as possible. Power plants are equipped with out-of-step protection, synchro-check equipment and synchronization equipment, amongst others. The standard [38] for generator circuit-breakers (see Section 10.1.2) takes out-of-phase switching into consideration and specifies test duties for an out-of-phase angle of 90°, corresponding to a TRV peak value of 3.18 p.u. Nevertheless, many users specify an out-of-phase angle of 180°, corresponding to 4.5 p.u. The RRRV is determined by the step-up transformer and is therefore comparable with the RRRV as specified for transformer-limited faults (see Section 1.4).

As shown in Figure 3.31, the total recovery voltage $U_{RV}$ is caused by the disappearance of the voltage drop across the reactances of the generator, the step-up transformer and the system:

$$U_{RV} = I_{oop} \cdot k_{pp} \cdot \left(X_d'' + X_{tr} + X_{sys}\right) \tag{3.22}$$

where

$I_{oop}$  is the out-of-phase fault current;
$k_{pp}$  is the overall first-pole-to-clear factor;
$X_d''$  is the sub-transient reactance of the generator;
$X_{tr}$  is the short-circuit reactance of the transformer;
$X_{sys}$  is the reactance of the system.

The overall first-pole-to-clear factor is a combination of the first-pole-to-clear factor of the systems at both sides of the circuit-breaker.

The largest voltage drop will generally be across the generator (sub)transient reactance. The transformer reactance is in the range 0.1 to 0.15 p.u., while many modern generators have a sub-transient reactance in the range 0.18 to 0.27 p.u. and even larger transient reactances; in old two-pole turbine generators, lower values of 0.12 to 0.15 p.u. are typical. The system reactance is normally at least five times lower than the (sub)transient reactance of the generator and the transformer together.

Further, the natural frequency of voltage oscillations due to the generator windings is two to three times lower than the natural frequency of the transformer windings. System transient's components usually show the lowest frequency, defined primarily by the travelling waves of the OH-lines. The source side transformer dominated transients are usually of a higher frequency.

The amplitude of the RV is determined by the natural frequency at each side of the circuit-breaker, and since the natural frequencies are substantially different, the components of the TRV at both sides of the circuit-breaker oscillate independently and their peaks probably do not coincide. At the generator side of the high voltage circuit-breaker the peak value of TRV is determined by a rather low first-pole-to-clear factor $k_{pp}$ (in systems with an earthed neutral somewhat smaller than 1.0) and a higher amplitude factor $k_{af}$. For low-loss generator/transformer units a value of $k_{af} = 1.8$ can be assumed.

In a system with a floating neutral, or equipped with Peterson coils, the first-pole-to-clear factor is 1.5 and the maximum RV will be 3.0 p.u. The total TRV for the same example can reach 4.55 p.u. at full phase-opposition. An out-of-phase angle of 75° gives a RV of 1.83 p.u. and a peak value of the TRV of 3.1 p.u., which is close to 3.13 p.u., as given in the standards for systems with first-pole-to-clear factor 1.5.

In Section 10.11.2 an analysis is given of the out-of-phase switching conditions that may arise in fault situations of wind generators coupled to the medium-voltage grid.

### 3.7.3  Switching between Two Systems

Switching between two power systems typically occurs in situations with power unbalance and system instability. Examples refer to large system disturbances, situations during system restoration, and due to the misoperation of protection systems. The more important transmission lines may be equipped with an out-of-phase blocking in their protection system and/or a special system-wide protection may be applied to prevent separation of the systems under severe out-of-phase conditions. Nevertheless, out-of-phase-current clearing will always lead to severe TRV stresses. Especially for radial transmission systems, utilities require circuit-breakers capable of switching out-of-phase currents under out-of-phase angles as large as 180°; that is, more than required by the IEC circuit-breaker standard [1].

Shortly before the system separation, the point of equilibrium can be on an overhead line or cable connection. The RRRV for the circuit-breaker that separates the systems will be determined by the equivalent surge impedance $Z_{eq}$ seen by the breaker and the rate-of-decline of the out-of-phase current $i_{oop}$:

$$\frac{du}{dt}\bigg|_{t = +0} = Z_{eq}\frac{di_{oop}}{dt}\bigg|_{t = -0} \tag{3.23}$$

At the line side, $Z_{eq}$ is equal to the surge impedance of the first-pole–to-clear, typically 350 to 400 $\Omega$, and the TRV builds up until the reflected travelling wave returns to the circuit-breaker from the substation at the other side of the overhead-line. The reflection coefficient at the remote substation depends on the equivalent surge impedance of a number of locally connected overhead lines. Similarly, $Z_{eq}$ at the busbar side of the circuit-breaker depends on the number of connected circuits to the busbar. The out-of-phase current is determined by the out-of-phase angle between both line ends, the busbar voltage and the line reactance. This implies that out-of-phase conditions on long lines are determined by travelling-wave phenomena and the system's natural frequencies.

Due to the travelling-wave effects, the TRV at the line side will exhibit a triangular shape (see Section 1.6.1) and its peak value can be calculated as twice the wave travelling time along the overhead line multiplied by the RRRV. The travelling time is proportional to the line length, but the RRRV decreases with increasing line length due to the decrease in the out-of-phase current $I_{oop}$. In specific cases, in addition travelling waves at the busbar side may also be encountered, leading to a further increase in the RRRV and the TRV peak value [11].

Calculations and simulations for real networks show out-of-phase TRV peak values as high as 3.3 to 3.5 p.u. [39] or even 3.9 p.u. [40] for very extended networks (hundreds of kilometres) and low currents, and 3.0 to 3.5 p.u. for meshed networks (hundred kilometres or less) and relatively high currents.

## 3.8   Fault-Current Making

### 3.8.1   Impact of Making a Short-Circuit Current on the Circuit-Breaker

Short-circuit currents impose severe stresses on circuit-breakers resulting in the following demands and challenges:

1. In spite of dynamic forces, the contacts must remain closed at the current passage. This is, especially for vacuum circuit-breakers, not a trivial requirement. Due to the *butt-type* construction (contacts that essentially touch head-on, and do not penetrate into each other), contacts in vacuum circuit-breakers tend to repel at high current, after which arcing at very small contact distance sometimes leads to irreversible welding. This is illustrated in Figure 3.32 showing the repelling forces that arise due to the radial currents in the butt contacts. A sufficient contact force must be applied [41]. In addition, the effective contact surface of vacuum interrupters is relatively small, and melting of contact bridges may appear as a consequence.

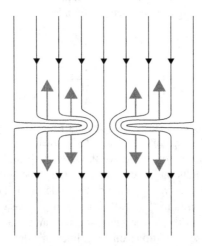

**Figure 3.32**   Butt-type contacts of a vacuum circuit-breaker with simplified indication of current distribution and repelling forces [41].

2. This is observed at short-time current tests, in which asymmetry plays only a minor role and the duration of current passage is crucial. Circuit-breakers with the finger-type contacts normally applied in $SF_6$ circuit-breakers are less prone to the phenomena discussed above since the passage of high currents tends to tighten the contacts by design.
3. The energization of a faulted circuit by a circuit-breaker represents another challenge for the circuit-breaker. During the closing operation with the contacts approaching each other, the withstand voltage of the gap decreases with time at a rate commonly called the *rate of decrease of dielectric strength* (RDDS). At a certain instant, the breakdown voltage ·of the gap becomes equal to the voltage imposed by the circuit across the contacts. At that moment, the gap will breakdown, the so-called *pre-strike*. An arc will be established lasting until the contacts touch mechanically. Due to this *pre-arcing*, the pressure rise in the interruption chamber of gas circuit-breakers challenges the so-called close-and-latch capability. In such cases, contacts may slow down, especially when a single mechanism is used to drive all three poles. It is required and has to be verified in testing that the travel of the contact system, stressed by a short-circuit making current, shall not deviate greatly from that of the no-load operation.
4. In vacuum circuit-breakers, the combination of high-temperature and high-contact forces involved in closing may lead to welding of the ultra-clean contacts.
5. During the arcing phase, lasting from the contact separation to the final extinction of the arc (*arcing time*), the contact system is stressed by the high-current arc. The arc energy is increased considerably at a higher asymmetry because of both the higher value of the current and the generally longer arcing time [42]. These quantities contribute to the arc energy, which must be dissipated in the interruption chamber.
6. Tests are defined to verify interruption capabilities under full asymmetrical conditions in the T100a test-duty specified in IEC 62271-100 [1], see Section 13.1.3.

Because short-circuit current starts to flow immediately after a pre-strike, the rate-of-rise of the short-circuit current determines the thermal stress during the *pre-arcing time*.

For the circuit-breaker, the most severe case regarding pre-arcing is the symmetrical current because:

- the rate-of-rise of the current is larger than that of the asymmetrical current, causing higher thermal and pneumatic arcing stresses during the pre-arcing time;
- symmetrical current arises with pre-strike at the voltage maximum, implying maximum pre-arcing time.

Both extremes are detailed in Figure 3.33.

To summarize: when the fault starts at voltage zero, the high asymmetrical current peak stresses mechanically all system components and increases arcing times.

When the fault is initiated at voltage maximum, the symmetrical current gives the highest thermal stresses to the circuit-breaker internal parts due to pre-arcing.

Both extreme making situations have to be tested by means of standardized making tests, see Section 14.2.3.3 [43].

Short-circuit making with a certain delay between poles (*pole discrepancy*) may, under the most severe conditions, lead to an asymmetrical-current peak factor much larger than the standardized value of $F_s = 2.5$ at 50 Hz or $F_s = 2.6$ at 60 Hz, see Section 2.2.3.

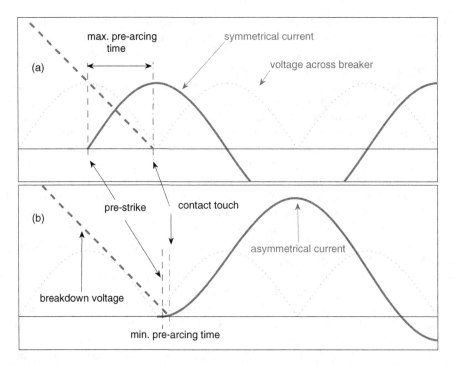

**Figure 3.33**  Pre-arc duration at making operations: (a) making operation with pre-strike near voltage maximum resulting in symmetrical current and maximum pre-arcing time; (b) making operation with pre-strike near voltage zero resulting in asymmetrical current and minimum pre-arcing time.

Another making operation that may have a major impact on circuit-breakers is the energization of capacitor banks in the *back-to-back* configuration, although this is considered a normal-load operation. Section 4.2.6 gives a detailed description.

## 3.8.2   Switching-Voltage Transients at Making in Three-Phase Systems

In three-phase systems, the voltage transients across each of the interrupter gaps are subject to phase-to-phase interaction.

Transients that accompany the pre-strike of switching devices can be described as the interaction of the decaying dielectric strength of the gap with the system voltage imposed across the gap.

For simplicity, the dielectric strength of the contact gap during contact approach is assumed to be a linear function of time, having a RDDS normalized against the maximum steepness of the system phase-to-phase voltage $U$:

$$\text{RDDS} = S_0 \omega U \sqrt{\frac{2}{3}}   \qquad (3.24)$$

with $S_0$ being a normalizing proportionality factor.

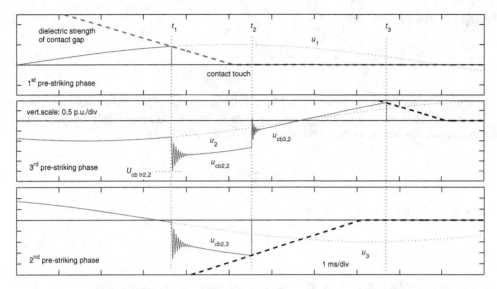

**Figure 3.34**  Voltage transients during a three-phase making operation.

In a three-phase system, the first pole will pre-strike when the momentary voltage across the contact gap exceeds the withstand voltage of the gap, see Figure 3.34.

As a complete example, Figure 3.34 can be used as an aid in understanding. This is a simulation of a system with $k_{pp} = 1.5$ and $k_{af} = 1.7$. The straight dashed lines represent the decreasing dielectric strength and the intersection with the phase voltage $u_1$ at $t_1$ leads to the first pre-strike. This results in transients in phases 2 and 3, where a shifted voltage of $u_{cb2,2}$ and $u_{cb2,3}$ in these phases is established across the contacts. Soon thereafter, phase 3 pre-strikes at $t_2$, leading to a new shifted voltage $u_{cb3,2}$ after the transient has decayed. The last pre-strike follows then in phase 2 at $t_3$, completing this making operation.

For clarity, a significant delay between the galvanic touch of the contacts has been introduced in the figure. In reality for a switch with simultaneously closing poles, the second pole will pre-strike immediately after the first pre-strike ($t_2 \approx t_1$), due to the transient across the second pole caused by the first pre-strike.

Depending on the earthing of the power system, this pre-strike will lead to transients and a temporary shift of the power-frequency voltages across the gaps in the two neighbouring phases.

The common characterization of the earthing situation is expressed by the ratio $k$ of zero-sequence and positive-sequence reactances ($k = X_0/X_1$) and the first-pole-to-clear factor (see Section 1.3.2.2) with $k_{pp} = 1.3$ for *effectively-earthed neutral systems*, and $k_{pp} = 1.5$ for *non-effectively-earthed neutral systems*.

After pre-strike of phase $i$ ($i = 1, 2, 3$) the new power-frequency voltage $u_{cb2,j}$ ($j \neq i$) across the other two gaps will be:

$$u_{cb2,j} = \frac{1-k}{k+2}u_i - u_j \qquad (3.25)$$

having a maximum value of:

$$\hat{U}_{cb2,j} = \frac{\sqrt{3}\sqrt{k^2 + k + 1}}{k + 2} \quad [\text{p.u.}] \tag{3.26}$$

where $u_i$ and $u_j$ are the phase voltages, and 1 p.u. is the amplitude of the phase-to-earth voltage.

A second pre-strike will occur in phase $j$ at time $t_2$, leading again to a shift of the power-frequency voltage across the last, still open gap $n$ (where $n \neq i \neq j$):

$$u_{cb3,n} = \frac{1-k}{2k+1}(u_i + u_j) + u_n \tag{3.27}$$

This voltage has a maximum value of:

$$\hat{U}_{cb3,n} = \frac{3k}{2k+1} = k_{pp} \quad [\text{p.u.}] \tag{3.28}$$

Finally, under the influence of this voltage, the third phase will pre-strike at $t_3$.

Only after this time, the full three-phase short-circuit current will develop.

It is interesting to note that making is the inverse process of the more often described breaking operation such that Equation 3.28 describes the recovery voltage of the first-pole-to-clear, and Equation 3.26 describes the recovery voltage across the second pole-to-clear (see Section 1.3.2.2).

The pre-strike events are accompanied by transients, the frequency and peak value of which are determined by the stray elements of the system.

Further investigation shows that the transient in the phase $j$, caused by the first pre-strike in phase $i$ at $t_1$, has a higher peak value than the transient occurring at the second or the third pre-strike. In Figure 3.34, the peak value of this transient in the given situation is marked as $U_{cb,tr2,2}$. Its value depends on the moment of the first pre-strike ($t_1$), and its maximum possible value is given by:

$$\max[u_{cb,tr2,j}(t_1)] = \hat{U}_{cb,tr2,j} = \sqrt{\left[\frac{2k_{af} - 2k_{af}k - k - 2}{2k+4}\right]^2 + \frac{3}{4}} \quad [\text{p.u.}] \tag{3.29}$$

where $k_{af}$ is the amplitude factor ($1 \leq k_{af} \leq 2$), depending on high-frequency damping (see Section 13.1.2).

In Table 3.3, values of the relevant transient voltages are given for various practical values of $k$, the first-pole-to-clear factor $k_{pp}$, and $k_{af} = 1.7$ where the voltages are in [p.u.], $T$ is the power-frequency cycle period, angles are in degrees relative to the positive-going phase voltage zero of the first pre-striking phase; the pre-striking sequence is $i \rightarrow j \rightarrow n$).

The case of the first pre-strike at the voltage maximum at $t_1 = T/4$ in a non-effectively-earthed system is illustrative.

Just before a pre-strike, the momentary p.u. voltages across the gaps are:

$$u_{cb1} = 1.0 \qquad u_{cb2} = u_{cb3} = -0.5$$

**Table 3.3** Transients during making ($k_{af} = 1.7$).

| Zero-sequence/Positive-sequence reactance ($X_0/X_1$) | $k = X_0/X_1$ | 1.00 | 1.38 | 3.25 | $\infty$ |
|---|---|---|---|---|---|
| First-pole-to-clear factor | $k_{pp}$ | 1.0 | 1.1 | 1.3 | 1.5 |
| Max. steady-state voltage, poles $j$ and $n$ after pre-strike of $i$ | $\hat{U}_{cb2,j}$ [p.u.] | 1.00 | 1.06 | 1.27 | 1.73 |
| Max. possible transient across poles $j$ and $n$ after pre-strike of $i$ | $\hat{U}_{cbtr2,j}$ [p.u.] | 1.00 | 1.11 | 1.50 | 2.36 |
| Phase angle of 1st pre-strike in pole $i$, giving max. transient in $j$ and $n$ | [el.deg.] | 30.00 | 39 | 55 | 69 |
| Transient across $j$ and $n$ when 1st pre-strike is at voltage max. of $i$ | $\hat{U}_{cbtr2,j}(T/4)$ [p.u.] | 0.50 | 0.69 | 1.23 | 2.20* |
| Max. steady-state voltage across pole $n$ after 2nd pre-strike of $j$ | $\hat{U}_{cb3,k}$ [p.u.] | 1.00 | 1.10 | 1.30 | 1.50 |

*Peak value of transients.

Since $u_{cb1}$ jumps to zero, this 1 p.u. jump is transferred to both $u_{cb2}$ and $u_{cb3}$, so they jump to $-1.5$ p.u.

The peak value of the transients in these phases with $k_{af} = 1.7$ is:

$$\hat{U}_{cb,tr2,2}(t = T/4) = \hat{U}_{cb,tr2,3}(t = T/4) = |-1.5[-(k_{af} - 1)(1.5 - 0.5)]| = 2.20 \text{ p.u.} \quad (3.30)$$

which is the value marked in Table 3.3 with the asterisk.

By the same reasoning it can be easily understood that for a pre-strike at 60°, that is, $t_1 = T/6$, with values in [p.u.]:

before pre-strike: $\quad u_{cb1} = 2\sqrt{3} \quad u_{cb2} = -2\sqrt{3} \quad u_{cb3} = 0$

after: $\quad u_{cb1} = 0 \quad u_{cb2} = -\sqrt{3} \quad u_{cb3} = -2\sqrt{3}$

thus: $\quad \hat{U}_{cb,tr2,2}(t = T/6) = |-\sqrt{3}[-(k_{af} - 1)(\sqrt{3} - 2\sqrt{3})]| = 2.34$

which is very close to the absolute maximum of $\hat{U}_{cb,tr2,2} = 2.36$ occurring at $t_1 = 0.19T = 69°$ (Table 3.3).

The importance of the analysis above lies in the influence that the transferred transient voltage at the moment of pre-strike exerts on the duration of the pre-arc. This duration of *pre-arcing time* $\Delta T_{pa}$ can be described as:

$$\Delta T_{pa} = \beta \frac{1}{S_0 \, \omega} \quad (3.31)$$

where $\beta$ is the per-unit value of the phase voltage at the moment of the pre-strike and $S_0$ a variable representing the closing velocity. Since $\beta$ can assume theoretically a value up to 2.36, the duration of the pre-arc between the approaching contacts can be considerably increased due to the phase-to-phase interaction.

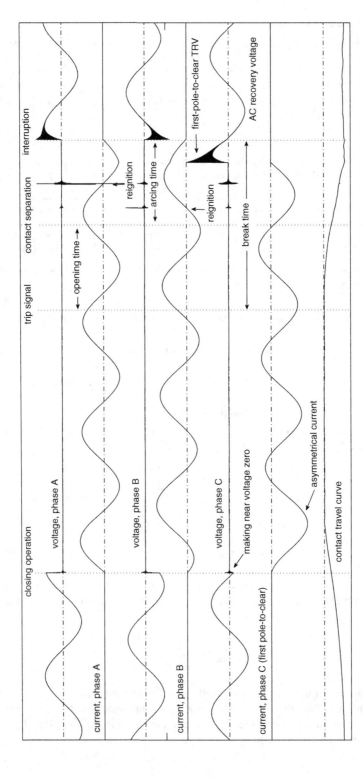

**Figure 3.35** Complete make-break cycle from a short-circuit test. Voltages are across the circuit-breaker.

The instant of the short-circuit initiation with respect to the voltage waveform determines the value of the asymmetrical short-circuit-current peak (see Section 2.1). The value of this peak is normally expressed as the asymmetrical-current peak factor $F_S$, which is equal to the maximum current peak divided by the symmetrical-current amplitude.

Pre-strike in general has a moderating influence on the probability of occurrence of very high values of the asymmetrical current peak. In principle, only fast closing circuit-breakers, having a value $S_0 > \sqrt{3}$ in systems having $k_{pp} = 1.5$, can initiate short-circuit current at voltage zero, thus providing the condition for maximum asymmetry.

In a simulation study [43], the reduction of the asymmetrical-current peak factor has been investigated for a slow circuit-breaker ($S_0 = 0.3$) and a fast one ($S_0 = 1.0$). It was concluded that the 90 percentile of the asymmetrical peak factors belonging to the fast circuit-breaker is 5% higher than that of the slow circuit-breaker. The theoretical asymmetrical-current peak value of 2.55 p.u. at 50 Hz and $\tau = 45$ ms is not reached at all due to the phase-to-phase interaction.

The following concluding remarks are justified:

1. The second phase to pre-strike can have the longest pre-arcing time if it faces the highest voltage transient of a theoretical maximum, which is 2.34 p.u. The practical value is 2.20 p.u. in a system with $k_{pp} = 1.5$. In switches without intentional pole delay, this value is naturally limited by the reduced contact distance.
2. Non-effectively earthed neutral networks tend to have longer pre-arcing times than those with an effectively earthed neutral.
3. A breaker with a deliberate pole delay at closing can generate higher asymmetrical-current peaks and has a higher probability of producing long pre-arcing times.
4. In general, pre-arcing increases the stress to the switching device but reduces, to a moderate extent, the electrodynamic stress of high asymmetrical-current peaks on the system components.

Apart from current, a number of voltage transients result in the system from energization. When a voltage is suddenly impressed on the system, voltage waves start to travel through it. These phenomena, however, fall outside the scope of this chapter. A detailed analysis can be found in [9].

The situation of energization of capacitor banks, which can be a major challenge in application and in testing due to severe inrush current, is treated in Section 4.2.6.

Overviewing the significant phenomena associated with fault current interruption, an oscillogram of a three-phase short-circuit test is shown in Figure 3.35, featuring a complete make–break cycle of a vacuum circuit-breaker in a non-effectively earthed circuit.

In this test, the pole in the lower phase (C) is the first pole-to-clear with the highest TRV and the shortest arcing time. Also it has the highest asymmetrical current peak because it makes near voltage zero. Note that poles A and B cannot interrupt at the first current zero during arcing and they show re-ignition. Having the longest arcing time, they face less severe TRV, compared with pole C.

# References

[1]  IEC 62271-100 (2012) High-voltage switchgear and controlgear – Part 100: Alternating-current circuit-breakers. Ed. 2.1.

[2] IEEE C37.09 (1999) IEEE Standard Test Procedure for AC High-Voltage Circuit-Breakers Rated on a Symmetrical Current Basis.

[3] Janssen, A.L.J., Knol, P. and van der Sluis, L. (1996) TRV-Networks for the Testing of High-Voltage Equipment. CIGRE Conference, Paper 13-205.

[4] Janssen, A.L.J., van Riet, M., Smeets, R.P.P. *et al.* (2012) Prospective Single and Multi-Phase Short-Circuit Current Levels in the Dutch Transmission, Sub-Transmission and Distribution Grid. CIGRE Conference, Paper A3.103.

[5] Harner, R.H. and Rodriguez, J. (1972) Transient recovery voltages associated with power system, three-phase transformer secondary faults. *IEEE Trans. Power Ap. Syst.*, **PAS-91**(5), 1887–1896.

[6] Horton, R., Dugan, R.C., Wallace, K. and Hallmark, D. (2012) Improved autotransformer model for transient recovery voltage (TRV) studies. *IEEE Trans. Power Deliver.*, **27**(2), 895–901.

[7] Steuer, M., Hribernik, W. and Brunke, J.H. (2004) Calculating the transient recovery voltage associated with clearing transformer determined faults by means of frequency response analysis. *IEEE Trans. Power Deliver.*, **19**(1), 168–173.

[8] Hribernik, W., Graber, L. and Brunke, J.H. (2006) Inherent transient recovery voltage of power transformers – a model-based determination procedure. *IEEE Trans. Power Deliver.*, **21**(1), 129–134.

[9] Greenwood, A. (1991) *Electrical Transients in Power Systems*, John Wiley & Sons Ltd, ISBN 0-471-62058-0.

[10] Dufournet, D., Janssen, A.L.J. and Hu, J. (2013) Transformer Limited Fault Transient Recovery Voltage for EHV and UHV Circuit-Breakers. Int. High Voltage Symp., Seoul.

[11] CIGRE Working Group A3.28 (2014) Switching Phenomena for EHV and UHV Equipment. CIGRE Technical Brochure 570.

[12] Janssen, A.L.J., Dufournet, D., Ito, H. *et al.* (2013) Transformer Limited Fault Duties for EHV and UHV. CIGRE UHV Colloquium, Session 2.1, paper A3 03, New Delhi.

[13] Kagawa, H., Maekawa, T., Yamagata, Y. *et al.* (2012) Measurement and Computation of Transient Recovery voltage of Transformer Limited Fault in 525 kV-1500 MVA Three-Phase Transformer. IEEE T&D Conference, Orlando.

[14] IEC 60076-6 (2007) Power transformers – Part 6: Reactors. Ed. 1.0.

[15] Li, Q., Liu, H. and Zou, L. (2008) Impact research of inductive FCL on the rate of rise of recovery voltage with circuit-breakers. *IEEE Trans. Power Deliver.*, **23**(4), 1978–1985.

[16] Peelo, D.F., Polovick, G.S., Sawada, J.H. *et al.* (1996) Mitigation of circuit-breaker transient recovery voltages associated with current limiting reactors. *IEEE Trans. Power Deliver.*, **11**(2), 865–871.

[17] Robert, D., Martin, F. and Taisne, J-P. (2007) Insertion of current limiting or load sharing reactors in a substation. Impact on specifications. CIGRE SC A2 Symposium, Bruges.

[18] IEEE Power Engineering Society, C37.011 (2006) IEEE Application Guide for Transient Recovery Voltage for AC High-Voltage Circuit Breakers. Section 4.4.2.

[19] Janssen, A.L.J., van Riet, M., Smeets, R.P.P. *et al.* (2014) Life extension of well-performing air-blast HV and MV circuit-breakers. CIGRE Conference.

[20] van der Sluis, L. (2001) *Transients in Power Systems*, John Wiley & Sons, ISBN 0-471-48639-6.

[21] Peelo, D.F. (2014) *Current Interruption Transients Calculation*, John Wiley & Sons, ISBN 978-1-118-70719-7.

[22] IEC 62271-306 (2012) High-voltage switchgear and controlgear – Part 306: Guide to IEC 62271-100, IEC 62271-1, and IEC standards related to alternating current circuit-breakers.

[23] Ahmethodžić, A., Smeets, R.P.P., Kertész, V. *et al.* (2010) Design improvement of a 245 kV $SF_6$ circuit-breaker with double speed mechanism through current zero analysis. *IEEE Trans. Power Deliver.*, **25**(4), 2496–2503.

[24] Smeets, R.P.P. and Kertész, V. (2000) Evaluation of high-voltage circuit-breaker performance with a new validated Arc model. *IEE Proc.-C*, **147**(2), 121–125.

[25] Smeets, R.P.P. and Kertész, V. (2006) A New Arc Parameter Database for Characterisation of Short-Line Fault Interruption Capability of High-Voltage Circuit Breakers. CIGRE Conference, Paper A3-110.

[26] CIGRE Working Group 07 (1974) Theoretical and Experimental Investigations of Compressed Gas Circuit-Breakers under Short-Line Fault Conditions, Report.

[27] Hoyaux, M.F. (1968) *Arc Physics*, Springer Verlag, Berlin.

[28] Lee, T.H. (1975) *Physics and Engineering of High Power Switching Devices*. MIT Press, ISBN 0-262-12069-0.

[29] Nakanishi, K. (1991) *Switching Phenomena in High-Voltage Circuit-Breakers*. Marcel Dekker, ISBN 0-8247-8543-6.

[30] Anders, A. (2008) *Cathodic Arcs. From Fractal Spot to Energetic Condensation*, Springer, ISBN 978-0-387-79107-4.

[31]  Greenwood, A. (1994) Vacuum Switchgear. the Institution of Electrical Engineers, ISBN 0 85296 855 8.
[32]  Lafferty, J.M. (1980) *Vacuum Arcs, Theory and Application* (ed. J.M. Lafferty), John Wiley & Sons, New York, ISBN 0-471-06506-4.
[33]  Slade, P.G. (2008) *The Vacuum Interrupter, Theory and Application*. CRC Press, ISBN 978-0-8493-9091-3.
[34]  Kapetanović, M. (2011) *High Voltage Circuit-Breakers*. ETF University of Sarajevo, ISBN 978-9958-629-39-6.
[35]  CIGRE Working Group 13.01 (1998) State of the art of circuit-breaker modelling. CIGRE Technical Brochure 135.
[36]  van Lanen, E.P.A. (2008) The Current Interruption Process in Vacuum. Ph.D. thesis Delft University of Technology, ISBN 978-90-5335-152-9.
[37]  CIGRE Working Group A3.19 (2010) Line fault phenomena and their implications for 3-phase short- and long-line fault clearing. CIGRE Technical Brochure 408.
[38]  ANSI/IEEE C37.013-1997 (1997) Standard for AC-High Voltage Generator Circuit-Breakers Rated on a Symmetrical Current Basis. IEEE New York. IEC 62271-37-013 (2014) High-voltage switchgear and controlgear. Alternating-current generator circuit-breakers".
[39]  Bui-Van, Q., Gallon, F., Iliceto, F. *et al.* (2006) Long Distance AC Power Transmission and Shunt/Series Compensation - Overview and Experiences. CIGRE Conference, Paper A3-206.
[40]  Amon, J. and Morais, S.A. (1995) Circuit-breaker Requirements for Alternative Configurations of a 500 kV Transmission System. CIGRE SC 13 Colloquium, Brazil, Report 2.3.
[41]  Slade, P.G. (1999) *Electrical Contacts*. Marcel Dekker, Inc., ISBN 0-8247-1934-4.
[42]  Smeets, R.P.P. and Lathouwers, A.G.A. (2001) Economy Motivated Increase of DC Time Constants in Power Systems and Consequences for Fault Current Interruption. IEEE PES Summer Meeting, 0-7803-7173-9/01.
[43]  Smeets, R.P.P. and van der Linden, W.A. (2001) Verification of the short-circuit current making capability of high-voltage switching devices. *IEEE Trans. Power Deliver.*, **16**(4), 611–618.

# 4

# Load Switching

## 4.1 Normal-Load Switching

Load-current switching is mainly characterized by currents up to the normal-load-current rating of the switching device. For this duty, the term *switching* is used, meaning energizing or de-energizing load circuits. This is in contrast with the terms *making* and *breaking operations*, reserved for a fault-current establishment and interruption.

Another fundamental difference is the frequency of operations: whereas a fault-current making/breaking is a relatively rare event, the switching of load currents is practised on a regular basis.

Switching of *normal load* has to be distinguished from switching of *capacitive* or *inductive loads*. Due to the reactive nature of these loads, they are considered "special" switching duties deserving particular attention – also in testing.

The main reason for this is that the TRV of normal-load switching is much less severe than in all other switching duties discussed so far. The circuit power factor is close to unity, the current and voltage have only a small phase difference, and current zeros occur far from the recovery voltage maximum. This is outlined in Figure 4.1.

The interruption of normal-load current is generally easy for any type of circuit-breaker that has proven its short-circuit making and breaking capabilities. Therefore, there is no test-duty defined for the verification of the capability to switch normal loads.

Two situations are compared in the figure:

- the dominantly inductive case, as in short-circuits discussed – but also in the case of inductive loads where current zero is close to the voltage maximum;
- the case where the phase lag of current with respect to voltage is significantly smaller than 90°.

As can be seen, the latter case has a much lower TRV peak value because of the relatively small voltage excursion caused by the dominant resistive nature of the circuit and its high power factor.

The reason that normal-load interruption is not critical is not the small current (as compared with the short-circuit duties). As will be demonstrated in Sections 4.2 and 4.3, even with much

*Switching in Electrical Transmission and Distribution Systems*, First Edition.
René Smeets, Lou van der Sluis, Mirsad Kapetanović, David F. Peelo, and Anton Janssen.
© 2015 John Wiley & Sons, Ltd. Published 2015 by John Wiley & Sons, Ltd.

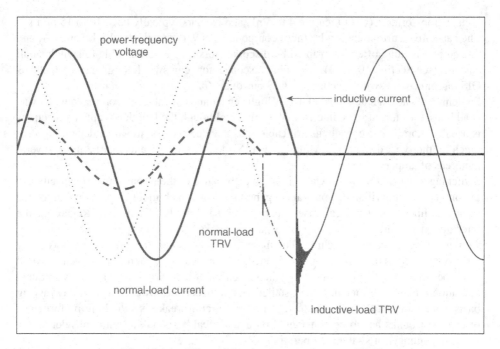

**Figure 4.1** Effect of the circuit power factor on TRV.

smaller current than a normal-load current, switching can become critical when current zero occurs near the voltage maximum. This is the case for capacitive and inductive load-current interruption due to the reactive nature of the respective loads.

## 4.2 Capacitive-Load Switching

### 4.2.1 Introduction

The interruption of capacitive currents is a very normal switching case, unlike the making and breaking of fault currents [1, 2].

The usual cases, in which a capacitive current is switched, are:

1. Switching of unloaded overhead transmission lines. In this case, the load is already switched off by a breaker at the remote end of the line, but due to the stray capacitance of the overhead-line, a small capacitive current is still flowing in the system, to be interrupted by the substation breaker. The current to be interrupted depends on the voltage level and the length of the line, and possibly on some station components. Typical capacitance values vary from 9.1 nF km$^{-1}$ per phase for single-conductor overhead lines to 14 nF km$^{-1}$ per phase for a four-conductor bundle [3]. The current to be interrupted is in the range of several amperes at medium voltage to several hundreds of amperes at high voltage, which corresponds to a line length of a few tens to hundreds of kilometres.

2. Switching of local substation components. Some substation components also draw capacitive current: current transformers typically have 1 to 1.5 nF of stray capacitance, capacitive

voltage transformers (CVT) feature 4 to 5 nF, and busbars typically have 10 to 15 pF m$^{-1}$. The capacitive current caused by these components is very small, usually below 1 A, and this current is quite often interrupted by disconnectors [4] generally able to switch small currents (see Section 10.3.2.3). Air-break disconnectors can switch long busbars, whereas GIS disconnectors switch sections of GIS bus ducts (*bus-charging current*) [5].

3. Switching of cables. Due to the relatively high capacitance of cables as compared with over-head lines, the current to be interrupted is higher, although the length of cable connections is usually shorter. The capacitance of cables has a wide variation in value, depending very much on the type, design, and voltage level. No-load cable currents can be up to several hundreds of amperes.

4. Switching of capacitor banks. Capacitor banks, because of their concentrated capacitance, in contrast with distributed capacitance, generally draw much more current than unloaded cables or lines – in practical cases up to several hundreds of amperes. Regarding the interruption of current, switching of capacitor banks is, in principle, a switching duty similar to line- or cable-switching. The main difference is the frequency of the switching operation: whereas the switching of unloaded lines and cables is an irregular event (several times per year to once in a few years), the switching of capacitor banks is a very frequent operation, since capacitor banks are installed to supply reactive power on a night/day varying basis, depending on the network load. Thus, the circuit-breaker's switching performance of capacitor banks has to be considered on a statistical basis, taking into consideration a very large number of switching operations.

Regarding the energization or making of capacitor-bank currents, the concentrated capacitance causes another onerous phenomenon for circuit-breakers: this is the inrush current, a very high transient current, drawn by the capacitor bank. Since the surge impedance of capacitor banks is far smaller than those of cables and lines, capacitor-bank inrush current management and its consequences is of considerable concern to users and developers of switchgear.

## 4.2.2   Single-Phase Capacitive-Load Switching

Current interruption in a single-phase capacitive circuit is relatively easy to understand and undemanding in practice. Such a circuit is outlined in Figure 4.2 showing a simple case of a single-oscillatory circuit at the source side, in agreement with all the earlier examples, and a lumped capacitance at the load side.

The resulting electrical transients are given in Figure 4.3. Since current leads the voltage by 90°, the voltage across the load is near its maximum at current zero. In fact, the load voltage

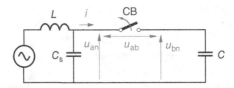

**Figure 4.2**   Basic circuit for single-phase capacitive-load switching.

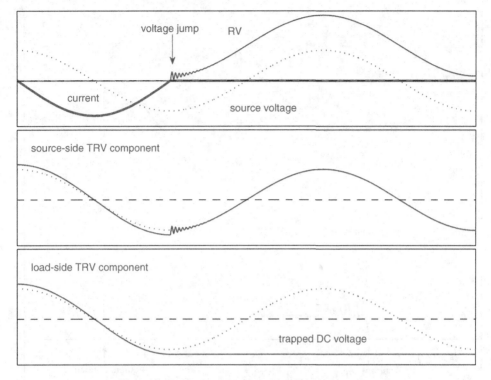

**Figure 4.3**  Single-phase capacitive-load interruption.

is even higher than the 1 p.u. of the source voltage due to driving a leading current through the source reactance. It can be easily derived that the voltage increase in [p.u.] is:

$$\Delta U = \frac{1}{1 - \omega^2 LC} - 1 \approx \frac{Q}{P} = \frac{I_{cap}}{I_{sc,loc}} \tag{4.1}$$

The symbols $L$ and $C$ refer to the inductance and capacitance as in Figure 4.2, $P$ is the local short-circuit active power (with a local short-circuit current $I_{sc,loc}$), and $Q$ is the reactive power rating of the capacitor bank corresponding to a capacitive current $I_{cap}$.

Therefore, the load capacitor is charged to a voltage $(1 + \Delta U)$ p.u. While this voltage remains at the load side, the source-side voltage first makes a transient excursion from $(1 + \Delta U)$ p.u. to 1 p.u., and then continues its power-frequency course. For the TRV, this implies initially a small "voltage jump" as a transient contribution, followed by a $(1 - \text{cosine})$ continuation of the recovery voltage (RV). After one half-cycle, a maximum recovery voltage of approximately 2 p.u. has developed across the circuit-breaker. Note that the rate-of-rise of this recovery voltage is much lower than in any fault situation discussed so far and, in effect, its frequency is equal to the power frequency.

Although minor in magnitude, the influence of the "voltage jump" on the interruption process should be considered, especially after short arcing times. Short arcing times are possible in this case, because the current to be interrupted is very small compared with the short-circuit current. The short arcing time can easily turn out to be advantageous, for example, with respect

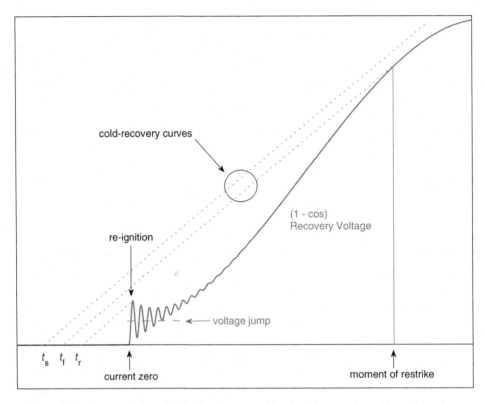

**Figure 4.4**  Time extension of a single-phase capacitive-load interruption with a voltage jump.

to wear of the circuit-breaker interruption chamber. The negative consequence of a short arcing time is the fact that at current zero, the contact gap is small and breakdown can follow shortly after current zero.

If breakdown follows at relatively low levels of voltage, such as that associated with the "voltage-jump" transient, the energy involved in the breakdown event is very limited and is generally of no concern. If breakdown occurs more than a quarter of the power-frequency cycle after current interruption, then the energy in the breakdown is significant and no longer harmless. In order to distinguish between the severities of these effects, IEC coined the term *re-ignition* for a breakdown occurring shorter than a quarter of the power-frequency cycle after interruption, and *restrike* for a breakdown occurring later.

The difference is outlined in Figure 4.4 where the dotted lines indicate the increasing dielectric strength of the opening contact gap – in simplification assumed linearly rising. Three different moments of contact separation lead to three essentially different interruption results:

- Contact separation at $t_s$ results in a normal interruption without breakdown. The arcing time is long enough to ensure sufficient contact spacing even at the maximum recovery voltage.
- Contact separation at $t_f$ results in restrike, at a relatively high breakdown voltage, generating a high transient voltage and current as the load side recovers to the source voltage.

- Contact separation at $t_r$ results in re-ignition, at a low breakdown voltage, normally harmless and considered part of the interruption process. Although the interruption in principle fails, this is generally not a problem: after another half-cycle of current, the contact gap is easily able to withstand the TRV and RV.

Since a restrike – and its consequences – is the main hazard of capacitive-current interruption, it needs adequately detailed treatment.

In Figure 4.5, the situation is given regarding a restrike near the TRV maximum.

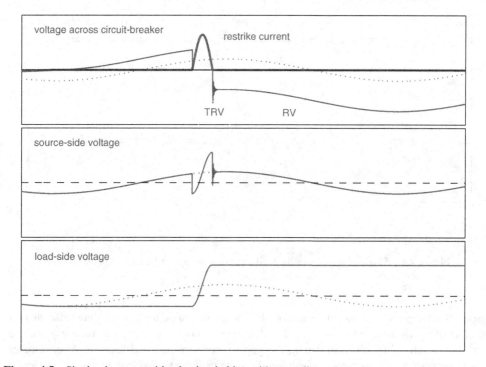

**Figure 4.5** Single-phase capacitive-load switching with a restrike and restrike current of a half-cycle.

Sudden conduction of the circuit-breaker gap leads to a discharge of the load capacitance. A restrike current will flow with the peak value $I_{res}$:

$$I_{res} = 2 U_r \sqrt{\frac{2}{3}} \sqrt{\frac{C}{L}}$$ (4.2)

where $U_r$ is the system voltage. Depending on the number of high-frequency half-cycles, during which the restrike current will flow, a new voltage pattern will be established in the system:

- An odd number of restrike-current half-cycles will increase the capacitor voltage. The maximum voltage shift (as illustrated in Figure 4.5) is from $-1$ to $+2$ p.u. if no damping is considered.

- An even number of restrike-current half-cycles will decrease the capacitor voltage. This case is shown in Figure 4.6. In the undamped case, there is no load-voltage change due to the restrike event.

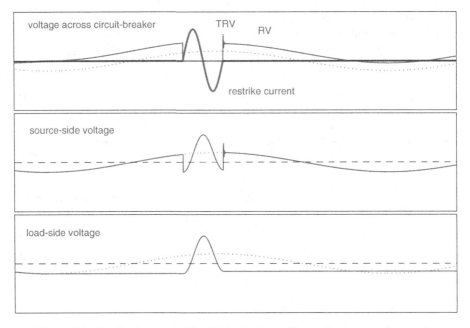

**Figure 4.6**  Single-phase capacitive-load switching with restrike current of one cycle.

Especially dangerous are *multiple restrikes*. Due to the increased load-side voltage after an odd number of restrike-current half-cycles and due to simultaneously increased recovery voltage that may reach values close to 4 p.u., the gap is prone to a second restrike. After another odd number of restrike-current half-cycles, the voltage at the load side may now theoretically escalate to 5 p.u. and the RV peak to 6 p.u. The gradual increase of the voltage by the multiple restrikes process is called capacitive *voltage escalation*.

This is visualized in Figure 4.7.

$SF_6$ circuit-breakers are able to interrupt capacitive currents after arcing times typically below 1 ms. Quite often, the current is chopped to zero before its natural current zero, because of the severe blast action of the $SF_6$ gas to the low-current arc. Although the phenomenon *current chopping* is associated with inductive-load interruption (see Section 4.3.1), capacitive-current chopping occurs, but without the consequences as in inductive-load current interruption.

Apart from dielectric stresses to the system components by the escalating voltage, the sudden release of energy in the circuit-breaker gap leads to very rapid pressure rise in the isolating gas. Restrike may damage the internal interrupter parts, holes in the nozzle being the most prominent damage observed. Traces of restrikes between the main contacts are occasionally observed. A proper dielectric coordination between arcing and main contacts should avoid this, as verified by tests [6].

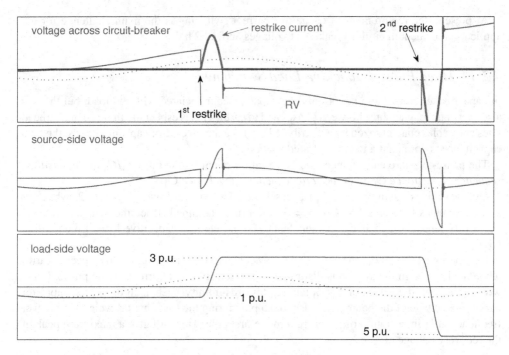

**Figure 4.7**    Single-phase capacitive-load switching with multiple restrikes.

The capacitive-current switching requirement tends to be the determining factor for the highest applicable voltage per break. Typically, for interrupters at the EHV ratings of 420 and 550 kV, the application of two moving contacts (*double-motion* principle) is normally needed to reach sufficient dielectric strength in time to withstand the capacitive-voltage stresses.

Vacuum circuit-breakers are, in general, able to interrupt current with a very high $di/dt$ at current zero. Therefore, they may interrupt pre-strike / re-ignition / restrike current at a high-frequency current zero. Under unfavourable system conditions, this effect may generate overvoltages [7].

Vacuum-interrupter contacts have a triple function: they carry the normal-load current, they have to cope with the arc under load- and fault-current conditions and they must isolate in open condition. In gas/oil circuit-breakers, these functions are assigned to dedicated contact systems, main and arcing contacts. In addition, the dielectric withstand in gas circuit-breakers is mainly determined by the gas, whereas in vacuum-interrupters it is mainly the contact material, surface, and geometry that determines its dielectric properties. This implies that the arcing history, the traces left over by the arc and the resulting changes in surface micro-topology, influences the dielectric properties of the gap. Even a contact separation under no-load conditions has an impact on the dielectric withstand strength. Cold welding of the ultra-clean metallic surfaces, pressed together in the closed position, causes micro-protrusions left over from the ruptured weld at opening. These may affect the dielectric withstand strength. A large current passage in the closed position may worsen this condition. Arcing after contact separation mitigates the effect: the arc melts local surface irregularities and conditions the contact surface in a positive way.

At present, the requirement of capacitive-current switching is the main challenge for the application of vacuum circuit-breakers at voltages above 52 kV.

## 4.2.3   Three-Phase Capacitive-Load Switching

In capacitive circuits, there is not only the capacitance associated with the load, but there is also (parasitic) capacitance between phases and to earth. This is especially the case in overhead lines and cables characterized by a distributed capacitance. In case of capacitor banks, the load capacitance is dominant and concentrated spatially.

The phase-to-phase capacitances are taken into account by the ratio $C_1/C_0$ of the *positive-sequence capacitance* $C_1$ over the *zero-sequence capacitance* $C_0$ [8,9].

Overhead lines having voltages in excess of 52 kV typically have $C_1/C_0 \approx 2$, while for system voltages below 52 kV, $C_1/C_0 = 3$ is normally assumed. Unearthed capacitor banks show very large $C_1/C_0$ values, greater than 100, and earthed capacitive loads by definition have $C_1/C_0 = 1$.

Depending on the $C_1/C_0$, the neutral of the system will rise in voltage (with respect to earth) when the balance in the system is disturbed after interruption of current in one phase. In the worst case, with the values of $C_1/C_0$ tending towards infinity, such as in unearthed capacitor banks, the voltage of the neutral will rise to 0.5 p.u. during the first quarter-cycle of the power frequency after interruption of the current in the first pole. This will give a maximum peak of the recovery voltage of:

> 1 p.u. (trapped charge on load capacitor) + 1 p.u. (reversed source voltage)
> + 0.5 p.u. (neutral shift) = 2.5 p.u.

The electrical phenomena in the case of an unearthed load are demonstrated in Figure 4.8. When the last poles do not clear a quarter-cycle after the first-pole-to-clear, the neutral shift continues to rise to 1 p.u., causing the recovery voltage to reach 3 p.u. This value may further increase to 4.1 p.u. in the case of non-simultaneity of the poles [1].

In order to prevent voltages larger than 2.5 p.u., it is recommendable that the non-simultaneity of poles is limited to a quarter power-frequency cycle.

## 4.2.4   Late Breakdown Phenomena

### 4.2.4.1   General

Occasionally, interrupters may break down after a relatively long time of up to 1 s after the interruption of current. Such an event is usually called *late breakdown*, and the events following immediately after the breakdown are of great importance for the assessment of the consequences.

There are two possibilities:

1. In the case of continued conduction of the contact gap, late breakdown results in *restrike* that may easily have consequences for the circuit and/or circuit-breaker; restrike is usually a reason for not passing the type test.
2. In the case where the insulation of the contact gap restores practically immediately after breakdown, such a late-breakdown event is called *non-sustained disruptive discharge*, or NSDD.

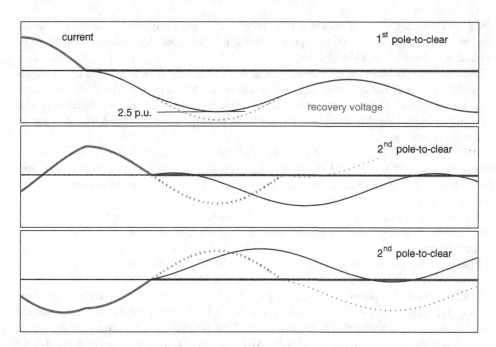

**Figure 4.8** Three-phase capacitive-current interruption. The drawn curves show current and recovery voltage when the current in the last two poles is cleared a quarter cycle after the first pole. The dotted curves show the interruption when the current in the last two poles is cleared three quarters of a cycle after the first pole.

The definition of NSDD in IEC 62271-100 [10], 3.1.126 is:

*"disruptive discharge associated with current interruption, that does not result in the resumption of power-frequency current or, in the case of capacitive-current interruption, does not result in current in the main load-circuit"*[1].

IEC [10], 6.102.8 states further:

*"NSDDs may occur during the recovery-voltage period following a breaking operation. However, their occurrence is not a sign of distress of the switching device under test. Therefore, their number is of no significance to interpreting the performance of the device under test. They shall be reported in the test report in order to differentiate them from restrikes. It is not the intent to require the installation of special measuring circuits to detect NSDDs. They should only be reported when seen on an oscillogram."*

NSDD events are observed frequently in testing [11], and ever since their discovery, their assessment has been a subject of much discussion in standardization committees. Though

---

[1] This is considered an improvement of the definition in IEC 60050 (441-17-46) where a *restrike* is defined as *"a resumption of current"*, not specifying the nature of the current, such as high-frequency, inrush-frequency, or power-frequency current.

NSDDs can no longer be a reason for rejecting a type-test certificate, it is often believed that they are a formidable issue to cope with in the development of vacuum switchgear to be applied at voltages of 36 kV and above.

Usually, *late breakdown* is associated with vacuum switching devices [12] (refer to Chapter 9), and it can occur not only during capacitive-current switching, but equally during short-circuit interruption. However, rare observations of late breakdown in $SF_6$ switchgear have also been made, indicating that this phenomenon is, in principle, not restricted only to vacuum switching technology.

The breakdown field strength of a vacuum gap, though for small gaps in principle higher than in $SF_6$, has a statistical probability distribution with a relatively large standard deviation. This is manifested by breakdowns that are possible under relatively low voltage stresses, such as during the AC recovery voltage.

A root cause of late breakdown may be metal particles in the vacuum gap. These particles, for example, solidified droplets formed after arcing, are sometimes loosely attached to the contact surface and they may become detached by vibrations originating from the opening mechanism [13]. Alternatively, a sudden increase of the field-emission current level, leading to breakdown, has been suggested as a root cause of late breakdown [14].

Under normal conditions, the vacuum gap is able to restore insulation capability very quickly, typically within a few (tens of) microseconds [15] because of its ability to interrupt the high-frequency currents that occur as a result of the breakdown. An example of this can be seen in Figure 4.9. Very high frequencies (>1MHz) of current can be observed after breakdown due to approximately 40 kV voltage stress. In spite of the very high value of $di/dt$ (several kA $\mu s^{-1}$), the vacuum gap can interrupt this current in this example already after 8 μs, keeping the duration of the conductive period very short. The high-frequency currents arise due to the discharge of stray capacitances through stray inductances in the immediate vicinity of the gap.

This self-restoring property is absent in $SF_6$ where interruption of current is only possible during a short-period of sufficient $SF_6$ pressure in the interruption chamber. Therefore, late breakdown occurring outside this "interruption window" can lead to significant arcing in $SF_6$ and possible destruction of the circuit-breaker.

### 4.2.4.2 Transients Initiated by Late Breakdown

The interpretation and assessment of NSDD have led to considerable discussion; particularly concerning the consequences NSDD could have in real circuits.

In order to make a fair assessment, it is important to make a clear distinction between the relevant breakdown events: re-ignition, restrike, and NSDD.

All three events can be observed in a single example taken from a real test of a vacuum circuit-breaker in a short-circuit test, as shown in Figure 4.10.

The following transients related to the breakdown of the switching gap can be distinguished in this example:

At first, re-ignition of the gap in the upper phase can be observed near current zero. This is related to a short arcing time being below the minimum arcing time in that phase; re-ignition is a natural part of the interruption process. Re-ignition is defined as breakdown not later than a quarter of the power-frequency cycle after interruption.

Then, *late breakdown* can be observed in the middle-phase gap, to be recognized by a collapse of voltage across this gap from a value $\Delta U_n$ to zero. Because of the very short

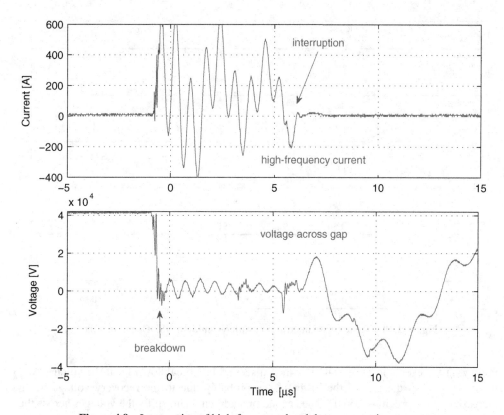

**Figure 4.9**   Interruption of high-frequency breakdown current in vacuum.

current duration, the event is identified as NSDD. Since the system's neutral is unearthed, the voltages across the healthy (open) gaps are increased by the same amount $\Delta U_n$. This leads to an essentially higher recovery voltage across these gaps, especially across the one in the upper phase. The increased recovery voltage may very well lead to an immediate breakdown in the most stressed gap, initiating another breakdown. If this does not happen, as in the present example, NSDD remains a single-phase event.

Generally, the relationship between the ratio ($k = X_0/X_1$, see Section 3.3.2.2) of the zero-sequence reactance to positive-sequence reactance (referring to the earthing of the capacitive load) and the voltage shift $\Delta U_f$ in the neighbouring phases (as a result of breakdown of the gap in the middle phase from voltage $\Delta U_n$ to zero) is given as:

$$\Delta U_f = \frac{k-1}{k+2}\Delta U_n \tag{4.3}$$

For the two standardized $k_{pp}$ values in IEC, this yields:

for non-effectively earthed loads: $\Delta U_f = \Delta U_n$
for effectively-earthed loads ($k = 3.25$): $\Delta U_f = 0.43\Delta U_n$

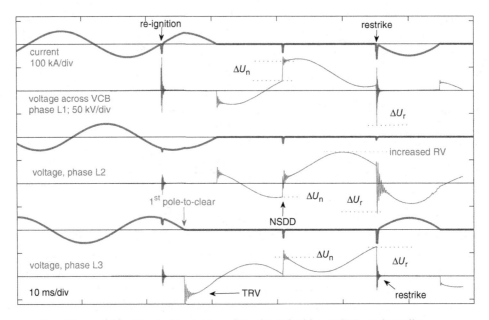

**Figure 4.10**   Three-phase interruption with re-ignition, NSDD, and restrike.

Following the events in Figure 4.10 further, a second late breakdown from the voltage $\Delta U_r$ to zero occurs in the gap of the lower phase, probably due to the recovery voltage being increased by the previous NSDD. The second breakdown again shifts the voltages across the neighbouring gaps by $\Delta U_r$, which immediately leads to a third breakdown in the upper phase gap. The upper and lower gaps are now conducting, thereby creating a two-phase path for current. Initially, this current consists of various frequency components and, depending on the capability of the gaps to interrupt this current, the late breakdown may develop into:

- NSDD in two phases – when the two-phase breakdown current is interrupted very soon after breakdown;
- restrike – when the breakdown current is not interrupted immediately and power-frequency current develops (in the case of short-circuit interruption) or the main capacitive load discharges during at least a half-cycle of capacitive restrike current (in the case of capacitive current interruption).

The latter event, restrike, has occurred in the test example of Figure 4.10.

From this example it can be understood that in three-phase unearthed neutral circuits, restrike needs simultaneous breakdown in at least two interrupter gaps. Most probably, the second breakdown is a result of the neutral voltage shift caused by the first breakdown. If the high-frequency component of the breakdown current is interrupted by one of the gaps, power-frequency or capacitive-load-discharge current will not develop, and the event is a two-phase NSDD. This development scheme is shown in Figure 4.11.

A typical example of two-phase and single-phase NSDD can be seen in Figure 4.12. The gap in the middle phase suffers late breakdown, immediately followed by the gap in the lower

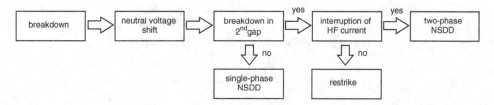

**Figure 4.11**   Scheme of development of late breakdown into either NSDD or restrike.

phase, after being stressed by increased voltage. Current starts to flow in two-phases but after several tens of microseconds the current is already interrupted, avoiding development of restrike current. A short time later, the middle-phase gap breaks down again, this time not followed by breakdown of a neighbouring gap. The upper trace shows the voltage across the gap that happens to be close to zero during the short time frame considered and no breakdown is to be expected. Some induced voltages can be seen during the flow of the NSDD current and, of course, the neutral shift $\Delta U_{\mathrm{f}} = 0.5$ p.u., first with the two-phase NSDD and later with the single-phase NSDD.

Single-phase NSDD is associated with high-frequency current flowing in the earth circuit, while two-phase NSDD causes two-phase high-frequency current in the involved phases.

To understand the *late breakdown* events, identification of the paths through which the currents flow is essential. This is outlined in the generalized three-phase capacitive circuit of Figure 4.13.

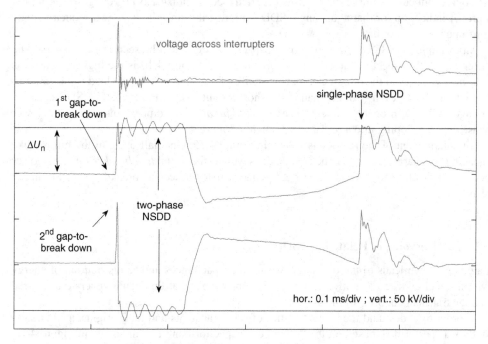

**Figure 4.12**   Two-phase and single-phase NSDD.

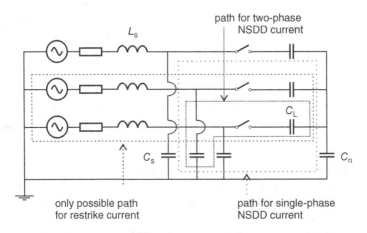

**Figure 4.13**   Equivalent three-phase circuit for analysis of restrikes and NSDDs.

In capacitive circuits, restrike causes the load capacitance(s) to discharge. NSDD in a single gap, with its very short duration, causes only a very small discharge of the capacitive load and only the stray neutral-to-earth capacitance $C_n$ will completely reverse polarity or discharge completely.

Restrikes and NSDDs can be clearly distinguished by the current; restrike must be identified by at least one half cycle of current determined by the main load capacitor discharging across the source inductance, the so-called *inrush current* (see Section 4.2.6.1). Voltage-shifts are not a proper indication for distinguishing NSDD from restrike since both show a sudden voltage shift in unearthed circuits.

An example of NSDD and restrike in a capacitive circuit can be seen in Figure 4.14. The difference with respect to the fault-current example of Figure 4.10 is the higher excursions that most voltages in the circuit can make because of the capability of the load capacitance to trap charge. Thus, even stronger than in the short-circuit case of Figure 4.10, NSDD causes a voltage shift, that probably initiates a second *late breakdown* some 30 ms later in the lower phase gap, developing into a restrike, as shown in Figure 4.14.

An enlargement of the restrike is shown in Figure 4.15. One half-cycle of the full restrike capacitor-discharge current can be observed, as well as the initial *late breakdown* in the lower phase gap, leading to a voltage shift $\Delta U_r$ that initiates a second breakdown in the upper phase gap.

### 4.2.4.3   Overvoltages Related to NSDD

Late breakdown leads to the discharge of various capacitances and re-distribution of energy within the circuit. In capacitive circuits, it is conceivable that this could generate undesired voltage transients in parts of the system.

In order to understand the characteristics of such transients and their origins, a simulation study was performed, based on the "reference" capacitor-bank circuit shown in Figure 4.16. Measurements provided input data for the model [16].

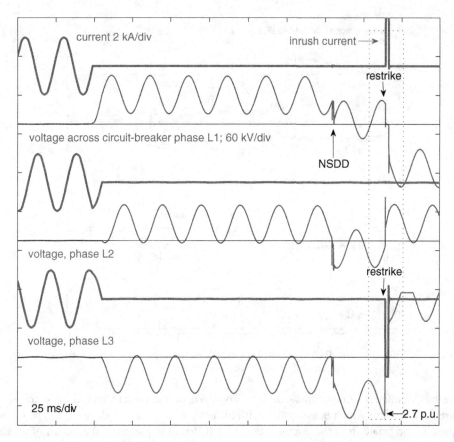

**Figure 4.14** NSDD and restrike in a vacuum circuit-breaker in a three-phase capacitive circuit. The restrike current is clipped and the voltage across the lower phase gap is limited to 100 kV.

The "reference" circuit represents a large 200 μF unearthed capacitor bank connected by a cable approximately 100 m long and having a lumped-element equivalent capacitance of 25 nF to the circuit-breaker. At the source-side of the circuit-breaker, a relatively large capacitance of 100 nF, representing multiple cables, is assumed.

The transient voltages that develop can now be calculated. The wave traces of the interrupter current and load-side voltage transients to earth, as indicated in Figure 4.16, are given in Figure 4.17. In a capacitive circuit, the most severe transients occur at breakdown of the circuit-breaker gap in the first-phase-to-clear at the recovery voltage peak, phase $L_1$ in this example. The voltage across the circuit-breaker is then at a level of 2.5 p.u.

It can be seen that the maximum load-side voltage excursion of approximately −5 p.u. (relative to earth) is reached after 33 μs in the healthy phase $L_3$. Note that the maximum voltage occurs in a phase that does not break down, reflecting the typical three-phase interaction that has been taken into account in this simulation.

In the simulation, it is assumed that the duration of the discharge is long enough to actually reach the above-mentioned value.

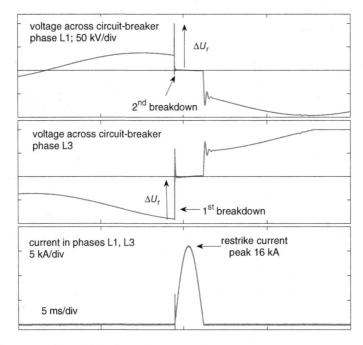

**Figure 4.15**   Time extension (boxed area of Figure 4.14) of restrike in a three-phase capacitive circuit.

Various frequency components can be recognized in the current and voltage traces in Figure 4.17, each of which is associated with different parts of the model circuit. The higher frequencies originate from the circuit-breaker itself, for example, from the circuit of $L_p$ and $C_p$ that generates oscillations in the tens of megahertz range and from circuit elements in the immediate vicinity of the circuit-breaker, that is, the cables and so on. The lower frequencies come from more distant circuit parts involving oscillations generated by the source circuit and earth return.

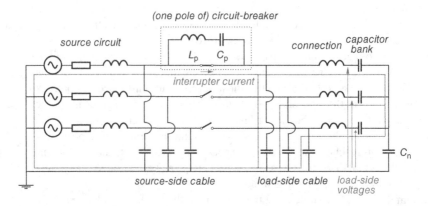

**Figure 4.16**   Circuit for simulation of NSDD-related voltage transients in a capacitive test-circuit.

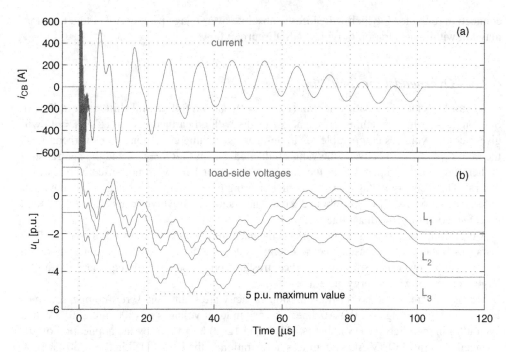

**Figure 4.17**   Result of simulation to estimate overvoltages by an NSDD in the circuit shown in Figure 4.16. (a) current through interrupter; (b) load-side voltages-to-earth in the phases $L_1$, $L_2$, and $L_3$.

From linking each of the oscillatory sub-circuits to the voltage and current transients they produce, it can be concluded that the major –5 p.u. voltage transient originates from the sub-circuit indicated in Figure 4.16 by the dotted-line segments. The voltage across the neutral capacitance $C_n$ will in this case make an excursion of 4 p.u. from (0.5 to –3.5) p.u. because the value of $C_n$ is so small that it will absorb the full voltage shift resulting from the breakdown. The voltage across the load capacitance in the lower phase remains at its initial value of $(-\frac{1}{2} - \frac{1}{2}\sqrt{3})$ p.u., making the load-side voltages reach $|(-\frac{1}{2} - \frac{1}{2}\sqrt{3} - 3.5)| \approx 5$ p.u.

As a conclusion of the present analysis, NSDD-related overvoltages in a capacitor-bank circuit can be potentially significant, and it is the duration of the NSDD current flow that will determine whether significant overvoltages actually develop.

Based on the observed very short duration of the NSDD current flow, which is much shorter than the period of the most hazardous oscillation [15], it may be concluded that an oscillation in the dotted reference circuit of Figure 4.16 will be disrupted before it can reach its maximum amplitude.

Apparently – assuming a very short conducting period – the effect of the NSDD remains localized in the circuit consisting of the circuit-breaker and its connecting cables. As a consequence, the overvoltage generating capability of very short duration NSDD current flow is limited.

However, in special cases, it is possible that even a very short-lasting NSDD current flow could cause significant overvoltages. Such a case is, for example, a circuit with a short cable connection to the load, which will shift the absolute overvoltage peak of 5 p.u. to a time that may fall within the duration of typical NSDD current flow.

## 4.2.5  Overhead-Line Switching

Capacitor banks with a floating neutral ($C_0/C_1 = 0$) show a full neutral shift, as described in Section 4.2.4, while switching off capacitor banks with an earthed neutral ($C_0/C_1 = 1$) will give recovery voltages comparable to single-phase switching conditions. For overhead lines the ratio $C_0$ to $C_1$ is in between the values indicated above: $C_0/C_1 = 0.6$, and the neutral shift is therefore small compared with the case of a capacitor bank with non-effectively-earthed neutral. The recovery voltage across the first clearing pole, assuming no clearing by the other poles, reaches $6/(2 + C_0/C_1)$, so about 2.3 p.u. The phenomena involved can best be understood by considering long overhead lines [22].

A clear trend in transmission networks is the need for large power transmission across long distances because of increased electric-energy trade. The drivers behind these needs are, on the one hand, power plants far from load centres and, on the other hand, the volatile nature of large amounts of wind-energy generators.

Means to increase the power-transmission capacity are shunt- and series-compensation of long extra-high-voltage (EHV) AC lines, for example, at voltages of (500 and 800) kV, and even ultra-high-voltage (UHV), that is, (1100 and 1200) kV, AC or by the application of DC connections. An 1150 kV AC system was in operation in the USSR [17] in the 1980s, a 1000 kV AC system was planned and partly realized in Japan [18], a 1050 kV AC system was put in commercial operation in 2009 in China [19], and a 1200 kV AC system is planned in India [20].

Heavy bundle-conductors are applied to reduce ohmic and corona losses. Flexible devices, such as thyristor-controlled series capacitor banks, *flexible AC transmission systems* (FACTS), and phase-shifting transformers are used to improve in an adaptive way the voltage profile along the overhead lines and the transmission capacity of the system.

Long transmission lines have a large phase angle between the voltage phasors at both ends of the transmission corridor associated with the transmission of active power. The lines also experience a voltage drop because of the line reactance. This voltage drop can be counteracted by series capacitor-banks that reduce, to a certain extent, the line overall reactance. Under low-load conditions, the line capacitance is dominating and the voltage will increase along the line due to the *Ferranti rise* [21], the voltage increase due to capacitance. Shunt reactors are applied at the line ends to compensate for the line capacitance. Preferably, the shunt reactors are connected to the overhead line and not to the substation busbars as the voltage rise is most severe when the line is unloaded, that is, when the far end is open and disconnected from the busbars.

The longer the line, the more pronounced the Ferranti rise becomes. The voltage along the overhead line due to the Ferranti effect can be calculated by Equation 4.4 and approximated by Equation 4.5.

$$U(x) = U_{\mathrm{p}} \cdot \cosh \left[ \beta(l - x) \right] \qquad (4.4)$$

$$U(x) = U_{\mathrm{p}} \cdot \cos \left[ \beta(l - x) \right] \qquad (4.5)$$

where

$U_p$     is the phase-to-earth voltage at the open receiving end of the line;
$U(x)$ is the phase-to-earth voltage along the line as a function of the distance $x$;
$x$       is the distance from the beginning of the line;
$l$        is the length of the line;
$\beta$       is the angle between the voltage phasors $U_p$ and $U(x)$; $\beta$ is a function of frequency:
$\beta = \omega/c = 2\pi f/c$ ($\beta = 0.001$ rad km$^{-1}$ for 50 Hz and $\beta = 0.0012$ rad km$^{-1}$ for 60 Hz).

Suppose now that the line is also to be de-energized at the sending end. After current inter-ruption, a considerable charge and voltage exists at the receiving end. The voltage $U_p$ and the 1 p.u. voltage $U_s$ at the sending end of the overhead line will be equalized by travelling waves to reach an average DC residual voltage $U_{dc}$ given by Equation 4.6.

$$U_{dc} = U_p \cdot \sin{(\beta \cdot l)}/\beta \cdot l = U_s \cdot \tan{(\beta \cdot l)}/\beta \cdot l \qquad (4.6)$$

Figure 4.18 shows the voltage rise at the line's receiving end $U_p$ due to the Ferranti rise with $U_s$ (sending end) $= U_s(x = 0)$ together with the average DC voltage $U_{dc}$.

Travelling waves will appear as a sinusoidal oscillation superimposed with double amplitude on the DC voltage and power-frequency components of the recovery voltage. The original voltage profile is reflected at both line ends.

Figure 4.19 shows the pattern that occurs when de-energizing long overhead lines without shunt reactors. For 440 km and 50 Hz, this gives a residual DC voltage of $U_{dc} = 1.07$ p.u.

Shown in Figure 4.19, the line-side recovery voltage $U_{RV}$ of the first pole interrupting the charging current of a 440 km long line (50 Hz, single circuit, fully transposed, and with ratio zero-sequence to positive-sequence capacitance $C_1/C_0 = 1.67$) experiences a typical waveform

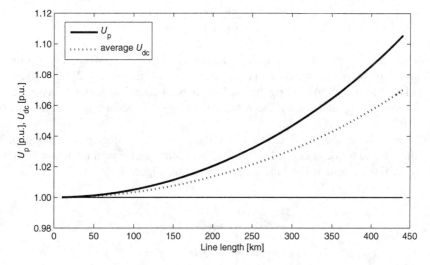

**Figure 4.18** Transmission line voltage $U_p$ at the open receiving end due to Ferranti rise and average residual DC voltage $U_{dc}$ after equalization of charges – both as a function of the line length.

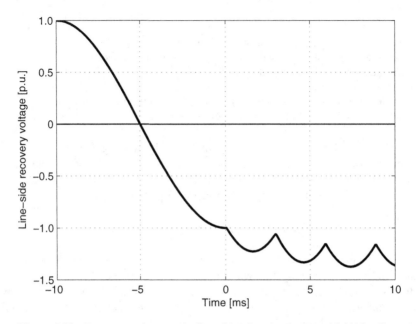

**Figure 4.19**  Recovery voltage at the line-side of an uncompensated 440 km line.

caused by the Ferranti rise. Apart from the Ferranti rise, the shift of the neutral voltage is 0.15 p.u., leading to a power-frequency RV of 2.15 p.u., as can be calculated with Equation 4.7 that is applicable when the other poles clear after about 90 electrical degrees:

$$U_{RV} = \frac{5 + C_0/C_1}{2 + C_0/C_1} \quad [\text{p.u.}] \tag{4.7}$$

Details of the calculation of the neutral shift as a function of $C_1/C_0$, considering the moments of interruption of the second and last pole-to-clear, can be found in [22]. In Figure 4.19, damping of the travelling waves along the line has not been taken into account. In practice, the waves will slowly damp and reach a DC voltage of a magnitude according to Equation 4.6, in this case 0.07 p.u.

When shunt reactors are connected to the overhead line, the voltage profile along the line will be different and the de-energized circuit will be discharged through the resonance circuit formed by the line capacitance and the shunt reactors, see Figure 4.20.

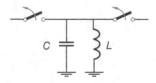

**Figure 4.20**  Simplified scheme of a shunt-reactor compensated line.

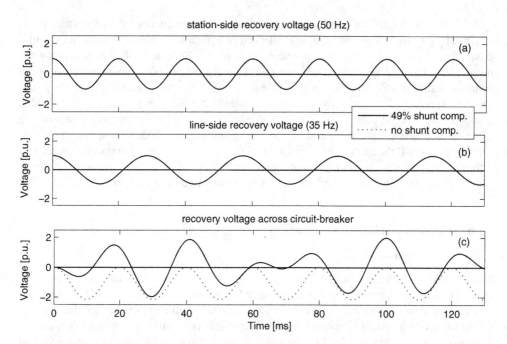

**Figure 4.21** Recovery voltage of unloaded-line switching. (a) station-side recovery voltage; (b) line-side recovery voltage; (c) recovery voltage across the circuit-breaker. Cases with 49% and zero compensation are shown.

Suppose, the degree of shunt compensation is $p$, where $(1/\omega L$ is $p$ % of $\omega C)$, then the resonance frequency $(1/\sqrt{LC})$ will be $0.1\sqrt{p}$ times the power angular frequency $\omega$. The recovery voltage will appear as a slowly rising low-frequency voltage pattern, showing a beat frequency of the difference between power- and resonance-frequency, that is, $0.1(10 - \sqrt{p})$ times the power frequency, as shown in Figure 4.21. Actual field traces are shown in Figure 11.14 in Section 11.4.5.2.

In the case of a single-phase-to-earth fault, the common procedure in transmission grids is to switch off the faulty phase at both line ends and to reclose within one to two seconds. This is known as a *single-pole auto-reclosure* (SPAR). Usually, the power arc disappears and the total three-phase power transfer is not interrupted. If the arc re-establishes at the fault location, a three-phase interruption is required and power transfer through the circuit is interrupted. The interruption of the capacitive-line current in the other two healthy phases is more severe than the interruption of a capacitive current without earth fault, as the healthy-phase voltage is higher by a factor $F_1$:

$$F_1 = \sqrt{3} \cdot \frac{\sqrt{k^2 + k + 1}}{2 + k} \tag{4.8}$$

where $k = X_0/X_1$.

With higher voltages and longer lines, SPAR may not be successful as the fault arc is sustained by the electrostatically coupled and electromagnetically induced voltage in the faulty phase. The sustained arc is called a *secondary arc* (see also Section 6.2.4). The secondary-arc current can be reduced by connecting the neutral point of the three-phase shunt reactors to earth through a neutral reactor, see Section 10.4.2. Instead of SPAR, some utilities apply *three-phase auto-reclosure* (TPAR), especially when two or three overhead lines run in parallel. Usually, by application of TPAR, the secondary arc will extinguish rapidly. An alternative measure to reduce the secondary-arc current and improve the reliability of SPAR is to apply *high-speed earthing switches* (HSES) that, in a very short time, connect the faulty conductor directly to earth, thus short-circuiting and eliminating the secondary arc [23, 24], see Section 10.4.2. Such measures facilitate successful auto-reclosure within one to two seconds [22, 25].

## 4.2.6   Capacitor-Bank Energization

### 4.2.6.1   Introduction

When a capacitive load is energized, it usually draws a certain *inrush current*. In the case where a lumped uncharged capacitor is connected to a voltage source, the sudden change of the capacitor voltage from zero to a certain source-voltage value involves a very large $du/dt$, leading to an inrush current. The inrush current is inversely proportional to the surge impedance of the load circuit. Capacitor banks have small surge impedance, far smaller than cables and lines. Because of their distributed nature, cables and lines have surge impedance from several tens (cables) to several hundreds of ohms (lines).

Due to the relatively high surge impedance, normally, the energization of cables and lines is not associated with inrush currents and related challenges to the circuit-breaker [26, 27]. Due to the low inductance of cable circuits, however, a very steep rate-of-rise of inrush current in cable systems is possible [28].

Energization of capacitor banks, however, having surge impedances of just a few ohms, may lead to very large inrush currents if no current-limiting measures have been taken.

The challenge of a capacitor-bank inrush current is twofold:

- For the switching device, the inrush current starts to flow at the moment of pre-strike, prior to contacts touching. Due to the high frequency of the inrush current, the peak value of the current appears during the pre-arc interval. This causes a high stress to the interrupter (see Section 4.2.6.2).
- For the system, depending on the topology of the capacitor bank(s), severe voltage transients can arise at the station bus, potentially causing *electromagnetic interference* (EMI), power-quality, and similar issues [29].

The severity of both transients depends greatly on the circuit topology of the capacitor banks. The situation is explained in Figure 4.22.

Two situations are distinguished:

- Single-bank topology (*one stand-alone capacitor bank*): a single bank (C) is energized with no other banks already connected to the busbar. In a simplified approach, it can be assumed that the inrush current is flowing mainly through the system's impedance

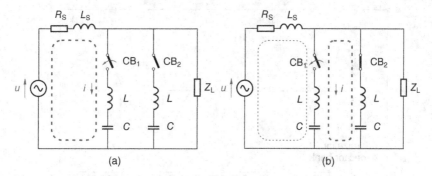

**Figure 4.22** Single-bank (a) and back-to-back (b) capacitor-bank switching topology. The dotted circuits indicate the inrush-current paths.

$X_S = \sqrt{R_S^2 + (\omega L_S)^2}$. This current path is indicated in Figure 4.22a. The advantage of this situation is that this impedance limits the inrush current. The drawback is that the busbar voltage is strongly affected by the switching operation, resulting in a severe transient, which is a power-quality issue. Electrically, the situation is shown in Figure 4.23.

Severe busbar transients can occur, though the peak inrush current is limited to several kiloamperes with a frequency of around several hundreds of hertz. Detailed calculations can be found in reference [27].

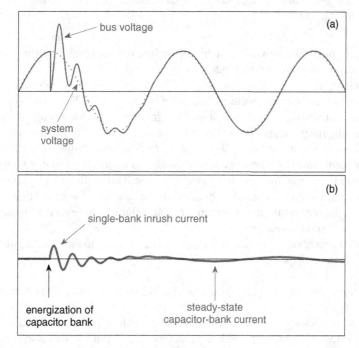

**Figure 4.23** Bus- and system-voltages (a) and breaker current (b) resulting from a single-bank capacitor-bank making operation.

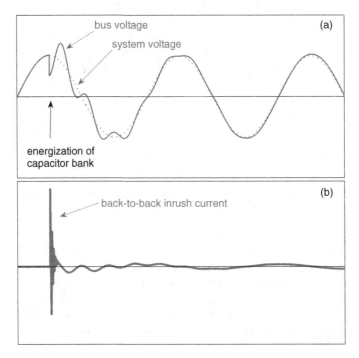

**Figure 4.24** Bus- and system-voltages (a) and breaker current (b) resulting from a back-to-back capacitor-bank making operation.

In the single-bank situation, the switching therefore imposes power-quality issues; and it stresses mainly the system and to a much lesser extent the circuit-breaker.

- Back-to-back topology (*parallel capacitor banks*): a single bank is energized while other bank(s) have already been connected to the same busbar before. This situation is shown in Figure 4.22b. The inrush current with a high magnitude may flow through the connection between neighbouring bank(s), limited only by the inductance of the busbar and leads or by current limiting elements, indicated in Figure 4.22b. The contribution of the source circuit is no longer dominant. The advantage here is an almost undisturbed busbar voltage, but the circuit-breaker is exposed to a very large inrush current. This is shown in Figure 4.24, where the peak value of the inrush current is 20 kA with a frequency of several kilohertz.

In practice, the inrush current can reach the peak values from several up to many tens of kiloamperes at several kilohertz.

In the back-to-back situation, the switching imposes stresses mainly on the circuit-breaker and much less so on the system.

In order to mitigate the effects of inrush current, the following measures are often considered:

1. Adding a series reactor: The intention is to reduce both the inrush current and its frequency. This is indicated in Figure 4.22 with the reactors *L*. The reactor will also limit potential restrike current upon de-energization and *outrush currents* as a result of faults near the capacitor-bank (see Section 10.7).

2. Application of controlled switching (see Section 11.4.5.1): In this case, the energization is chosen to coincide with the relevant voltage-zero crossings in each of the phases, resulting in virtually no inrush current. Although this method is widely applied, it does not reduce the restrike current, which is normally larger than the inrush current.
3. Insertion of non-linear elements, such as damping resistors into the circuit during the inrush period [30].

#### 4.2.6.2 Impact on Circuit-Breakers

Inrush current starts to flow at the moment when the contact gap of the circuit-breaker pre-strikes. From that moment on, the pre-arc will start and the inrush current supplies the pre-arc until the contacts establish galvanic connection. Depending on the frequency of the inrush current and the duration of the pre-arcing period, very high current can flow through the pre-arc.

Figure 4.25 shows a comparison of the arc energy, assumed to be proportional to the integral of the current, for the following cases:

- closing operation of a capacitor bank in a back-to-back topology with the peak value of the inrush current 20 kA (as specified in IEC 62271-100);
- a *symmetrical short-circuit current* of 50 kA;
- a fully asymmetrical short-circuit current of 50 kA.

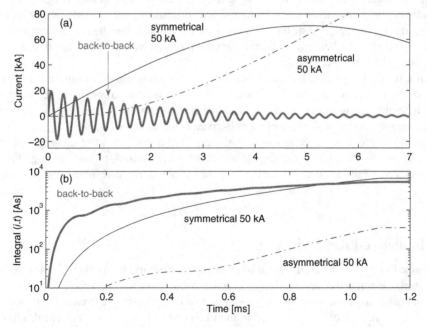

**Figure 4.25**   Currents (a) and integrated currents (b) for the making of a back-to-back capacitor bank, a symmetrical and an asymmetrical 50 kA, 50 Hz fault current.

From this, it is clear that the arc energy – and especially the rate of the energy input – is very large under back-to-back inrush conditions.

$SF_6$ circuit-breakers are exposed to major stresses to their contact system upon pre-strike when it is followed by inrush current having a rate-of-rise of hundreds of amperes per microsecond in the back-to-back case, as compared with the several tens of amperes per microsecond during closing operation with an ordinary power-frequency fault current. The steep rising current leads to extremely rapid heating and gas expansion in the inter-contact gap. The dielectric coordination between the main contacts and the arcing contacts during making must be such that pre-strikes will occur under all circumstances only between the arcing contacts. Due to the high frequency of the transient current of a capacitor-bank discharge, the skin-effect will force the arc foot-points to burn near the circumference of the stationary arcing contact rather than causing homogeneous and uniform erosion over the contact surface area. As a result, after many switching operations, the stationary contact takes on a conical structure instead of a rather hemi-spherically rounded form, which appears in the case of fault-current making [31]. The conical structure, in turn, has been observed to increase the probability of a pre-strike/restrike between the main contacts, leading to malfunction of the circuit-breaker [32].

In several cases, during the required visual contact inspection after back-to-back testing, punctures were found in the nozzle of HV circuit-breakers, even without restrikes [33]. Such punctures are detrimental for the pressure build-up, necessary for fault-current interruption.

Since vacuum circuit-breakers do not have separate main and arcing contacts, pre-arcing is on the same contacts that have to withstand the voltage in the open position. Due to the back-to-back pre-arc, the very high current causes local contact melting and, when the contacts touch each other, local welding of them may occur. The contact mechanism should be designed to break this weld, but remnants of the weld may still cause local surface irregularities that act as electric-field enhancing sites. If these micro-protrusions are not sufficiently removed by arcing during the opening of the contacts, they may impair the dielectric strength of the contact gap. Thus, a higher current during switching off, as well as longer arc duration, will reduce the effect of the weld remnants.

A full test of a back-to-back capacitor-bank energization is shown in Figure 4.26. Frame (a) shows the inrush current and its interruption; (b) shows the power-frequency capacitive current including the recovery voltage with a late restrike; and in (c), the restrike current of twice the inrush-current peak value can be observed (see Equation 4.2).

A common failure mode in testing is welding of the contacts after closing and stuck contacts. Also, the observed probability of late restrike during back-to-back testing [34] may be explained by an impaired dielectric integrity due to pre-arcing and subsequent welding.


## 4.3  Inductive-Load Switching

As discussed in Section 4.2.2, the main issue in capacitive-current switching is the release of energy stored as trapped charge in the load capacitor-bank due to restrike.

An analogous situation arises in the interruption of inductive-load currents. Here, however, the issue is the release of energy stored in the magnetic field due to the current through the inductive load.

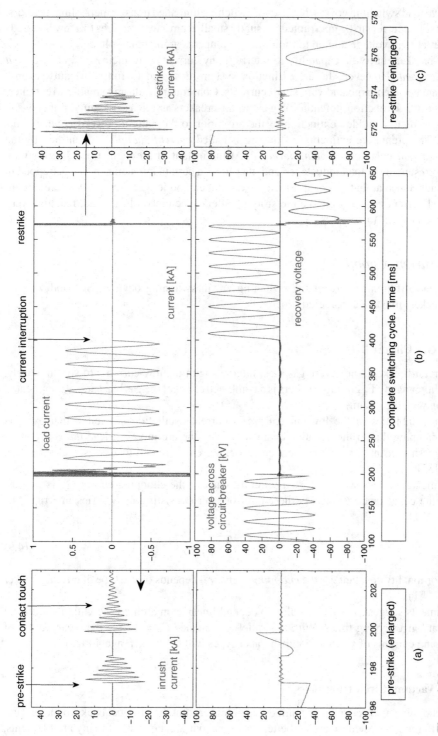

**Figure 4.26** Example of a test of a vacuum circuit-breaker in a back-to-back configuration (b), including enlarged views of inrush current (a) and restrike (c).

Inductive load switching covers the cases of switching unloaded transformers, shunt reactors, and motors. The current to be interrupted is usually small, from a few amperes to a few hundreds of amperes at high voltage and up to thousands of amperes at medium voltage.

Circuit-breakers have the capability to interrupt any current up to their *rated short-circuit breaking current*. However, the arc-extinction system, designed to fulfil this duty, cannot discriminate between large and very small currents. Consequently, during small-current interruption, the full extinction potential of gas circuit-breakers is released against rather weak arcs. This leads to a sudden extinction of the arc, prior to the natural power-frequency current zero. The premature extinction of the arc is called *current chopping*, quantified by the *chopping-current value* or *chopping level* $i_{ch}$. At the instant of arc extinction, the chopping current is present in the load inductance and, therefore, a certain magnetic energy is trapped in the load inductance at that moment. Since the current can no longer flow in the main circuit, the trapped current can only charge the stray parallel capacitance of the load, reaching very high overvoltages in some cases.

## 4.3.1   Current Chopping

Due the very different nature of the switching arc in switching devices, $SF_6$ and vacuum circuit-breakers must be considered separately.

### 4.3.1.1   Gas Circuit-Breakers

In $SF_6$ circuit-breakers, the severe gas blast interacts with the low-current arc very intensely. Current chopping in $SF_6$ circuit-breakers is a combined effect of the gas blast and a complicated interaction with the circuit.

Due to instabilities in the low-current arc, a current oscillation is excited that exhibits negative damping due to the negative characteristic of the arc in gases, leading to exponentially increasing instability and, consequently, the current is forced to zero (see also Section 10.10.3) [35].

Experiments have shown that for gas circuit-breakers, the chopping current $i_{ch}$ is proportional to the capacitance $C_p$ across the circuit-breaker consisting of $N$ series interrupters:

$$i_{ch} = \lambda \sqrt{NC_p} \qquad (4.9)$$

The proportionality constant $\lambda$ – the *chopping number* – depends on the type of circuit-breaker (see Table 4.1).

Experiments suggest that $\lambda$ tends also to depend linearly on arcing time with a significant increase at longer arcing times. With a parallel capacitance $C_p = 10$ nF, the mean values of the chopping currents of various circuit-breaker types will be in the range 4 to 20 A.

### 4.3.1.2   Vacuum Circuit-Breakers

Current chopping also occurs in vacuum circuit-breakers, but the physical background is completely different. The vacuum arc or, better, metal-vapour arc, consists of many arc filaments,

**Table 4.1**   Gas circuit-breaker chopping numbers.

| Circuit-breaker type | Chopping number $\lambda$ [A F$^{-0.5}$] |
| --- | --- |
| Minimum-oil | (5.8 to 10) $\times$ 10$^4$ |
| Air-blast | (15 to 20) $\times$ 10$^4$ |
| SF$_6$ puffer | (4 to 19) $\times$ 10$^4$ |
| SF$_6$ self-blast | (3 to 10) $\times$ 10$^4$ |
| SF$_6$ rotating arc | (0.4 to 0.8) $\times$ 10$^4$ |

with foot-points at the negative contact, the cathode, called *cathode spots* [36]. Each cathode spot carries a fraction of the current at about 30 to 50 A (see Section 8.3.3). Due to the appearance of the metal-vapour arc in a discrete number of vapour-emitting cathode spots proportional to the current magnitude, their number declines with the falling slope of the power-frequency current sine-wave. Near current zero, the arcing medium is supplied by only a single cathode spot, which cannot sustain itself fully down to current zero. Due to lack of supply of ionized material by the single, weakly emitting low-current cathode spot to the inter-contact gap, space charges build up that cause a very short-lasting temporary arc voltage increase. These arc voltage spikes interact with the stray circuit elements and force the arc current to zero prior to the natural zero [37].

The chopping current is to a high degree determined by the contact material supplying the arcing medium and is typically 2 to 10 A for circuit-breakers [38], but can be as small as 0.1 A for special-duty interrupters, for example, for contactor functions, having contact material optimized for low chopping currents. Chopping current in vacuum interrupters is dependent on the circuit parameters, however, not on simple equations. The chopping current increases slightly with parallel capacitance [39], and also with the circuit's surge impedance [40].

In the early days of the development of vacuum circuit-breakers, *oxygen-free high conductivity* (OFHC) copper was used as contact material. This resulted in rather high current chopping levels and current chopping was a serious problem. For many years, because of the application of special contact material (see Section 9.2), the chopping current levels of vacuum circuit-breakers have not differed from those of SF$_6$ circuit-breakers.

### 4.3.2   Implication of Current Chopping

At interruption, the presence of a large load inductance will result in a small inductive current and subsequent chopping. The inductive-load current is much smaller than the normal-load current, and thus easily interrupted by circuit-breakers. The energy $E_m$ trapped in the load with inductance $L_L$ at the moment of chopping is:

$$E_m = \frac{1}{2}L_L\, i_{ch}^2 \tag{4.10}$$

Inductive-load switching is a special case because it is a highly interactive event between the circuit-breaker and the circuit in which it is applied. First, the chopping process itself is the result of a high-frequency interaction with the arc as a physical entity and the stray circuit elements.

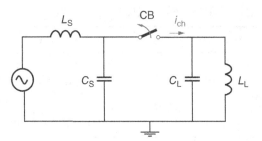

**Figure 4.27**   Basic circuit for inductive-load switching.

Even more important for the user of switchgear is the fact that the consequences of current chopping are also directly dependent on the circuit-breaker and the circuit parameters. The circuit is shown schematically in Figure 4.27, where $C_L$ denotes the (normally) small stray capacitance of the load.

As regards the energy in the load at the moment of current chopping, a simple equation can be set up:

$$E_{\text{load}} = E_e + E_m = \frac{1}{2}C_L \hat{U}^2 + \frac{1}{2}L_L\, i_{\text{ch}}^2 \tag{4.11}$$

assuming single-phase current and chopping at the peak value $\hat{U}$ of the system voltage. The load circuit starts to oscillate immediately at the current chopping, exchanging energy between $L_L$ and $C_L$. The resulting voltage oscillation has its maximum value when the full load energy $E_{\text{load}}$ is stored in the capacitance:

$$\frac{1}{2}C_L\, u_m^2 = E_{\text{load}} \quad \Rightarrow \quad u_m = \sqrt{\hat{U}^2 + \frac{L_L}{C_L}\, i_{\text{ch}}^2} = \sqrt{\hat{U}^2 + Z_0^2\, i_{\text{ch}}^2} \tag{4.12}$$

From this equation, it is clear that the maximum voltage across the load reaches a value that is dependent on a circuit-breaker characteristic, the chopping-current $i_{\text{ch}}$ level, and additionally on a circuit characteristic, namely the load surge impedance $Z_0 = \sqrt{L_L/C_L}$. The absolute maximum value of the TRV, without damping, is:

$$U_{\text{TRVmax}} = \hat{U} + \sqrt{\hat{U}^2 + Z_0^2\, i_{\text{ch}}^2} \tag{4.13}$$

From this, it is clear that in inductive-load switching, the TRV maximum can reach very high values. Of all the switching cases studied so far, the potentially highest TRV is attributed to inductive-load switching. From the equations above it is clear that the generated overvoltages are higher at lower current with the same system voltage, because the load inductance is higher at lower current. Also, small stray capacitances lead to higher overvoltages. On the other hand, chopping overvoltages can be reduced by adding capacitance across the load, although higher capacitance on the other hand may increase the value of the chopping current itself.

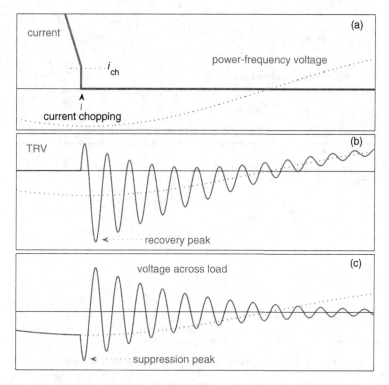

**Figure 4.28** Example of current chopping (a) with associated TRV (b) and load-side voltage (c).

The electrical switching process is outlined in Figure 4.28. As soon as the current chops to zero, the TRV starts to rise, but in this switching duty, it rises with the same polarity as the previous arc voltage before it crosses zero. The maximum load-side voltage during this time is called the *suppression peak*. The absolute maximum of the TRV without re-ignition is called the *recovery peak*.

In Figure 4.29, a comparison is plotted between current interruption with and without current chopping. The impact on TRV can be clearly seen.

## 4.3.3  Inductive-Load Switching Duties

In practice, the interruption of inductive load can occur in the following switching duties:

- unloaded transformer switching;
- shunt-reactor switching; and
- MV-motor switching.

A vast volume of literature is available on the switching of inductive currents and its repercussions for the circuit. An extensive overview can be found in the legacy of CIGRE Working Group 13.02 [41], summarized in the application guide [42] to the IEC circuit-breaker standard [10].

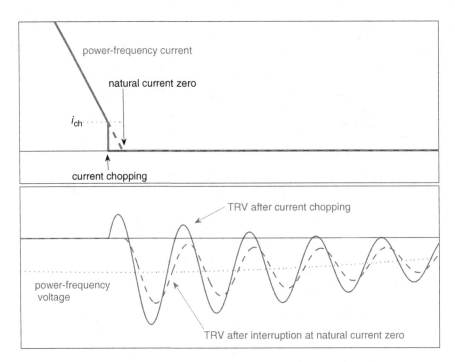

**Figure 4.29** Comparison of TRVs: with current chopping (drawn line) and without current chopping (dashed line).

### 4.3.3.1   No-Load Transformer Switching

The current to be interrupted in this switching duty is the transformer magnetizing current, having a very small value of a few amperes at most but more often less than one ampere. This implies that even the current peak value remains below the chopping-current level, which means that the current is interrupted immediately at the contact separation of the circuit-breaker. Re-ignition, even with repetitive occurrences, will follow, but at a very low voltage level.

Experience shows that significant overvoltages are not typically generated by the interruption of the steady-state no-load current of a modern transformer. However, in the case of interruption of the transformer inrush current, overvoltage generation may be considered [41]. In this case, currents up to 10 to 15 times the rated normal current may arise. It is common practice to avoid the effects of transformer inrush current by blocking the protective relay when the ratio of inrush current to fault current, measured by harmonic currents, is too large. Alternatively, the transformer can be protected at the high-voltage side phase-to-phase and phase-to-earth with surge arresters [43].

HV transformers are not switched frequently in practice, and the interruption of inrush current in particular is rare, since it would mean interruption very soon after energization.

The possibility of excitation of internal resonances due to repetitive re-ignitions has been discussed [21] but the probability is considered negligible [44]. A case where voltage escalation due to multiple re-ignitions might occur is the switching of dry-type transformers with vacuum circuit-breakers or switches. However, in this case, the excitation of resonating overvoltages seems to be very unlikely due to the statistical variation of the re-ignition rate of occurrence.

As a general conclusion, the current interruption of modern transformers in no-load condition with modern circuit-breakers will usually give overvoltages below the insulation levels specified by the standards. For HV transformers ($\geq$ 72.5 kV), the maximum overvoltages are reported [41] to be generally below 1.5 p.u.; even in the most onerous condition of very high power and high magnetizing current, the maximum overvoltage values remain below 4 p.u.

For MV transformers, the maximum overvoltages will normally be below 2.5 p.u. and in rare cases for dry-type transformers below 4 p.u.

For the special cases with older types of transformers with hot rolled steel cores and/or heavy duty air-blast circuit-breakers, overvoltages can be estimated with sufficient accuracy by calculation methods [41], the chopping characteristics being provided by reactor switching tests.

### 4.3.3.2 Shunt-Reactor Switching

Shunt-reactor switching is by far the most common practice of the inductive-load switching. Shunt reactors are installed for overhead-line capacitance compensation, and switched in or out depending on the momentary line load. This implies very frequent switching operations (similar to capacitor-bank switching, see Section 4.2.6) and thus, merely from the number of required switching operations, this duty needs special attention, not only electrically, but also mechanically.

Since the shunt reactor can be treated as a lumped circuit element having a stray capacitance, the equivalent load circuit can be simplified to a straightforward $LC$ circuit.

At interruption, actually current chopping, the $LC$ circuit produces voltage oscillations, reaching a maximum voltage $U_{mp}$, which is the 1 p.u. of the system voltage augmented with the additional current-chopping contribution. Generally, the single-frequency oscillatory TRV is of high frequency, standardized by IEC 62271-110 [45] to values between 6.8 kHz at the rated voltage of 72.5 kV and 1.5 kHz at 800 kV.

Similar to the capacitive-current switching, the reactor current is low enough that interruption can occur after a very short arcing time. This implies that the circuit-breaker gap might have not yet reached a sufficient spacing at current zero to withstand the TRV and, if so, a breakdown is the consequence. In this case, the breakdown is termed a *re-ignition* because the high-frequency TRV makes it happen within a quarter of a power-frequency period after interruption.

Unlike a restrike in capacitive circuits, the energy delivered to the inductive re-ignition discharge is relatively low, being the discharge of the stray capacitance; a high-frequency re-ignition current will flow briefly and the gap may or may not recover from the event.

During the flow of re-ignition current, the opening gap reaches only a slightly higher breakdown voltage and the subsequent higher TRV, after re-ignition current interruption, will again lead to re-ignition. This is more likely to occur, because, during the brief conducting period, the power-frequency current in the reactor increases slightly, causing the second TRV to be steeper and potentially higher than the previous one. The sequence of re-ignitions is called *multiple re-ignitions* (MR) or *repetitive re-ignition* and the gradual increase in the re-ignition voltage value is called *(inductive) voltage escalation*.

As a result of this process, a number of re-ignitions will have occurred before the gap reaches sufficient withstand strength. In this case, the interruption is successful but accompanied with multiple re-ignitions. The other alternative, especially after very short arcing times, is that, after a few re-ignitions, another loop of power-frequency current will flow and potentially clear at the end of its half cycle when the gap has gained significantly higher dielectric strength.

Multiple re-ignitions may be onerous for gas and oil circuit-breakers, and this is the reason why the shunt-reactor switching is sometimes termed "a circuit-breaker's nightmare" – also and not least because it is a daily switching operation [46].

Extremely high switching overvoltage values above 3 p.u. have been observed after multiple re-ignitions and the internal parts of circuit-breakers were regularly found damaged [47]. Mitigation of overvoltages that result from shunt-reactor switching is discussed in Section 11.5.2.

For vacuum circuit-breakers, the re-ignition current is unlikely to damage the contact system, but due to its capability of interrupting a current of very high frequency, the conducting period of the re-ignition current is very short, making the number of multiple re-ignitions significantly higher than in $SF_6$ circuit-breakers. Damage due to multiple re-ignitions is sometimes observed [48].

An example from a $SF_6$ circuit-breaker test is presented in Figure 4.30. Seven re-ignitions can be observed before recovery is achieved. Immediately after re-ignition, a re-ignition current of very high frequency keeps the gap conducting during approximately 100 µs. The maximum voltage reached across the load reactor is 2.3 p.u. Without re-ignitions it would have been 1.08 p.u. due to the very small chopping current. The peak value of TRV is 3.3 p.u.

In this case, despite the very small chopping current, the load voltage reached after multiple re-ignitions escalates to a very high value.

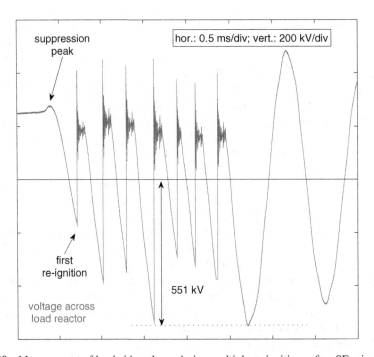

**Figure 4.30**  Measurement of load-side voltage during multiple re-ignitions of an $SF_6$ circuit-breaker.

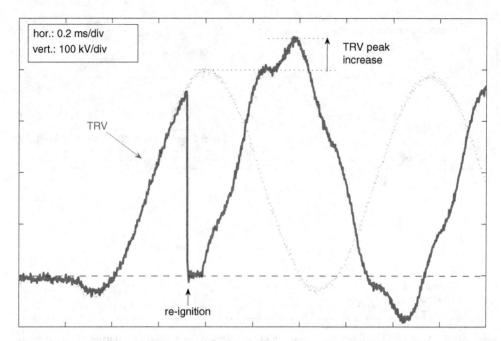

hor.: 0.2 ms/div
vert.: 100 kV/div

TRV peak
increase

TRV

re-ignition

**Figure 4.31** Comparison of TRV in the case of re-ignition (drawn line) and without re-ignition (dotted line).

This demonstrates that, contrary to common belief, it is the multiple re-ignitions process, rather than the current chopping itself, that makes reactor switching potentially hazardous.

In Figure 4.31, a detail of the measured TRV oscillation of two tests is superimposed, one with a single re-ignition and the other without, clearly demonstrating the voltage-increasing effect of a single re-ignition.

In Figure 4.32, interruption of inductive-load current by a vacuum circuit-breaker is shown. The upper part shows the first pole-to-re-ignite (also the last-to-clear) whereas the lower part shows the enlargement of the multiple re-ignitions, seen in the upper part after 0.5 ms of arcing time. Again, significant overvoltages, up to 3.3 p.u., with steep fronts are produced.

Critical $SF_6$ chamber parts, like the nozzle, can be damaged by the very high and high-frequency re-ignition current that excites shock waves in the interruption chamber [49]. In addition, the internal voltage coordination in the interruption chamber may change because of the extremely high d$i$/d$t$ associated with the re-ignition current, causing the re-ignition to take place between other parts of the contact system than the arcing contacts. Punctures of the nozzle have been observed from time to time, as well as discharge traces on the main contacts.

The influence of shunt-reactor switching on the circuit-breaker condition has been the subject of several studies. Practical application guidelines are provided in the relevant IEC [42] and IEEE application guides [50].

Multiple re-ignitions of breakers during shunt-reactor switching can be avoided by controlled switching. Using controlled, instead of random contact separation well in advance of current zero, avoids short arcing times and interruption follows when the gap has reached sufficient spacing. This technology is discussed in Section 11.4.5.1.

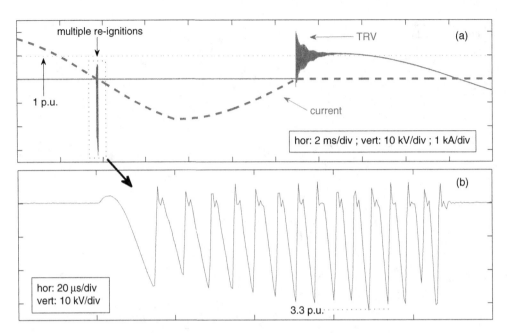

**Figure 4.32**  Multiple re-ignitions in vacuum circuit-breaker. (a) current and TRV during the interruption attempt and interruption; (b) enlarged multiple re-ignitions at the first current zero.

In non-controlled applications, it is generally accepted that re-ignition is an integral part of inductive-load switching, and tests are only intended to demonstrate that the occurrence of (multiple) re-ignitions is limited to one current-zero region only. Circuit-breakers have to be designed to withstand these very frequently occurring electrical shocks. This is a difference from capacitive-current switching tests that are intended to demonstrate a low or very low probability of restrikes for a particular circuit-breaker design.

Due to the breakdown processes in switching gaps, that take place in tens of nanoseconds, the $du/dt$ values are very high, causing extremely steep-fronted waves. This poses a threat to the neighbouring equipment in substations – notably transformers.

The voltage across the windings in transformers is equally distributed across each turn as long as the frequency content of the voltage does not extend to values above 1 MHz. However, as soon as impulse voltages become so steep that frequency components above 1 MHz become dominant, the voltage distribution inside the winding becomes non-homogeneous. There will be disproportional stresses across the terminal turns, where the voltage wave enters the winding.

Although HV transformers are electrically reinforced at their terminals, stresses due to steep-fronted waves, initiated by re-ignitions, may be a challenge. Any capacitance in the system (such as a bushing capacitance in metal-enclosed switchgear and transformers) mitigates these stresses.

In addition, there are *electromagnetic compatibility* (EMC) related concerns due to the high penetration potential of steep-fronted waves in secondary equipment for control, measurement, diagnostics, and so on [5].

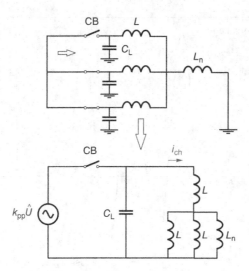

**Figure 4.33** Circuit simplification (for calculation) from a three-phase circuit to a single-phase circuit.

### 4.3.3.3 Calculation of Shunt-Reactor Switching Overvoltages in Three-Phase Circuits

A large body of literature exists on the calculation of overvoltages at shunt-reactor switching in three-phase systems [41, 42].

The initial approach is as follows:

In the first step, the general three-phase circuit is reduced to a single-phase circuit (see Figure 4.33) having a source voltage $k_{pp}\hat{U}$ and an effective load-inductance $L'$:

$$L' = L\left(1 + \frac{1}{2 + \dfrac{L}{L_n}}\right) = k_{pp}L \tag{4.14}$$

For solidly earthed shunt reactors: $L_n \rightarrow 0, \ k_{pp} = 1.0 \ L' = L$

For non-effectively earthed shunt reactors: $L_n \rightarrow \infty, k_{pp} = 1.5 \ L' = 1.5 \ L$

As explained in Section 4.3.2, the energy stored in the load network at the moment of current chopping will transfer completely at a certain moment into the capacitor $C_L$, charging it to a voltage $u_m$:

$$\frac{1}{2}C_L u_m^2 = \frac{1}{2}C_L \left(k_{pp}\hat{U}\right)^2 + \frac{1}{2}L' i_{ch}^2 \qquad \Rightarrow$$

$$\frac{u_m}{\hat{U}} = k_{pp}\sqrt{1 + \frac{1}{k_{pp}}\left(\frac{i_{ch}}{\hat{U}}\right)^2\left(\frac{L}{C_L}\right)} \equiv k_a + k_{pp} - 1 \tag{4.15}$$

From this, the suppression peak overvoltage $k_a$ can be calculated:

$$k_a = k_{pp}\sqrt{1 + \frac{1}{k_{pp}}\left(\frac{i_{ch}}{\hat{U}}\right)^2\left(\frac{L}{C_L}\right)} - k_{pp} + 1 \tag{4.16}$$

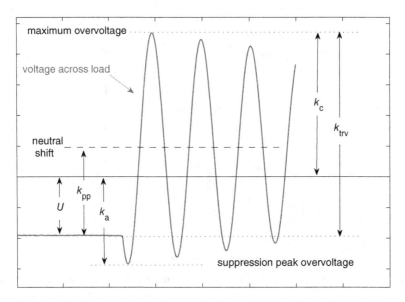

**Figure 4.34**  Symbols for the characterization of TRV in inductive-load testing.

The recovery-voltage peak $k_c$ (the highest voltage across the reactor) is then:

$$k_c = 2\,k_{pp} + k_a - 2 \tag{4.17}$$

Whereas the TRV peak across the circuit-breaker $k_{trv}$ (ignoring damping) is given as:

$$k_{trv} = 2\,k_{pp} + k_a - 1 \tag{4.18}$$

The relevant symbols are entered in Figure 4.34, taken from the first-pole-to-clear in an interruption test with a vacuum circuit-breaker ($U_r = 12$ kV, current 500 A) in a non-solidly earthed neutral three-phase system ($k_{pp} = 1.5$).

The characteristic overvoltage parameters in this example are:

- $k_a = 1.45$ p.u.;
- $k_c = 2.4$ p.u. (compared with a prospective value of 1.85 without current chopping);
- $k_{trv} = 3.4$ p.u. (compared with a prospective value of 2.85 p.u. without current chopping).

As a consequence of current chopping, the load voltage is increased by 0.55 p.u., or 30%. The frequency of the TRV was 19.2 kHz, corresponding to a load capacitance $C_L = 1$ nF.

### 4.3.3.4   High-Voltage Motor Switching; Virtual Current Chopping

A third major application of inductive-load switching is the de-energization of MV motors during their start-up or during stall. It is only during these operation modes that motors are inductive loads, given that, in steady-state running, the power factor is close to one and interruption is not associated with significant TRV. Since the mechanical time constants are much larger than the electrical ones, running motors hardly lose speed during the TRV period and maintain a load-side electromotive force during the slowing down of the rotation.

In most cases, vacuum circuit-breakers or contactors are used for this switching duty, due to the typical MV range in which they are applied. A prominent property of vacuum switching devices is that they are capable of interrupting high-frequency re-ignition currents much better than $SF_6$ circuit-breakers [51]. Values of $di/dt$ well above several hundreds of amperes per microsecond have been reported to be in the interruptable range [52]. In addition, the TRV frequencies are much higher in distribution systems compared with transmission systems. Therefore, the duration of the recovery periods as well as the duration of periods of conduction are much shorter than in HV $SF_6$ applications (as shown in Figure 4.35) and re-ignitions in MV systems succeed each other at a much faster rate. Therefore, the number and the repetition rate of re-ignitions is higher, and voltage can escalate faster and to higher per unit levels.

Another difference with HV shunt-reactor switching is the compactness of circuits at medium voltage. The phase distance between conductors, for example busbars and conductors in belted three-phase cables, is small. This enables a significant electromagnetic coupling between phases, whereby transients in one phase are coupled to the neighbouring phases through mutual inductance and phase-to-phase capacitance [53].

This may have severe consequences, as outlined schematically in Figure 4.35. Note that the re-ignition current and frequency are not to scale. Part of the re-ignition current in the re-igniting

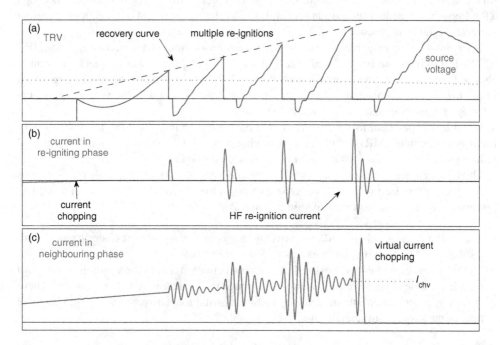

**Figure 4.35** Virtual current chopping (schematically, from a simulation). (a) TRV with multiple re-ignitions; (b) re-ignition current in re-igniting pole; (c) current in neighbouring arcing pole.

phase is coupled inductively into the neighbouring-phase conductors and is superimposed upon the power-frequency current. When neighbouring pole is still arcing – the re-igniting pole being the first-pole-to-clear – and the induced high-frequency current is large enough to force the power-frequency current to zero in that pole, a vacuum interrupter may interrupt at this high-frequency current zero. In this situation, the power-frequency current is chopped to zero from a very high value $I_{chv}$, much higher than the conventional current-chopping level. The forced chopping in the non-re-igniting pole, as illustrated here, is called *virtual current chopping* (VCC). It is a three-phase phenomenon and is practically only observed in vacuum switching devices. Due to the high virtual current-chopping level, extreme overvoltages can develop.

Virtual current chopping is always initiated by multiple re-ignitions, but only occasionally multiple re-ignitions events will lead to virtual chopping. No virtual current chopping occurs for example:

- if the power-frequency current is too high and high-frequency current zero cannot be reached in the neighbouring pole; or
- if the mutual inductance/capacitance between phases is too small and induced re-ignition current is too small; or
- if the surge impedance of the re-ignition circuit is too high and the re-ignition current itself is too small.

In all these cases, multiple re-ignitions is still possible because this is determined by arcing time (since short arcing time enhances the re-ignition probability), TRV frequency (since high TRV frequency enhances re-ignition probability) and chopping current (high chopping current increases re-ignition probability) [54].

In Figure 4.36, an example is given of a MV-motor switching test in accordance with IEC 62271-110 [45]. It can be observed that in the middle pole, the power-frequency current is chopped to zero from an instantaneous value of 237 A by escalating multiple re-ignitions in the lower pole, in this example, the first-pole-to-clear. As a result, the voltage across the interrupter rises to almost 100 kV, nearly 10 p.u.

In Table 4.2, practical factors are given that have a mitigating effect on the probability of multiple re-ignitions (MR) and virtual current chopping (VCC). The factors marked with "+" reduce probability, factors marked with "−" increase probability.

Though, for a number of reasons, virtual chopping is a rather remote possibility, re-ignitions are very common, having a possible impact on apparatus in the vicinity of the breaker. Re-ignitions may stress the motor insulation in two ways:

- The escalating peak voltage and/or the virtual-chopping overvoltage stress the phase-to-earth insulation [55]. Overvoltages in excess of 5 p.u. are reported [41].
- The steepness of the surges stresses the insulation between turns of the winding. The hazard is, as in the case of shunt-reactor switching, through the breakdown due to re-ignition. These steep surges lead to an inhomogeneous voltage distribution across the winding, if the rise time of the surge is sufficiently short [56].

As with transformers (see Section 4.3.3.2), the impulse withstand level of motors is greatly reduced by voltages having a rise time shorter than 1 μs. A manufacturer enquiry on 6.6 kV

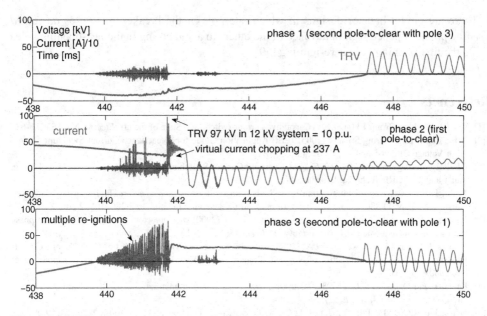

**Figure 4.36**   Three-phase measurement of virtual chopping in a 12 kV motor-switching test circuit.

motors shows a reduction of the impulse-voltage withstand level from 6 p.u. at 1.5 μs rise time of the transient to 2 p.u. at 0.1 μs rise time [41]. Unlike transformers, testing of the ability of motor insulation to withstand steep-fronted voltage surges is not standardized [57].

Protection is recommended [58, 59], especially when currents are below 600 A. Various methods are available: surge capacitors or *RC* snubbers reduce the steepness but do not protect against high overvoltages [60]. Surge arresters may limit the phase-to-earth overvoltages but do not protect against steep fronts (see Section 11.3.3.2).

**Table 4.2**   Factors influencing multiple re-ignitions MR and virtual current chopping VCC.

| Circuit-breaker parameter | MR | VCC | Interaction mechanism |
|---|---|---|---|
| High motor power | – | + | lower probability of current zero in neighbouring phase |
| Long load cables | + | + | lower TRV frequency |
| Screened cables | | + | weak inter-phase capacitive coupling |
| Belted cables | | – | strong inter-phase capacitive coupling |
| Long busbar | | – | strong inductive inter-phase coupling |
| Short phase distance | | – | strong inductive inter-phase coupling |
| Short arcing time | – | – | dielectric withstand developed insufficiently |
| SF$_6$ interruption medium | | + | no or late interruption of re-ignition current |
| High chopping current | – | – | higher TRV peak value |
| Fast opening breaker | + | + | faster recovery |
| Delayed pole opening | | + | arcing in neighbouring pole during re-ignition not possible |

There are circuit-breaker designs in which one pole of the breaker, known as *advanced opening pole*, is opened well in advance of the others to avoid arcing in the last-opening poles while the first-opening pole is re-igniting [59].

# References

[1] CIGRE Working Group 13.04 (1994) Capacitive current switching – State of the art, *Electra*, No. **155**, 33–63.
[2] IEEE Power Engineering Society (2005) IEEE Application Guide for Capacitance Current Switching for AC High-Voltage Circuit-Breakers, IEEE Std. C37.012.
[3] CIGRE Working Group 13.04 (1996) Line-Charging Current Switching of HV Lines – Stresses and Testing, Part 1 and 2, CIGRE Technical Brochure 47.
[4] Peelo, D.F. (2004) Current interruption using high-voltage air-break disconnectors. Ph.D. Thesis Eindhoven University, ISBN 90-386-1533-7 (available through http://alexandria.tue.nl/extra2/200410772.pdf).
[5] Smeets, R.P.P., van der Linden, W.A., Achterkamp, M. *et al.* (2000) Disconnector switching in GIS: three-phase testing and – phenomena. *IEEE Trans. Power Deliver.*, **15**(1), 122–127.
[6] Smeets, R.P.P. and Lathouwers, A.G.A. (2000) Capacitive Current Switching Duties of High-Voltage Circuit-Breakers: Background and Practice of New IEC Requirements. IEEE PES Winter Meeting, Singapore, Paper No. 2000WM-690.
[7] Fu, Y.H. and Damstra, G.C. (1993) Prestriking overvoltages during energizing capacitive load by a vacuum circuit-breaker. *IEEE Trans. Electr. Insul.*, **28**(4), 657–666.
[8] van Sickle, R.C. and Zaborszky, J. (1951) Capacitor switching phenomena. *AIEE Trans.*, **70**, 151–159.
[9] Johnson, I.B., Schultz, A.J., Schultz, N.R. and Shores, R.B. (1955) Some fundamentals on capacitive switching. *Proc. AIEE*, **74** (Pt III), 727–736.
[10] IEC 62271-100 (2012) High-voltage switchgear and controlgear – Part 100: Alternating-current circuit-breakers, Ed. 2.1.
[11] Smeets, R.P.P. and Lathouwers, A.G.A. (2002) Non-sustained disruptive discharges: test experiences, standardization status and network consequences. *IEEE Trans. Diel. Electr. Insul.*, **9**, 194–200.
[12] Bernauer, C., Knie, E. and Rieder, W. (1994) Restrikes in vacuum circuit-breakers within 9s after current interruption. *Eur. Trans. on Electr. Power*, **4**(6), 551.
[13] Gebel, R. and Falkenberg, D. (1993) Mechanical shocks as cause of late discharges in vacuum circuit-breakers. *IEEE Trans. Dielct. Electr. Insul.*, **28**, 468–72.
[14] Jüttner, B., Lindmayer, M. and Düning, G. (1999) Instabilities of prebreakdown currents in vacuum I: late breakdowns. *J. Phys. D: Appl. Phys.*, **32**, 2537–2543.
[15] Smeets, R.P.P., Thielens, D.W. and Kerkenaar, R.P.W. (2005) The duration of arcing following late breakdown in vacuum circuit-breakers. *IEEE Trans. Plasma Sci.*, **33**(5), 1582–1587.
[16] Smeets, R.P.P., Lathouwers, A.G.A. and Falkingham, L. (2004) Assessment of Non-Sustained Disruptive Discharges (NSDD) in Switchgear. CIGRE Conference, Paris.
[17] Bortnik, I.M., *et al.* (1988) 1 200 kV Transmission line in the USSR. The first results of operation. CIGRE Conference, Paper 38-09, Paris.
[18] Zaima, E., Shindo, T. and Ishii, M. (2007) System Aspects of 1 100 kV AC Transmission Technologies in Japan: Solutions for Network Problems Specific to UHV AC Transmission System and Insulation Coordination. Int. IEC/CIGRE Symp. on Standards for Ultra High Voltage Transmission, Beijing.
[19] Shu, Y. (2005) Development of Ultra High Voltage Transmission Technology in China. Proc. of the XIVth Int. Symp. on High Voltage Eng., Paper K-01.
[20] Nayak, R.N., Bhatnagar, M.C., De Bhowmick, P.L. and Tyagi, R.K. (2009) 1 200 kV Transmission System and Status of Development of Substation Equipment/Transmission Line Material in India. Second Int. IEC/CIGRE Symp. on Standards for Ultra High Voltage Transmission, New Delhi.
[21] Greenwood, A. (1991) *Electrical Transients in Power Systems*, John Wiley & Sons Ltd, ISBN 0-471-62058-0.
[22] CIGRE Working Group A3.22 (2011) Background of Technical Specifications for Substation Equipment exceeding 800 kV AC, CIGRE Technical Brochure 456.
[23] Agafonov, G.E., Babkin, I.V., Berlin, B.E. *et al.* (2004) High-Speed Grounding Switch for Extra High Voltage Lines. CIGRE Conference, Paper A3-308, Paris.

[24] Toyoda, M. *et al.* (2011) Considerations for the standardization of high-speed earthing switches for secondary arc extinction on transmission lines. CIGRE Colloquium, Paper A3-103, Bologna.

[25] CIGRE Working Group A3.22 (2008) Technical Requirements for Substation Equipment exceeding 800 kV, CIGRE Technical Brochure 362.

[26] CIGRE SC 13 (1979) Circuit-breaker Stresses when switching Back-to-Back Capacitor Banks, *Electra*, No. 62, pp. 21–45.

[27] IEC 62271-306 (2012) High-voltage switchgear and controlgear – Part 306: Guide to IEC 62271-100, IEC 62271-1, and IEC standards related to alternating current circuit-breakers.

[28] Kalyuzhny, A. (2013) Switching capacitor bank back-to-back to underground cables. *IEEE Trans. Power Deliver.*, **28**(2), 1128–1137

[29] CIGRE Working Group 36.05/CIRED 2 CC02 (2001) Capacitor Switching and its Impact on Power Quality, *Electra*, No. 195, pp. 27–37.

[30] Sabot, A., Morin, C., Guilliaume, C. *et al.* (1993) A unique multiporpose damping circuit for shunt capacitor bank switching. *IEEE Trans. Power Deliver.*, **8**(3), 1173–1183.

[31] da Silva, F., Bak, C. and Hansen, M. (2010) Back-to-back energization of a 60 kV cable network – inrush currents phenomenon. IEEE PES General Meeting, pp. 1–6.

[32] Kasztenny, B., Voloh, I., Depew, A. *et al.* (2008) Re-strike and Breaker Failure Conditions for Circuit-Breakers Connecting Capacitor Banks. 61st Ann. Conf. for Prot. Relay Eng., pp. 180–195.

[33] Smeets, R.P.P. and Lathouwers, A.G.A. (2000) Capacitive Current Switching Duties of High-Voltage Circuit-Breakers: Background and Practice of New IEC Requirements. IEEE PES Winter Meeting, Singapore, Paper No. 2000 WM-690.

[34] Smeets, R.P.P., Wiggers, R., Bannink, H. *et al.* (2012) The Impact of Switching Capacitor Banks with Very High Inrush Current on Switchgear. CIGRE Conference, Paper A3-201.

[35] van den Heuvel, W.M.C. (1980) Current chopping induced by arc collapse. *IEEE Trans. Plasma Sci.*, **8**(4), 326–331.

[36] Anders, A. (2008) *Cathodic Arcs. From Fractal Spot to Energetic Condensation*, Springer, ISBN 978-0-387-79107-4.

[37] Smeets, R.P.P. (1986) Stability of low-current vacuum arcs. *J. Phys. D: Appl. Phys.*, **19**, 575–587.

[38] Holmes, F.A. (1974) An Empirical Study of Current Chopping by Vacuum Arcs, IEEE Paper C-74-088-11.

[39] Damstra, G.C. (1976) Influence of Circuit Parameters on Current Chopping and Overvoltages in Inductive MV Circuits. CIGRE Conference, Paper 13-8.

[40] Slade, P.G. (2008) *The Vacuum Interrupter. Theory, Design, and Application*, CRC Press, ISBN 978-0-8493-9091-3.

[41] CIGRE Working Group 13.02 (1995) Interruption of Small Inductive Currents, CIGRE Technical Brochure 50 (ed. S. Berneryd).

[42] IEC 62271-306 (2012) High-voltage switchgear and controlgear – Part 306: Guide to IEC 62271-100, IEC 62271-1, and IEC standards related to alternating current circuit-breakers.

[43] Liljestrand, L., Lindell, L., Bormann, D. *et al.* (2013) Vacuum Circuit Breaker and Transformer Interaction in a Cable System. 22nd Int. Conf. on Electr. Distr., CIRED, Paper 0412.

[44] CIGRE Working Group 12.07 (1984) Resonance Behaviour of High-Voltage Transformers. CIGRE Conference, Report 12–14.

[45] IEC 62271-110 (2012) High-voltage switchgear and controlgear – Part 110: Inductive load switching.

[46] Peelo, D.F., Avent, B.L., Drakos, J.E. *et al.* (1988) Shunt reactor switching tests in BC hydro's 500 kV system. *IEE Proc.-C*, **135**, (5), 420–434.

[47] Bachiller, J.A., Cavero, E., Salamanca, F. and Rodriguez, J. (1994) The Operation of Shunt Reactors in the Spanish 400 kV Network – Study of the Suitability of Different Circuit-Breakers and Possible Solutions to Observed Problems. CIGRE Conference, Paper 23–106.

[48] Du, N., Guan, Y., Zhang, J. *et al.* (2011) 'Phenomena and mechanism analysis on overvoltage caused by 40.5-kV vacuum circuit-breakers switching off shunt reactors. *IEEE Trans. Power Deliver.*, **26**(4), 2102–2110.

[49] Tanae, H., Matsuzaka, E., Nishida, I. *et al.* (2004) High-frequency reignition current and its influence on electrical durability of circuit-breakers associated with shunt-reactor current switching. *IEEE Trans. Power Deliver.*, **19**(3), 1105–1111.

[50] IEEE Switchgear Committee (2009) IEEE Guide for the Application of Shunt Reactor Switching, IEEE Std C37.15.

[51] Smeets, R.P.P., Funahashi, T., Kaneko, E. and Ohshima, I. (1993) Types of reignition following high-frequency current zero in vacuum interrupters with two types of contact material. *IEEE Trans. Plasma Sci.*, **21**(5), 478–483.

[52] Greenwood, A. (1994) *Vacuum Switchgear*, The Institution of Electrical Engineers, ISBN 0 85296 855 8.

[53] Damstra, G.C. (1976) Influence of Circuit Parameters on Current Chopping and Overvoltages in Inductive MV circuits. CIGRE Conference, Paper 13-08.

[54] Smeets, R.P.P., Kardos, R.C.M., Oostveen, J.P. *et al.* (1993) Essential parameters of vacuum interrupters and circuit related to the occurrence of virtual current chopping in motor circuits. IEE Japan Power & Energy, Int. Session, Sapporo, Japan.

[55] Murano, M., Fujii, T., Nishikawa, H. *et al.* (1974) Voltage escalation in interrupting inductive current by vacuum switches. *IEEE Trans.*, **PAS-93**(1), 264–271.

[56] Cornick, K.J. and Thompson, T.R. (1982) Steep-fronted switching transients and their distribution in motor windings. Part 1: system measurements of steep-fronted switching transients; Part 2: distribution of steep fronted switching voltage transients in motor windings. *IEE Proc.-B*, **129**, (2), 45–63.

[57] IEEE Std. C57.142-2010 (2011) IEEE Guide to Describe the Occurrence and Mitigation of Switching Transients Induced by Transformers, Switching Device, and System Interaction.

[58] Nailen, R.L. (1979) Transient surges and motor protection. *IEEE Trans. Ind. Appl.*, **15**(6), 606–610.

[59] Schoonenberg, G.C. and Menheere, W.M.M. (1989) Switching Overvoltages in M.V. Networks. CIRED Conference, Paper 2.17.

[60] Murai, Y., Nitti, T., Takami, T. and Itoh, T. (1974) Protection of motor from switching surge by vacuum switch. *IEEE Trans.*, **PAS-93**, 1472–1477.

# 5

# Calculation of Switching Transients

## 5.1 Analytical Calculation

### 5.1.1 Introduction

Switching operations, short-circuits, and disturbances during normal operation often cause temporary overvoltages and high-frequency oscillations. The power system must be able to withstand the overvoltages as well as lightning strokes without damage to the system components. The calculation and simulation of transient voltages and currents is of great importance for the insulation coordination, correct operation, and adequate functioning of the system protection.

Transient phenomena can occur in different time frames, such as:

- microseconds – in the case of the initial rate-of-rise of the transient recovery voltages and short-line faults;
- milliseconds – when looking at transient recovery voltages caused by switching actions; or even
- seconds – for instance, in the case of ferroresonance (see Section 10.6).

Calculation of transients in power systems is treated in many books [1–9]. A practical approach, taking into account the application of switchgear can be found in application guides, related to the relevant standards [10, 11]. Calculation of transients generated by current interruption is treated extensively in a new book [12].

The following sections give examples of analytical solutions to the relevant differential equations and numerical simulation of electrical transients. The mathematical expression for asymmetrical current described in Section 2.1 is derived in Section 5.1.2. Section 5.1.3 provides the derivation of the expression for TRV originating from two oscillating circuits separated by the switching device – the source-side circuit and the load-side circuit, described earlier in Section 1.6.2.

*Switching in Electrical Transmission and Distribution Systems*, First Edition.
René Smeets, Lou van der Sluis, Mirsad Kapetanović, David F. Peelo, and Anton Janssen.
© 2015 John Wiley & Sons, Ltd. Published 2015 by John Wiley & Sons, Ltd.

Transients depend on the voltage and topology of the network and on the short-circuit power of the system. They usually comprise components caused by travelling waves on HV transmission lines and underground cables, oscillations due to lumped network elements, generators, transformers, and so on.

## 5.1.2  Switching LR Circuits

A sinusoidal voltage is switched on to a series connection of an inductance and a resistance, see Figure 5.1. This is in fact the simplest single-phase representation of a circuit-breaker closing into a short-circuited transmission line or a short-circuited underground cable. The voltage source $u(t)$ represents the electromotive forces of all connected synchronous generators. The inductance $L$ comprises a combination of the inherent inductances of these generators, the leakage inductances of the power transformers, and the inductance of the busbars, cables, and transmission lines. The resistive losses in the circuit are represented by the resistance $R$.

All these elements are supposed to be:

- *linear*, that is, the value of the element is independent of the magnitude of the current in or the voltage across it;
- *time invariant* (constant); and
- *bilateral*, that is, the value of the element is independent of the direction of the current in or the polarity of the voltage across it.

Therefore, the *principle of superposition* applies to this network, and consequently, during the *transient period*, the total short-circuit current $i(t)$ flowing in the circuit after closing the switch is the superposition of a transient DC component $i_{dc}$ and a steady-state symmetrical AC component $i_{ac}$:

$$i(t) = i_{dc} + i_{ac} \tag{5.1}$$

The *total current $i(t)$* after the instant of switching is determined by the inductance and the resistance, by the sources in the network (in this case by the voltage source $u(t)$), by the instant of switching, and the initial condition of the current through the inductance at the moment of switching. The usual procedure of computation is to start the time scale at the instant of switching (that is whenever the process starts, $t = 0$ at this moment). The instant of switching is characterized by the initial phase angle of the voltage source at this moment. The initial condition of $i(t)$ at $t = 0$ is zero in the present case.

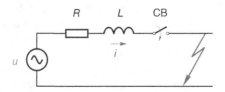

**Figure 5.1**  A sinusoidal voltage source is switched on to an *LR* series circuit.

Applying Kirchhoff's voltage law to the circuit in Figure 5.1 yields a first-order inhomogeneous differential equation with constant coefficients:

$$Ri + L\frac{di}{dt} = u(t) \tag{5.2}$$

where (see Equation 2.1)

$$u(t) = \hat{U}\cos(\omega t + \psi) \tag{5.3}$$

$\hat{U}$ is the amplitude of the particular voltage source $u(t)$;
$\psi$ is the *initial phase angle* (also known as the *actuating or energizing angle*) of the voltage at the short-circuit initiation, that is, the angle on the voltage wave at which the switch is closed ($t = 0$).

The *complete solution* of the inhomogeneous Equation 5.2 describes the *total current* that is the sum of two components:

- The *transient (DC) component* $i_{dc}$, which is exponential as results from a solution of the related *homogeneous differential equation*, that is, the voltage equilibrium equation with $u(t) \equiv 0$. It is, therefore, also referenced as a *force-free* or *natural component* in order to point out that it describes the general behaviour of the circuit free from any driving voltage. The magnitude of the DC component must fit the initial state of excitation of the circuit and, hence, the initial value of the DC component is equal to the instantaneous value of the forced symmetrical component but of the opposite polarity.
- The *steady-state (AC) component* $i_{ac}$ of the current is sinusoidal, being forced by a particular driving voltage source $u(t)$. It is given by the *particular harmonic solution*, or *particular integral* of the *inhomogeneous* differential equation, the right-hand side of which equals $u(t)$ as given in Equation 5.3. It is also referenced as the *forced component* because it describes the forced current that is proportional to the driving voltage. The steady-state component continues as long as $u(t)$ is applied and it is finally free from the DC contribution.

The DC component thus does not by itself satisfy the inhomogeneous differential equation of the circuit. It is rather a *complementary function* to the AC steady-state current, providing it with an initial offset – the offset just enough to comply with the physically inevitable condition that the total current has to start from the zero (or initial) value because the magnetic flux in the inductance cannot change instantly. In this way, the DC component bridges the gap between the initial and steady-state current.

In general, the switch may close the circuit at any instant and, consequently, the initial phase angle $\psi$ can have any value which can be normalized between 0 and $2\pi$. To find the natural solution of the differential equation, the following *characteristic equation* has to be solved for $\lambda$:

$$R + L\lambda = 0 \tag{5.4}$$

This algebraic equation is obtained by substituting the standard form of the transient component $i_{dc} = K \cdot \exp(\lambda t)$ into the homogeneous differential equation:

$$Ri + L\frac{di}{dt} = 0 \tag{5.5}$$

The scalar $\lambda$ is the eigenvalue of the characteristic equation: $\lambda = -(R/L)$, and thus the natural solution is:

$$i_{dc}(t) = K \cdot \exp\left(-\frac{t}{\tau}\right) \tag{5.6}$$

where

$t$  is the time from the initiation of the short circuit;
$\tau$  is the DC time constant of the circuit: $\tau = -1/\lambda = L/R$; and
$K$  is an arbitrary constant considering the homogeneous Equation 5.5. Considering the inhomogeneous Equation 5.2 it is determined by the initial condition.

The steady-state current, that is, the particular solution of the inhomogeneous differential equation, must be a harmonic function of the same frequency as the forcing voltage function $u(t)$, but possibly with a different initial phase. Therefore, the general expression for such a harmonic particular solution (with amplitude and phase to be determined) is:

$$i_{ac}(t) = I_1 \cdot \cos(\omega t + \psi) + I_2 \cdot \sin(\omega t + \psi) \tag{5.7}$$

Since the steady-state component by itself must satisfy the inhomogeneous equilibrium equation of the circuit, it is substituted into Equation 5.2:

$$R[I_1 \cos(\omega t + \psi) + I_2 \sin(\omega t + \psi)] - \omega L[I_1 \sin(\omega t + \psi) - I_2 \cos(\omega t + \psi)] = \hat{U} \cos(\omega t + \psi) \tag{5.8}$$

After rearranging the terms:

$$(RI_1 + \omega L I_2) \cos(\omega t + \psi) + (RI_2 - \omega L I_1) \sin(\omega t + \psi) = \hat{U} \cos(\omega t + \psi) \tag{5.9}$$

$$(RI_1 + \omega L I_2 - \hat{U}) \cos(\omega t + \psi) + (RI_2 - \omega L I_1) \sin(\omega t + \psi) = 0 \tag{5.10}$$

This equation must be satisfied at all instants of time, which is possible only if:

$$(RI_1 + \omega L I_2 - \hat{U}) = 0 \qquad \text{and} \qquad (RI_2 - \omega L I_1) = 0 \tag{5.11}$$

Solving for $I_1$ and $I_2$ yields:

$$I_1 = \frac{\hat{U} \cdot R}{R^2 + \omega^2 L^2} \qquad \text{and} \qquad I_2 = \frac{\hat{U} \omega L}{R^2 + \omega^2 L^2} \tag{5.12}$$

Substitution into the equation 5.7 for $i_{ac}$ results in:

$$i_{ac}(t) = \frac{\hat{U}}{R^2 + \omega^2 L^2}[R \cdot \cos(\omega t + \psi) + \omega L \cdot \sin(\omega t + \psi)] \quad (5.13)$$

Introducing the following well-known terms can be introduced, (assuming $Z = R + j\omega L$ is the complex impedance and $Z = |Z|$):

$Z = \sqrt{(R^2 + \omega^2 L^2)}$ is the *magnitude of impedance*;
$\phi = \tan^{-1}(\omega L/R)$ is the *impedance angle*;
$R = Z \cos \phi$ *and* $\omega L = Z \sin \phi$.

The Equation 5.13 then becomes:

$$i_{ac}(t) = \frac{\hat{U}}{Z}\left[\frac{R}{Z} \cdot \cos(\omega t + \psi) + \frac{\omega L}{Z} \cdot \sin(\omega t + \psi)\right]$$

$$= \frac{\hat{U}}{Z}[\cos \phi \cdot \cos(\omega t + \psi) + \sin \phi \cdot \sin(\omega t + \psi)] \quad (5.14)$$

The cosine and sine terms can be combined into a single term:

$$i_{ac}(t) = \frac{\hat{U}}{Z}[\cos(\omega t + \psi - \phi)] = \hat{I}[\cos(\omega t + \psi - \phi)] \quad (5.15)$$

Here, the sinusoidal current amplitude $\hat{I}$ is introduced as $\hat{I} = \hat{U}/Z$.
  The complete expression for the total current is then:

$$i(t) = i_{dc} + i_{ac} = K \cdot \exp\left(-\frac{t}{\tau}\right) + \hat{I}[\cos(\omega t + \psi - \phi)] \quad (5.16)$$

It remains to evaluate the free constant $K$ from the initial conditions of the circuit. At the instant of $t = 0$, the switch closes, but the current immediately after $t = 0$ remains zero ($i(0) = 0$) because the inductance prevents discontinuity of current. Therefore:

$$i(0) = K \cdot \exp\left(-\frac{t}{\tau}\right)\Big|_{t=0} + \hat{I}[\cos(\omega t + \psi - \phi)]\Big|_{t=0} = 0 \quad (5.17)$$

$$K = -\hat{I}[\cos(\psi - \phi)] \quad (5.18)$$

The final equation for the total current is then:

$$i(t) = i_{dc} + i_{ac} = \hat{I}\left[\cos(\omega t + \psi - \phi) - \cos(\psi - \phi)\exp\left(-\frac{t}{\tau}\right)\right] \quad (5.19)$$

The second part of this equation is the DC component of the total current; it contains the term $\exp(-t/\tau)$ and therefore it exponentially damps out. The expression $\cos(\psi - \phi)$ is a constant and its value is determined by the voltage phase angle at the instant the switch is closed and by the angle of the circuit impedance $\mathbf{Z}$.

Three conditions can now be identified:

1. For $\psi - \phi = \pi/2 \pm n\pi$ ($n$ is an integer), the DC (transient) component is zero, and the current is immediately in its steady state. In this case, where no transient occurs, we speak of a *symmetrical current*. This case is shown in Figure 2.2.
2. For $\psi - \phi = \pm n\pi$. This refers to the switch closing the circuit 90° (electrical degrees) earlier or later than described by condition 1. In this case, the DC component will reach its maximum value, which does not imply a maximum possible peak value of the total current.
3. The maximum possible peak value of the total current is reached when the following conditions are satisfied:

$$\frac{\partial i(t, \psi)}{\partial t} = 0 \quad \text{and} \quad \frac{\partial i(t, \psi)}{\partial \psi} = 0 \qquad (5.20)$$

This yields:

$$\frac{1}{\tau} \exp\left(-\frac{t}{\tau}\right) \cos(\psi - \phi) - \omega \sin(\omega t + \psi - \phi) = 0 \quad \text{and}$$

$$\exp\left(-\frac{t}{\tau}\right) \sin(\psi - \phi) - \sin(\omega t + \psi - \phi) = 0 \qquad (5.21)$$

After some manipulation and using the identity: $1/\tan(x) = \tan(\pi/2 \pm n\pi - x)$ yields:

$$\psi = \pi/2 \pm n\pi$$

This simply implies closing the switch at voltage zero, irrespective of the circuit parameters $L$ and $R$. Thus although the DC component is not (necessarily) at maximum, the maximum asymmetrical peak is obtained at short-circuit initiation at voltage zero. This case is shown in Figure 2.2.

In all cases, the total current comprising the non-zero DC component is called an *asymmetrical current*.

When the time constant of the source circuit is rather high, which is the case for short-circuit faults close to the generator terminals, the *transient reactance* and *subtransient reactance* of the synchronous generator cause an extra-high first peak of the short-circuit current. After approximately 20 ms, when the influence of the subtransient reactance is no longer present, the transient reactance dominates, and seconds later, the synchronous reactance takes over, thus further reducing the value of the steady-state component of the short-circuit current. Under these circumstances, with a current determined by the generator subtransient reactance and an AC component determined by the generator transient reactance, a current *without* current zeros can flow in one of the phases for several periods due to the large DC component, see also Section 10.1.2. This current cannot be interrupted because a current zero, necessary for current interruption in an AC circuit-breaker, is not available. Note that this case mathematically can be represented by a time-varying $L$ instead of a constant in the original Equation 5.2.

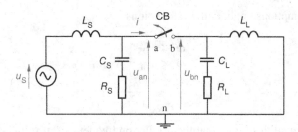

**Figure 5.2**   Basic circuit for the calculation of oscillatory switching transients.

## 5.1.3   Switching RLC Circuits

In Section 5.1.2, a first-order differential equation, comprising trigonometric forms of the time-dependent variables, has been solved via *ordinary calculus*, which can be rather straight-forward in the case of similar simple circuits. However, calculation of switching transients in more complicated circuits becomes lengthy and cumbersome when instantaneous values of trigonometric real functions of time are used.

This section highlights the calculation of the switching transients that occur when separating two discrete *RLC* circuits. This simplification is the basis for understanding of most switching duties of power switching devices. The switch separates the source circuit (modelled as an ideal voltage source $u_S(t)$ and a supply grid represented by $L_S$, $C_S$, and $R_S$) and the load circuit (represented by $L_L$, $C_L$, and $R_L$). The circuit arrangement is shown in Figure 5.2.

Current interruption occurs at $t = 0$. For $t < 0$, the circuit-breaker is closed; for $t \geq 0$, the circuit-breaker is open. The circuit-breaker is assumed to be ideal: if closed it has zero impedance, if open its impedance becomes infinite instantaneously.

### 5.1.3.1   Notations

$u_S(t) = \hat{U}\cos(\omega t)$               the source voltage with amplitude $\hat{U}$ and angular power frequency $\omega$,

$\omega_{0S} = \dfrac{1}{\sqrt{C_S L_S}}$               angular frequency of the source side (undamped case, neglecting $R_S$),

$\beta_S = \dfrac{R_S}{2L_S}$               damping factor of the source side,

$\omega_{dS} = \sqrt{\omega_{0S}^2 - \beta_S^2}$               angular frequency of the source side with damping,

$\omega_{0L} = \dfrac{1}{\sqrt{C_L L_L}}$               angular frequency of the load side (undamped case, neglecting $R_L$),

$\beta_L = \dfrac{R_L}{2L_L}$               damping factor of the load side,

$\omega_{dL} = \sqrt{\omega_{0L}^2 - \beta_L^2}$               angular frequency of the load side with damping,

$u_{an}(t) = u_{an,s}(t) + u_{an,t}(t)$               voltage at the source side, consisting of stationary component ($u_{an,s}$) and transient component ($u_{an,t}$),

$u_{bn}(t) = u_{bn,s}(t) + u_{bn,t}(t)$               voltage at the load side, consisting of stationary component ($u_{bn,s}$) and transient component ($u_{bn,t}$).

The following assumptions apply in the calculations:

$$\omega \ll \omega_{0S}$$
$$\omega \ll \omega_{0L}$$
$$\frac{L_L}{L_S} \ll \frac{\omega_{0S}^2}{\omega^2}$$

(5.22)

These last three inequalities imply that the oscillation frequencies of source-side ($L_S$ and $C_S$) and load-side ($L_L$ and $C_L$) circuits, and that of the oscillatory circuit comprising $L_L$ and $C_S$ are well above the power frequency.

In addition, the following assumptions are made regarding damping (the oscillations at source- and load-side encounter little damping):

$$\beta_S \ll \omega_{0S}$$
$$\beta_L \ll \omega_{0L}$$

(5.23)

Lastly, it is assumed that all transients have already been decayed before the moment of switching. Note that only the case when the switching occurs at voltage maximum is studied here.

In the subsequent calculations the complex-impedance method will be used for the stationary solutions and the complex form of the transient voltages will be applied. Further, making use of the assumptions above for obtaining approximations valid in practical cases, essential simplifications will be achieved. For more complicated cases, the Laplace transform method can be used [1, 13].

To obtain the desired results the calculation is carried out in the following steps:

First, the initial conditions that exist at the moment of switching will be determined (Section 5.1.3.2). Then the stationary components will be derived (Section 5.1.3.3). Next, the transient solutions will be computed (Section 5.1.3.4). Lastly, the voltages $u_{an}$ and $u_{bn}$ will be computed by adding together the stationary and transient solutions and taking into consideration the initial conditions (Section 5.1.3.5).

### 5.1.3.2   Calculation of the Initial Conditions

The initial conditions are the values of the currents and voltages at the moment of switching. First, the complex impedance $Z_C$ of the parallel branch in Figure 5.3 is calculated (symbols in bold denote complex variables).

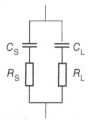

**Figure 5.3**   Sub-circuit 1 for the calculation of transients.

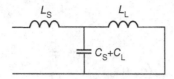

**Figure 5.4**    Sub-circuit 2 for the calculation of transients.

To obtain a good approximation for $\mathbf{Z}_C$, assumptions (5.22) and (5.23) are used. Namely:

$$R_S + \frac{1}{j\omega C_S} = 2L_S\left(\frac{R_S}{2L_S} + \frac{1}{2j\omega C_S L_S}\right) = 2L_S\omega_{0S}\left(\frac{\beta_S}{\omega_{0S}} + \frac{\omega_{0S}}{2j\omega}\right) \approx \frac{1}{j\omega C_S} \quad (5.24)$$

An analogous approximation can be obtained for the load side. It results in:

$$R_S + \frac{1}{j\omega C_S} \approx \frac{1}{j\omega C_S} \quad \text{and} \quad R_L + \frac{1}{j\omega C_L} \approx \frac{1}{j\omega C_L} \quad (5.25)$$

Making use of these two approximations, a simple approximate expression for $\mathbf{Z}_C$ can be found:

$$\mathbf{Z}_C = \frac{\left(R_S + \dfrac{1}{j\omega C_S}\right)\left(R_L + \dfrac{1}{j\omega C_L}\right)}{R_S + \dfrac{1}{j\omega C_S} + R_L + \dfrac{1}{j\omega C_L}} \approx \frac{1}{j\omega(C_S + C_L)} \quad (5.26)$$

The next step is to calculate the impedance $\mathbf{Z}$ in Figure 5.4. This is the approximate impedance seen from the source of the circuit in Figure 5.2 with a closed circuit-breaker. It is the impedance to be seen, not the circuit-breaker. Assumptions (5.22) and (5.23) are used again:

$$\mathbf{Z} \approx j\omega L_S + \frac{j\omega L_L}{1 - \omega^2 L_L(C_S + C_L)} = j\omega L_S + \frac{j\omega L_L}{1 - \dfrac{\omega^2}{\omega_{0L}^2} - \dfrac{\omega^2}{\omega_{0S}^2}\dfrac{L_L}{L_S}} \approx j\omega(L_S + L_L) \quad (5.27)$$

From the calculated impedance $\mathbf{Z}$, it can be seen that the current is purely inductive. Consequently the current at $t = 0$ is zero. The voltages $u_{an}$ and $u_{bn}$ at $t = 0$ are equal and their value is:

$$u_{an}(0) = u_{bn}(0) = \hat{U}\frac{L_L}{L_S + L_L} \quad (5.28)$$

### 5.1.3.3    Calculation of the Stationary Voltages for $t \geq 0$

The source and the load sides are behaving independently now. The source side can be calculated using Figure 5.5.

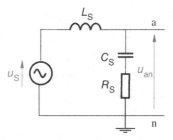

**Figure 5.5**   Sub-circuit 3 for the calculation of transients.

The complex-impedance method will be used again:

$$u_{an,s}(t) = u_S(t) \frac{R_S + \dfrac{1}{j\omega C_S}}{j\omega L_S + R_S + \dfrac{1}{j\omega C_S}} \approx u_S(t) \frac{\dfrac{1}{j\omega C_S}}{j\omega L_S + \dfrac{1}{j\omega C_S}} = u_S(t) \frac{1}{1 - \dfrac{\omega^2}{\omega_{0S}^2}} \approx u_S(t) = \hat{U}\cos(\omega t)$$

$$(5.29)$$

At the load side (see Figure 5.6) there is no source, therefore the stationary voltage is zero.

$$u_{bn,s}(t) = 0 \qquad\qquad (5.30)$$

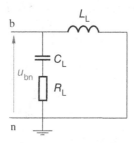

**Figure 5.6**   Sub-circuit 4 for the calculation of transients.

### 5.1.3.4   Calculation of the Transient Voltages for $t \geq 0$

The source-side transients can be easily calculated because the related circuit is the well-known *LRC* series circuit, see Figure 5.7.

   For this circuit, the Kirchhoff-loop equation is:

$$u_{CS}(t) + R_S i(t) + L_S \frac{di(t)}{dt} = 0 \qquad\qquad (5.31)$$

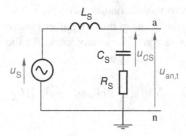

**Figure 5.7**  Sub-circuit 5 for the calculation of transients.

The relation between the current and voltage across the capacitance is:

$$i(t) = C_S \frac{du_{CS}(t)}{dt} \tag{5.32}$$

From these two equations, the following second-order differential equation can be obtained for $u_{CS}$:

$$u_{CS}(t) + C_S R_S \frac{du_{CS}(t)}{dt} + L_S C_S \frac{d^2 u_{CS}(t)}{dt^2} = 0 \tag{5.33}$$

The related characteristic equation is:

$$1 + C_S R_S \lambda + L_S C_S \lambda^2 = 0 \tag{5.34}$$

Solving this second-order algebraic equation for the (complex) eigenvalue $\lambda$ ($\lambda = j\omega_{dS} - \beta_S$), the (complex) time function $u_{CS}(t)$ can be obtained:

$$u_{CS}(t) = (\hat{U}_{CS} + j\hat{V}_{CS}) \exp\{(j\omega_{dS} - \beta_S)t\} \tag{5.35}$$

It can be easily verified that the Kirchhoff loop equation is satisfied. Note that the amplitude is a complex constant.

Knowing $u_{CS}$, one can compute $u_{an,t}$ as follows:

$$u_{an,t}(t) = u_{CS}(t) + R_S i(t) = u_{CS}(t)\{1 + (j\omega_{dS} - \beta_S)C_S R_S\} = u_{CS}(t)\left(1 + j2\frac{\omega_{dS}\beta_S}{\omega_{0S}^2} - 2\frac{\beta_S^2}{\omega_{0S}^2}\right)$$

$$\approx u_{CS}(t) = (\hat{U}_{CS} + j\hat{V}_{CS}) \exp\{(j\omega_{dS} - \beta_S)t\} \tag{5.36}$$

In an analogous way, the transient component of $u_{bn,t}$ can be calculated:

$$u_{bn,t}(t) = (\hat{U}_{CL} + j\hat{V}_{CL}) \exp\{(j\omega_{dL} - \beta_L)t\} \tag{5.37}$$

To obtain the real forms, the zero initial conditions of the load- and source-side currents can provide the necessary information. Namely, the real parts of the time derivatives of the voltages across the capacitances have to be zero. The computation for the source side is shown below. The computation for the load side is analogous.

$$\text{Re}\left[\frac{d}{dt}\{(\hat{U}_{CS}+j\hat{V}_{CS})\exp[(j\omega_{dS}-\beta_S)t]\}\right]=\text{Re}[(j\omega_{dS}-\beta_S)\{(\hat{U}_{CS}+j\hat{V}_{CS})\exp[(j\omega_{dS}-\beta_S)t]\}]$$

(5.38)

The right-hand side at $t = 0$ has to be zero:

$$0=\text{Re}[(j\omega_{dS}-\beta_S)(\hat{U}_{CS}+j\hat{V}_{CS})]=-(\hat{U}_{CS}\beta_S+\hat{V}_{CS}\omega_{dS})$$

(5.39)

This last equation gives a relationship between the real and imaginary parts of the voltage amplitude:

$$\hat{V}_{CS}=-\frac{\beta_S}{\omega_{dS}}\hat{U}_{CS}$$

(5.40)

With this relation the complex amplitude is:

$$\hat{U}_{CS}+j\hat{V}_{CS}=\hat{U}_{CS}\left(1-j\frac{\beta_S}{\omega_{dS}}\right)\approx\hat{U}_{CS}$$

(5.41)

Lastly, the real forms $u_{\text{an,t}}$, $u_{\text{bn,t}}$ of these voltages obtained from their complex expressions $u_{\text{an,t}}$, $u_{\text{bn,t}}$ are:

$$u_{\text{an,t}}(t)=\hat{U}_{CS}\exp(-\beta_S t)\cos(\omega_{dS}t)$$

(5.42)

$$u_{\text{bn,t}}(t)=\hat{U}_{CL}\exp(-\beta_L t)\cos(\omega_{dL}t)$$

(5.43)

### 5.1.3.5   Calculation of the Voltages $u_{\text{an}}$ and $u_{\text{bn}}$

Using the results of Sections 5.1.3.3 and 5.1.3.4 the voltages $u_{\text{an}}$ and $u_{\text{bn}}$ are as follows:

$$u_{\text{an}}(t)=\hat{U}\cos(\omega t)+\hat{U}_{CS}\exp(-\beta_S t)\cos(\omega_{dS}t)$$

(5.44)

$$u_{\text{bn}}(t)=\hat{U}_{CL}\exp(-\beta_L t)\cos(\omega_{dL}t)$$

(5.45)

The constants $\hat{U}_{CS}$ and $\hat{U}_{CL}$ can be determined from the initial conditions, see Section 5.1.3.2.

$$\hat{U}_{CS}=-\hat{U}\frac{L_S}{L_S+L_L}$$

$$\hat{U}_{CL}=\hat{U}\frac{L_L}{L_S+L_L}$$

(5.46)

Note that according to the assumptions in Section 5.1.3.1, the angular frequencies $\omega_{dS}$ and $\omega_{dL}$ can be substituted by $\omega_{0S}$ and $\omega_{0L}$, respectively.

The voltage $u_{ab}$ across the circuit-breaker after interruption (the TRV) is simply the difference between the voltages $u_{an}$ and $u_{bn}$:

$$u_{ab}(t) = \hat{U}\left[\cos(\omega t) - \frac{L_S}{L_S + L_L}\exp(-\beta_S t)\cos(\omega_{0S}t) - \frac{L_L}{L_S + L_L}\exp(-\beta_L t)\cos(\omega_{0L}t)\right] \quad (5.47)$$

This is the Equation 1.14 in Section 1.6.2.

Note that the present equation is an approximation and is valid only if the assumptions in Section 5.1.3.1 are fulfilled. If, for example, the oscillation is close to the critically damped case, that is $\beta \approx \omega_0$, the derived equation is incorrect.

An encyclopaedic collection of closed analytical solutions of switching problems in a variety of networks can be found in [13].

## 5.2 Numerical Simulation of Transients

### 5.2.1 Historical Overview

Simulation of transient phenomena, based on an ideal model of the electrical network, is an important tool for the design of the system protection. Manual analytical calculation of transients is laborious, cumbersome, or even impossible, and from the early days, analogue scale models have been developed – the so-called *transient network analysers* or TNA. The TNA consists of analogue building blocks, and transmission lines are built from lumped $LC$ pi-sections. The first large TNA was built in the 1930s and even today, the TNA is used for large system studies. The availability of cheap computer power (at first mainframes, later workstations, and presently personal computers) had a great influence on the development of numerical simulation techniques. Sometimes it can still be convenient to make use of the TNA, but in the majority of the cases, computer programs, such as the widespread and well-known *electromagnetic transients program* (EMTP), are used. These computer programs are often more accurate and cheaper than a TNA, but not always easy to use.

The computer programs that were developed first were based on the techniques to compute the propagation, refraction and reflection of travelling waves on lossless transmission lines. For each node, the reflection and refraction coefficients are computed from the values of the characteristic impedances of the connected transmission-line segments. The bookkeeping of the reflected and refracted waves was done and visualized by means of a *lattice diagram*, first published by Bewley in 1931 [14]. Another fruitful development was the application of the Bergeron method. This method was developed by O. Schnyder in 1929 in Switzerland and L. Bergeron in 1931 in France for solving pressure-wave problems in hydraulic piping systems [15].

The Bergeron method applied to electrical networks represents lumped network elements, like $L$ or $C$, by short transmission lines. An inductance becomes a stub lossless line with a characteristic impedance $Z = L/\tau$ and a travel time $\tau$. In the same manner, a parallel capacitance becomes a stub transmission line with a characteristic impedance $Z = \tau/C$ and a travel time $\tau$.

Almost all programs for the computation of electrical transients solve the network equations in the time domain, but some programs apply the frequency domain, such as the *frequency-domain transient program* (FTP), or use the Laplace domain to solve the network equations. A great advantage of the frequency domain is that the frequency-dependent effects of HV lines and underground cables are automatically included. The advantage of using the Laplace transform is that inverse transformation into the time domain results in a closed analytical expression. When a certain time parameter is substituted in that equation, the currents and voltages can be calculated directly for different circuit parameters. For a program with a solution directly in the time domain, the computation has to be repeated when any circuit parameter changes. The Laplace transform has, however, its drawbacks. The method cannot cope with nonlinear elements, such as surge arresters and arc models.

### 5.2.2   The Electromagnetic Transients Program

The *electromagnetic transients program* (EMTP) is the creation of H.W. Dommel, who started to work on the program at the Munich Institute of Technology in the early 1960s [9]. He continued his work at BPA (Bonneville Power Administration) in the United States. The EMTP became popular for the calculation of power-system transients when Dommel and Scott-Meyer, his collaborator in those days, made the source-code public domain. This became both the strength and weakness of EMTP; many people spent time on the program development but their actions were not always as concerted as they should have been. This resulted in a large amount of computer code for every conceivable power-system component but very often without much documentation. This problem has been overcome in the commercial version of the program, the so-called EPRI/EMTP version [16]. The Electric-Power Research Institute (EPRI) has recoded, tested, and extended most parts of the program in a concerted effort and this has improved the reliability and functionality of the transient program [17].

The EMTP method is based on *Kirchhoff's current law* (KCL). This means that all voltage sources $e(t)$ have to be transformed into equivalent current sources $j(t)$ based on the Thévenin-Norton theorem. For example, in the electrical circuit of Figure 5.8, the source $e(t)$ with the resistance $R_2$ needs to be replaced by its Norton current-source equivalent shown in Figure 5.9. For simplification of equations, all resistances are expressed as conductances $G_i = 1/R_i$ , where $i$ is the identification number of a resistive element.

After transformation of the voltage source(s) into equivalent current sources, the KCL is applied to each node. For node no. 1, the KCL reads:

$$-j(t) + G_1(u_{10} - u_{20}) + G_2(u_{10} - u_{20}) + G_2 e(t) = 0 \qquad (5.48)$$

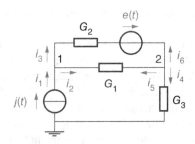

**Figure 5.8**   Circuit diagram with an equivalent voltage source $e(t)$.

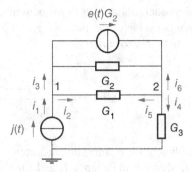

**Figure 5.9**   Circuit diagram with Norton current-source equivalent $G_2\,e(t)$.

A similar equation for node no. 2 gives:

$$G_3 u_{20} + G_1(u_{20} - u_{10}) + G_2(u_{20} - u_{10}) - G_2 e(t) = 0 \qquad (5.49)$$

These two equations can be written in the matrix notation:

$$\begin{bmatrix} G_1 + G_2 & -(G_1 + G_2) \\ -(G_1 + G_2) & G_1 + G_2 + G_3 \end{bmatrix} \cdot \begin{bmatrix} u_{10} \\ u_{20} \end{bmatrix} = \begin{bmatrix} j(t) - G_2 e(t) \\ G_2 e(t) \end{bmatrix} \qquad (5.50)$$

Using conventional symbols for matrices and vectors, the following equation holds in general:

$$Y\underline{u} = \underline{I}$$

$Y$ is the nodal admittance matrix;
$\underline{u}$ is the vector with unknown node voltages;
$\underline{I}$ is the vector with current sources.

Inductances or capacitances cause difficulty for this method because they impose differential equations. However, the EMTP transforms differential equations of the network into algebraic equations. The principle of transformation can be derived from the trapezoidal rule of numerical integration of a time-dependent function $y(t)$ within a certain integration interval. Let the differential equation to be integrated be:

$$\frac{dy(t)}{dt} = f(t) \qquad (5.51)$$

$$dy(t) = f(t)dt \qquad (5.52)$$

$$\int_0^t dy(t) = y(t) = y(0) + \int_0^t f(t)dt \qquad (5.53)$$

If this interval is divided into $n$ equal subintervals of a time step $t$, the following approximate relation can be written to obtain the value of the function $y(t)$ at a time $t_i + \Delta t$, derived from the function value at a time $t_i$ one time step earlier:

$$y(t_i + \Delta t) = y(t_i) + \int_{t_i}^{t_i + \Delta t} f(t)dt \approx y(t_i) + \frac{f(t_i) + f(t_i + \Delta t)}{2} \Delta t \qquad (5.54)$$

This approach is demonstrated below for elements of inductance $L$ and capacitance $C$. The relation between the current $i_L$ through inductance $L$ and the voltage $u_L$ across it, can be written as:

$$\frac{di_L(t)}{dt} = \frac{1}{L} u_L(t) \qquad (5.55)$$

$$i_L(t_i + \Delta t) = i_L(t_i) + \frac{u_L(t_i) + u_L(t_i + \Delta t)}{2L} \Delta t = I_L(t_i) + \frac{u_L(t_i + \Delta t)}{2L} \Delta t \qquad (5.56)$$

where

$$I_L(t_i) = i_L(t_i) + \frac{u_L(t_i)}{2L} \Delta t \qquad (5.57)$$

Similar equations can be written for the current $i_C$ through capacitance $C$ and the voltage $u_C$ across it:

$$\frac{du_C(t)}{dt} = \frac{1}{C} i_C(t) \qquad (5.58)$$

$$u_C(t_i + \Delta t) = u_C(t_i) + \frac{\Delta t}{2C}(i_C(t_i) + i_C(t_i + \Delta t)) \qquad (5.59)$$

$$i_C(t_i + \Delta t) = -i_C(t_i) + \frac{2C}{\Delta t}(u_C(t_i + \Delta t) - u_C(t_i)) = I_C(t_i) + \frac{2C}{\Delta t} u_C(t_i + \Delta t) \qquad (5.60)$$

where

$$I_C(t_i) = -i_C(t_i) - \frac{2C}{\Delta t} u_C(t_i) \qquad (5.61)$$

Based on these equations, an inductance and capacitance can be represented by a current source with a parallel resistance:

$$R_L = \frac{2L}{\Delta t} \quad \text{and} \quad R_C = \frac{\Delta t}{2C} \qquad (5.62)$$

This means that any network can be described by current sources and resistances using the equivalent circuits shown in Figure 5.10.

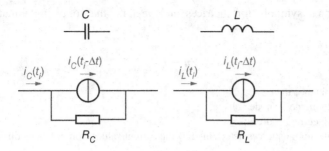

**Figure 5.10**   EMTP representation of inductance and capacitance by a current source and a parallel resistor.

This approach will be demonstrated on a sample *RLC* network shown in Figure 5.11. By means of the equivalent models for the inductance and the capacitance, as depicted in Figure 5.10, and the replacement of the voltage source and the series resistance by a current source with a parallel resistance, the *RLC* circuit can be converted into the equivalent circuit shown in Figure 5.12. To compute the unknown voltages of the nodes, a set of equations is formulated by applying the *nodal analysis* (NA) method.

$$
\begin{bmatrix} \dfrac{1}{R} + \dfrac{1}{R_L} & \dfrac{-1}{R_L} \\[2ex] \dfrac{-1}{R_L} & \dfrac{1}{R_L} + \dfrac{1}{R_C} \end{bmatrix} \begin{bmatrix} u_1(t_i) \\[2ex] u_2(t_i) \end{bmatrix} = \begin{bmatrix} \dfrac{e(t_i)}{R} \\[2ex] 0 \end{bmatrix} - \begin{bmatrix} I_L(t_i - \Delta t) \\[2ex] I_C(t_i - \Delta t) - I_L(t_i - \Delta t) \end{bmatrix} \qquad (5.63)
$$

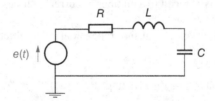

**Figure 5.11**   Sample *RLC* circuit.

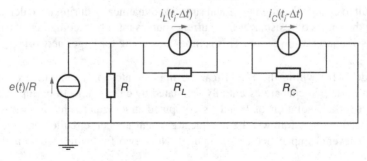

**Figure 5.12**   Equivalent EMTP circuit.

Using conventional symbols for matrices and vectors, the following equation holds in general:

$$\boldsymbol{Y}\underline{u} = \underline{i} - \underline{I}$$ (5.64)

$Y$ is the nodal admittance matrix;
$\underline{u}$ is the vector with unknown node voltages;
$\underline{i}$ is the vector with current sources;
$\underline{I}$ is the vector with current sources determined by the current values from previous time steps.

The actual computation procedure is as follows:

- Building up and inversion of the $Y$-matrix. This step has to be taken only once for a given network configuration. However, any switching action in the network requires repetition of this step because of the changed topology.
- The time-step calculation loop is entered and the resulting vector at the right-hand side of Equation 5.64 is computed regarding a time step $\Delta t$. The set of linear equations is solved using the inverted matrix $Y^{-1}$ and the vector with the nodal voltages $u$ becomes known.
- The resulting vector at the right-hand side of Equation 5.64 is computed for the next time step $\Delta t$. This procedure is repeated until the end of the given total time interval is reached.

The advantages of the 'Dommel–EMTP' method are, amongst others:

- Simplicity – the network is reduced to a number of current sources and resistances of which the $Y$-matrix is easy to construct.
- Robustness – the EMTP makes use of the trapezoidal rule, which is a numerically stable and robust integration routine.

However, the method also has some disadvantages:

- A voltage source poses a problem. This is clear from the sample $RLC$ circuit; a small series resistance will result in an ill-conditioned $Y$-matrix.
- It is difficult to change the computational step size dynamically during the calculation. This is because the values of resistances and current sources must be recomputed at each change that entails $Y$-matrix re-inversion. This is time-consuming for larger networks.

The arc models [18] within the EMTP can be implemented by means of a compensation method. Nonlinear elements are essentially simulated by current injections superimposed on the linear network, a solution of which is computed first without the nonlinear elements [19]. The procedure is as follows: the nonlinear element is open-circuited and the Thévenin voltage and Thévenin impedance are computed. Now, two following equations have to be satisfied:

1. The equation of the linear part of the network, that is, the instantaneous Thévenin equivalent circuit as seen from the arc model:

$$V_{th} - i \cdot R_{th} = i \cdot R_a \qquad (5.65)$$

$V_{th}$ is the Thevenin (open-circuit) voltage;
$R_{th}$ is the Thevenin impedance;
$i$ is the current through the arc;
$R_a$ is the arc resistance.

2. The equation of the nonlinear element itself. Application of the trapezoidal method of integration yields for the arc resistance at the simulation time $t_i$:

$$R_a(t_i) = R_a(t_i - \Delta t) + \frac{\Delta t}{2} \left( \left. \frac{dR_a}{dt} \right|_{t_i} + \left. \frac{dR_a}{dt} \right|_{(t_i - \Delta t)} \right) \qquad (5.66)$$

The $dR_a/dt$ is described by the differential equation of the arc model. To find a simultaneous solution of Equations 5.64 and 5.65, the equations have to be solved by means of an iterative process (e.g. with the Newton–Raphson approach).

Therefore the solution process is as follows:

- The node voltages are computed without the nonlinear branch;
- Equations 5.64 and 5.65 are solved iteratively; and
- The final solution is found by superimposing the response to the current injection $i$.

### 5.2.3 Overview of Electrical Programs for Transient Simulation

The development of the electrical programs for transient simulation has been improved to avoid the most important inconvenience of the EMTP which is the time step $\Delta t$. A system has two statuses, the steady state and the transient part. During the transient part, a small $\Delta t$ is needed, however, during the steady state, a large $\Delta t$ can be applied and changing the $\Delta t$ with EMTP means recalculation of the matrix $Y$ and its inverse.

A new method of modelling electrical networks appeared in the 1970s, the modified nodal analysis (MNA) developed by Ho, Ruehli and Brennan [20]. This method made it possible to include voltage-sources directly without transformation. The MNA is based on Kirchhoff's laws. The set of equations resulting from this method is a *differential algebraic equation* (DAE) system and has to be solved by a DAE solver. The advantage of using DAE solvers is that the $\Delta t$ can change as a function of the stiffness of the system.

The third way to represent an electrical network by a set of equations is to write them as *ordinary differential equations* (ODE), also called the space-state representation. The set of equations is solved by an ODE solver. As for DAE solvers, variation of the step size is possible.

## 5.3    Representation of Network Elements when Calculating Transients

Power-system analysis is a broad subject. An interesting aspect of power systems is that the modelling of the system depends on the time scale that is being viewed. Accordingly, the models for the power-system components that are used for 50 or 60 Hz steady-state analysis have a limited validity, they are only valid for low-frequency phenomena.

The simulation of transient phenomena may require a representation of network components valid for a frequency range that varies from DC to the megahertz range. The objective is usually to provide wideband models for power-system components, but an acceptable representation of each component throughout a wide frequency range is not easy and for most components it is not practically possible, either because it results in computational inefficiency or requires more complex data, which are not readily available.

Each range of frequencies usually corresponds to some particular transient phenomena. In one of the most accepted classifications, proposed by the IEC and CIGRE, frequency ranges are classified into four groups:

- low-frequency oscillations (from 0.1 Hz to 3 kHz);
- slow-front surges (from 50 Hz or 60 Hz to 20 kHz);
- fast-front surges (from 10 kHz to 3 MHz); and
- very-fast-front surges (from 100 kHz to 50 MHz).

Several reports on modelling guidelines for time-domain digital simulations have been published:

- The Technical Brochure *Guidelines for Representation of Network Elements when Calculating Transients* published by CIGRE WG (Working Group) 33-02 covers the most important power components and proposes the representation of each component, taking into account the frequency range of the transient phenomena to be simulated [21].
  CIGRE WG C4.501 studied the application of various simulation methods for surge phenomena [22].
- The documents produced by the IEEE WG on "Modelling and Analysis of System Transients Using Digital Programs", and its task forces present modelling guidelines for several particular types of studies.
- The fourth part of the IEC standard 60071 (TR 60071-4) provides modelling guidelines for insulation-coordination studies when using numerical simulation, for example, EMTP-like tools [23].

The part of the system to be modelled depends on the frequency range of the transients – the higher the frequencies, the smaller the range. One should try to keep the part of the system to be represented as small as possible. Although computer programs allow representation of very large networks through advanced graphical user interfaces, an increased number of components does not necessarily result in higher precision and, in addition, a rather detailed representation of a part of the system will often result in a longer simulation time.

Losses are difficult to model. Although there are cases where losses do not play a critical role, there are also situations for which losses are critical, for instance in defining the magnitude of overvoltages. Cases where losses are particularly important include ferroresonance, dynamic overvoltage conditions involving harmonic resonance, and capacitor-bank switching. Losses can be caused in windings, cores, or in insulating materials. Other sources of losses are corona in overhead lines and the losses in screens and sheaths of cables.

Losses are commonly represented using a circuit analysis approach. In some situations, losses cannot be separated from electromagnetic fields, for instance:

- the *skin effect* is caused by the magnetic field constrained in windings/conductors and it produces frequency-dependent winding losses;
- the *magnetic core losses* depend on the peak magnetic flux and on the frequency of the magnetic field;
- the *corona losses* play a role when the electric field exceeds the corona inception voltage; and
- the *insulation losses* are caused by the electric field and show almost linear behaviour.

Engineers and researchers spend only a small amount of their time running simulations; most of the time is spent obtaining parameters for component models. In the case where one or more parameters cannot be determined accurately enough, a sensitivity study could be very practical since this shows whether these parameters are of concern or if their influence is of secondary importance.

A large volume of research-output is available on models of the switching arc in $SF_6$ gas as a network element. A good survey is given by CIGRE WG 13.01 [24], augmented by updated overviews [25, 26].

For simulation, a wide choice of "black-box models" is available. In all cases, the basis of these models, sometimes referred to as "P-T models" – after the main parameters the arc cooling and arc time constant – are:

the Cassie [27] model

$$\frac{dg}{dt} = \frac{g}{\tau_c} \left[ \left( \frac{u_a}{U_m} \right)^2 - 1 \right] \qquad (5.67)$$

and the Mayr [28] arc model

$$\frac{dg}{dt} = \frac{g}{\tau_m} \left( \frac{u_a i_a}{P} - 1 \right) \qquad (5.68)$$

where $g$, $i_a$, $u_a$ are the arc conductivity, current, and voltage; $\tau_m$, $\tau_c$ are the arc time constants, $U_m$ is the arc-voltage parameter, and $P$ is the cooling power.

Both models describe the rate of change of arc conductivity as a function of current and voltage in a simple ordinary first-order differential equation. The Cassie model is well suited for studying the behaviour of the arc conductivity in the high-current interval whereas the Mayr model describes the arc conductivity around current zero.

Many variants of these equations, their combination, with parameters taken as a function of current and so on, have been formulated and applied in simulation programs [25]. In only a few cases, a large number of high-power tests of real HV circuit-breakers forms the basis for validation of such a model and for the supply of arc parameters [29].

Similar models for vacuum switching arcs do not exist.

# References

[1] Greenwood, A. (1991) *Electrical Transients in Power Systems*, 2nd edn, Chapters 1–4, John Wiley & Sons Inc., New York, ISBN 0-471-62058.
[2] van der Sluis, L. (2001) *Transients in Power Systems*, Chapter 1, John Wiley & Sons Ltd, Chichester, England, ISBN 0-471-48639-6.
[3] Boyce, W.E. and DiPrima, R.C. (1977) *Elementary Differential Equations and Boundary Value Problems*, 3rd edn, Chapter 3, John Wiley & Sons Inc., New York, ISBN 978-0-470-03940-3.
[4] Edminister, J.A. (1983) *Electric Circuits*, 2nd edn, Chapter 5, McGraw-Hill, New York, ISBN: 978-0-070-21233-6.
[5] Happoldt, H. and Oeding, D. (1978) *Elektrische Kraftwerke und Netze*, 5th edn, Chapter 7, Springer-Verlag, Berlin, ISBN 3-540-00863-2.
[6] Rüdenberg, R. (1950) *Transient Performance of Electric Power Systems: Phenomena in Lumped Networks*, Chapter 3, McGraw-Hill, New York.
[7] Rüdenberg, R. (1974) in *Elektrische Schaltvorgänge*, 5th edn, Vol. I (eds H. Dorsch and P. Jacottet), Chapters 2 and 3, Springer-Verlag, Berlin, ISBN 978-3-642-50334-4.
[8] Chowdhuri, P. (1996) *Electromagnetic Transients in Power Systems*, John Wiley & Sons, ISBN 0-471-95746 1.
[9] Ametani, A., Nagaoka, N., Baba, Y. and Ohno, T. (2014) *Power System Transients. Theory and Calculation*, CRC Press, Boca Raton, ISBN 978-1-4665-7784-9.
[10] IEC 62271-306 (2012) High-voltage switchgear and controlgear – Part 306: Guide to IEC 62271-100, IEC 62271-1, and IEC standards related to alternating current circuit-breakers.
[11] IEEE Power Engineering Society, C37.011 (2006) IEEE Application Guide for Transient Recovery Voltage for AC High-Voltage Circuit Breakers".
[12] Peelo, D.F. (2014) *Current Interruption Transients Calculation*, John Wiley & Sons, ISBN 978-1-118-70719-7.
[13] Slamecka, E. and Waterschek, W. (1972) *Schaltvorgänge in Hoch- und Niederspannungsnetzen*, Siemens Aktiengesellschaft, Berlin.
[14] Bewley, L.V. (1951) *Travelling Waves on Transmission Systems*, 2nd edn, John Wiley & Sons, New York.
[15] Bergeron, L. (1961) *Water Hammer in Hydraulics and Wave Surges in Electricity*, ASME Committee, John iley & Sons, New York.
[16] Phadke, A.G., Scott Meyer, W. and Dommel, H.W. (1981) Digital simulation of electrical transient phenomena, IEEE tutorial course, EHO 173-5-PWR, IEEE Service Center, Piscataway, New York.
[17] Dommel, H.W. (1971) Nonlinear and time-varying elements in digital simulation of electromagnetic transients. *IEEE Trans. Power Ap. Syst.*, **90**(6), 2561–2567.
[18] van der Sluis, L., Rutgers, W.R. and Koreman, C.G.A. (1992) A physical arc model for the simulation of current zero behaviour of high-voltage circuit-breakers. *IEEE Trans. Power Deliv.*, **7**(2), 1016–1022.
[19] Phaniraj, V. and Phadke, A.G. (1987) Modelling of circuit-breakers in the electromagnetic transients program. Proceedings of PICA, pp. 476–482.
[20] Ho, C.-W., Ruehlti, A.E. and Brennan, P.A. (1975) The modified nodal approach to network analysis. *IEEE Trans. Circuits Syst.*, **CAS-22**(6), 504–509
[21] CIGRE Working Group 33.02 (1990) Guidelines for Representation of Network Elements when Calculating Transients. CIGRE Technical Brochure 39.
[22] CIGRE Working Group C4.501 (2013) Guide for numerical electromagnetic analysis methods: application to surge phenomena. CIGRE Technical Brochure 543.
[23] IEC TR 60071-4 (2004) Insulation Co-ordination – Part 4: Computational Guide to Insulation Co-ordination and Modeling of Electrical Networks, IEC.
[24] CIGRE Working Group 13.01 (1998) State of the art of circuit-breaker modelling, CIGRE Technical Brochure 135.

[25] Kapetanović, M. (2011) *High Voltage Circuit-Breakers*, ETF – Faculty of Electrotechnical Engineering, Sarajevo, ISBN 978-9958-629-39-6.
[26] de Lange, A.J.P. (2000) High Voltage Circuit Breaker Testing with a Focus on Three Phases in One Enclosure Gas Insulated Type Breakers. Ph.D. Thesis, Delft University of Technology, ISBN 90-9014004-2.
[27] Cassie, A.M. (1939) Arc rupture and circuit severity: A new theory, Report No. 102, CIGRE.
[28] Mayr, O. (1943) Beiträge zur Theorie des statischen und dynamischen Lichtbogens. *Archiv für Elektrotechnik*, **37**(H12), s. 588–608.
[29] Smeets, R.P.P. and Kertész, V. (2006) A New Arc Parameter Database for Characterisation of Short-Line Fault Interruption Capability of High-Voltage Circuit Breakers. CIGRE Conference, Paper A3-110.

# 6

# Current Interruption in Gaseous Media

## 6.1 Introduction

When a HV circuit-breaker receives a command to separate its contacts and to interrupt a short-circuit current, the contact separation can generally take place anywhere within the sine wave of the alternating current. After that, regardless of the arc-extinction medium, the current will flow through an arc between the contacts. The arc core comprises extremely hot gas (temperature above 15 000 K), which is fully ionized and has an electrical conductivity comparable to graphite.

As long as the current is high, the arc has negligible influence on the current flow, and the arc voltage is usually only a few hundred volts. Cooling the arc by a gas flow at this stage will only result in a slight increase in the arc voltage, just to compensate for the increased arc cooling.

To interrupt the current, the circuit-breaker has to wait for a natural current zero. As the current drops towards zero, the arc cross-section decreases. When the current reaches zero, the channel has reduced to a thin thread of ionized gas. At the very moment of current zero, no energy is put into the arc. If it were possible for the arc to disappear at this instant, the current would be successfully interrupted. The arc, however, has a thermal inertia quantified by the arc time constant. Success of interruption depends on how much thermal energy is stored in the arc and how quickly this energy can be removed by cooling. This means that, immediate after current zero, the arc column as such still has a certain conductivity that enables the network to inject energy into this channel by the transient recovery voltage. For a successful current interruption, the cooling of the conductive path needs to be more effective than the heating after current zero. In this way, the arc-column temperature could drop rapidly, turning the gaseous medium between the contacts into an insulator.

Figure 6.1 shows that cooling a gas (plasma) by less than only one order of magnitude (from 5000 to 1500 K) causes a drop in conductivity by more than 12 orders of magnitude. In this, way the gas changes from being a good conductor (like graphite) to an insulator.

Therefore, the arc, and especially the remnants after arc extinction, needs intensive cooling. Cooling takes place by different processes, like radiation, thermal conduction, convection, and

*Switching in Electrical Transmission and Distribution Systems*, First Edition.
René Smeets, Lou van der Sluis, Mirsad Kapetanović, David F. Peelo, and Anton Janssen.
© 2015 John Wiley & Sons, Ltd. Published 2015 by John Wiley & Sons, Ltd.

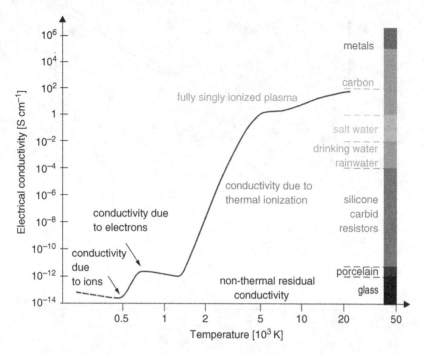

**Figure 6.1**  The electrical conductivity of a gas versus temperature [1].

turbulent mixing of cold gas with the arc plasma. Radiation is dominant at high currents and high arc temperatures, whereas turbulent mixing is generally considered the most important cooling mechanism near current zero. Different arc-extinction media have different properties in this respect.

As an example, Figure 6.2 shows the dependence of thermal conductivity on the temperature for the most commonly used media for arc extinction[1] in HV circuit-breakers:

- Nitrogen ($N_2$), the main component of the gas in air-blast circuit-breakers;
- Hydrogen ($H_2$), dominating in the gas bubble in which the arc burns in oil circuit-breakers;
- Sulfur-hexafluoride ($SF_6$), the gas used in $SF_6$ circuit-breakers.

From the point of view of arc-cooling efficiency in the interval around current zero, obviously it is favourable to use the gas that has a high thermal conductivity in the temperature range where the transition from conductor to insulator takes place. According to this criterion it may be noted that nitrogen has a high thermal conductivity but at temperatures too high to be effective. This could explain the inferior short-line-fault interrupting capability of air-blast circuit-breakers.

Hydrogen has very high thermal conductivity exactly in the region of interest, which explains the very good short-line-fault interrupting capability of oil circuit-breakers.

---

[1] The mechanism for arc extinction in vacuum circuit-breakers is completely different; therefore this consideration does not apply to vacuum. Interruption in vacuum is discussed in Chapter 8.

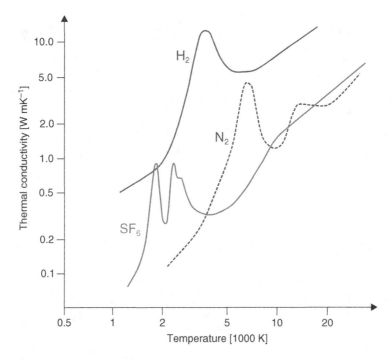

**Figure 6.2**   The dependence of the thermal conductivity of $H_2$, $N_2$, and $SF_6$ on temperature [1].

$SF_6$ has two distinct peaks on the thermal conductivity curve in the temperature interval where the conductivity should drop dramatically. This would explain the good short-line-fault interruption performance of $SF_6$ circuit-breakers (see Section 3.6.1).

The dielectric characteristics of the gaseous medium play a crucial role just after successfully passing the thermal interval of current interruption. This is illustrated in Figure 6.3 that shows the breakdown voltage as a function of pressure for air, $SF_6$ and mineral oil. The breakdown voltage of mineral oil, which is independent of the pressure, is given as a reference value. The breakdown voltage of cold $SF_6$ is about twice as high as that of air. It is independent of frequency, and already at a pressure of 0.3 MPa, the breakdown voltage of $SF_6$ is comparable to that of mineral oil. This diagram clearly shows that $SF_6$ has superior dielectric properties. This is because the $SF_6$ molecule captures free electrons, thus blocking them from starting an avalanche-type of breakdown. It is, therefore, an ideal insulating and arc-extinction medium for HV circuit-breakers. With $SF_6$ as insulating medium, the highest rated voltage per interrupting unit of a circuit-breaker of all known media can be achieved.

## 6.2   Air as an Interrupting Medium

### 6.2.1   General

Dry air is a mixture of gases and consists of 78.08% nitrogen, 20.95% oxygen, 0.93% argon, 0.038% carbon dioxide and traces of other gases. Air also contains a variable amount of water

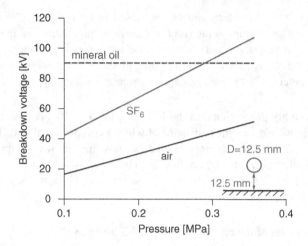

**Figure 6.3**   Breakdown voltage as a function of pressure for air, SF$_6$ and mineral oil [2].

vapour, on average around 1%. The humidity of air depends on atmospheric conditions and temperature, and saturation is present only on foggy and rainy days.

Having in mind that air has good insulation properties under normal conditions and that it is freely available and widely accessible, it is obvious that it has been immediately proposed as an insulating and extinction medium for switching devices. It can be dried, compressed, and stored in high-pressure tanks for all kinds of use.

As an insulating medium, air is applied for transmission lines, for air-insulated indoor and for outdoor switchgear, and for many other applications in electrical power systems.

Air can also be used as an arc extinction medium for current interruption. However, the interrupting capability of air at atmospheric pressure is rather low, and, therefore, such use of air is limited to low-voltage (LV) and medium-voltage (MV) applications.

### 6.2.2   Fault-Current Interruption by Arc Elongation

Current interruption by simple arc elongation in air was probably the first known arc-extinction technology and the simplest principle to apply. This principle is explained in Figure 6.4 that shows a circuit consisting of an ideal voltage source, the source-side impedance ($Z_S = \omega L_S$), and a circuit-breaker being short-circuited at its output terminals.

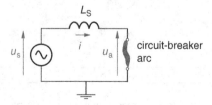

**Figure 6.4**   Basic circuit of AC short-circuit current limitation and interruption by arc elongation in air.

When the circuit-breaker contacts are in the closed position, there is a current flowing through the circuit. In an attempt to interrupt this current, the contacts of the circuit-breaker get separated. At the instant of contact separation, the arc comes into being in the contact gap. The voltage across the arc $u_a(t)$ is a function of time given by the arc development. Generally, the following parameters affect the arc voltage in gaseous media:

- Arc voltage is roughly proportional to the length of the arc. This is basically because the arc body is a resistive plasma, the resistance of which is proportional to its length.
- The arc voltage increases when the arc is cooled, because the only way for the arc to maintain itself, when its energy is drained by cooling, is to increase its voltage.
- The arc voltage increases when the current approaches zero. The same physical mechanism, as stated above, applies also here.

The equilibrium equation of the circuit in Figure 6.4 is given by:

$$u_S(t) = L_S \frac{di}{dt} + u_a(t) \Rightarrow \frac{di}{dt} = \frac{1}{L_S}[u_S(t) - u_a(t)] \tag{6.1}$$

From this equation it can be seen that the short-circuit current cannot increase ($di/dt = 0$) when the instantaneous value of the arc voltage becomes equal to the instantaneous value of the source voltage. The more important consequence is:

$$u_a(t) > u_S(t) \Rightarrow \frac{di}{dt} < 0 \tag{6.2}$$

Expressed in words: when a situation is reached in which the arc voltage exceeds the source voltage, the short-circuit current can no longer increase but will reduce its value until current zero is reached and the arc is extinguished. Such a situation is shown in Figure 6.5.

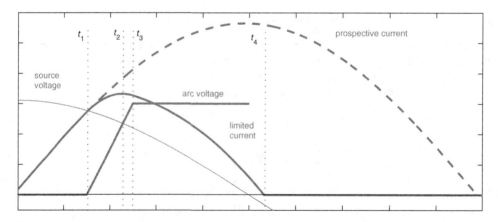

**Figure 6.5**   Principle of AC short-circuit current limitation and interruption by arc elongation in air. For symbols see text.

In the interruption process outlined schematically in Figure 6.5, the following instants can be identified:

$t_1$          the arc voltage appears (the arc starts at contact separation) and current begins to deviate from its *prospective*[2] course;

$t_1$ to $t_3$    elongation of the arc, resulting in arc-voltage increase, which then reduces the current in its steepness (between $t_1$ and $t_2$) and even in its peak value starting from $t_2$;

$t_2$          the arc voltage is equal to the instantaneous value of the supply voltage; from this moment on, $di/dt$ is negative and the current decreases towards an early current zero;

$t_3$          the arc voltage, simply presented by straight lines, reaches its maximum value;

$t_4$          the current reaches zero well before the natural power-frequency current zero and is interrupted.

If the circuit-breaker is able to build up the arc voltage sufficiently fast and maintain it at a sufficiently high level, it can strongly reduce the peak value as well as the duration of the fault current compared to the prospective current. This is very beneficial for both the interruption chamber and the circuit components that experience lower thermal and electrodynamic stresses.

Circuit-breakers designed to act as outlined in Figure 6.5 are called *current-limiting circuit-breakers*. Technically, there are several ways to create the conditions described above:

- Contact separation should occur as soon as possible after fault initiation ($t_1$ must be kept short, which is accomplished by an automatic and fast mechanism). The arc voltage must rise sufficiently fast; this can be achieved by using the Lorentz force, generated by the current loop, to "blow" the arc into a set of diverging arc-runners or by establishment of many arcs in series, such as in fuses, see Section 10.12.2. A design challenge is to minimize the *arc stagnation period*, a certain time needed to move the arc away from its location of origin, the contact gap, into the diverging arc-runners. In certain designs, special coils are used that create an extra magnetic field ("*magnet blast*") in addition to the field generated by the regular current-carrying parts of the interruption chamber, see Section 6.2.3.
- The arc voltage must reach a sufficiently high level. In practice, this is realized by segmentation of the arc, that is, by chopping the arc into many shorter arcs in series. Segmentation of the arc works in two ways: each partial arc has its own cathode and anode voltage-drop, added to the voltage across the initial arc body. In addition, the arc is brought into contact with the cold arc-splitter plates in the arc chute and is cooled and this also raises the arc voltage.
- Interruption takes place in the interruption chamber (*arc chute*, see Section 6.2.3) at current zero. This is called the *deion principle* [3]. Each arc segment should become almost instantly deionized when the current drops to zero. Normally, the walls of the interruption chamber are made of material that releases gas in order to provide additional cooling of the arc. The release of wall material in order to assist the interruption process is called *ablation*.

---

[2] *Prospective current* (of a circuit and with respect to a switching device or a fuse): "*The current that would flow in the circuit if each pole of the switching device or the fuse were replaced by a conductor of negligible impedance.*" (IEC 60050, IEV, #441-17-01).

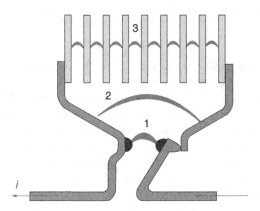

**Figure 6.6** Consecutive arc positions in deion chamber: (1) arc at location of initiation; (2) arc on divergent runners; (3) segmented arc between the metal plates in the interruption chamber.

The described current-interruption strategy is used for interrupting devices in LV networks. Household installations are equipped with the so-called *miniature moulded-case circuit-breakers* (MMCB) with an automatic coil-triggered contact-opening mechanism designed to perform the desired fast operation. In this way, it is possible to clear faulty circuits even before the first prospective peak current could have been reached. Figure 6.6 shows this principle schematically. In *fuses*, see Section 10.12.2, special low-melting temperature spots (*M-spots*) are built-in to establish many series arcs as quickly as possible after short-circuit initiation.

LV circuit-breakers in industrial installations – the so-called *moulded-case circuit-breakers* (MCB) – often cannot manage to provide such limitation of the short-circuit current before the first prospective peak.

In MV systems, the principle of current limitation is difficult to realize – in fact, only fuses can act fast enough. However, certain applications exist that can create very high arc voltages; examples are provided in Section 6.2.3.

In HV networks, it is not possible to bring this principle to fruition because the arc voltage cannot be raised to even anywhere near the system voltage, let alone above it.

For most MV and all HV applications, circuit-breakers have to wait for the natural current zero. Consequently, both the networks and the circuit-breakers have to be designed to withstand the stresses of the full prospective short-circuit current, including asymmetry.

In case of DC-current interruption, the arc-elongation principle is an often practised solution since DC current will not have current zeros. The situation is shown in Figure 6.7 where a short-circuit current (rising with the time constant of the circuit to a very high prospective value) is interrupted shortly after its initiation by forcing the current to zero.

Of course, making use of this principle is realistic only for a limited voltage range – usually up to the lower level of medium voltage, that is, a few kilovolts. See Section 10.10 for alternative switching strategies for DC

The principle of current interruption by arc elongation in air was widely applied in the pioneering days of electrical engineering when both voltage and current levels were relatively low [4].

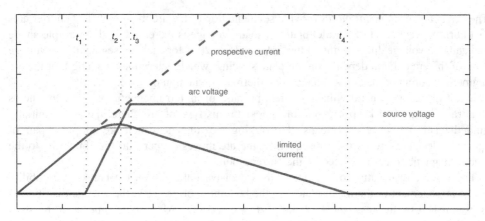

**Figure 6.7** Principle of DC short-circuit current limitation and interruption by increasing arc voltage.

At that time, circuit-breakers were thought of as simple devices. However, using air at atmospheric pressure, as a medium for arc-extinction in power circuit-breakers, is not as simple as it may seem at first sight. Once a power arc has been drawn in air at atmospheric pressure, it soon becomes very long. A considerable volume of air surrounding the arc is raised to high temperatures, which makes cooling of the arc difficult. A risk of a flashover to earth or to neighbouring conductors through the wide reach of ionized air also increases. A design of a HV circuit-breaker using the principle of arc elongation in air at atmospheric pressure for a system that operates with hundreds of kilovolts and tenths of kiloamperes would need a contact system with a gap of tens of metres.

### 6.2.3 Arc Chutes

An effective method for increasing the voltage of the arc is the inclusion of *arc chutes*. They provide a more effective control of the arc. The primary functions of the arc chutes are:

- arc motion, elongation, and cooling by convection;
- squeezing the arc between insulating side walls;
- cooling the arc by gassing wall materials;
- splitting up the arc into a number of shorter arcs in series by metal plates (arc segmentation); or
- increasing the arc length by using isolating plates as barriers.

In most arc chutes, at least two of the above functions are included.

There are two basic types of arc chutes; each type is characterized primarily by the material of the arc plates:

- arc chutes with metal plates; and
- arc chutes with insulating plates.

The *arc chutes with metal plates* provide segmentation of the arc in a number of shorter arcs that burn between a set of parallel plates. The arc voltage is increased by the multiple anode and cathode voltage drops, being of the order of 15 to 20 V for each arc segment. The voltage drop of the arc column depends on the plate spacing, which determines the length of the arc segment. Cooling of the arc is enhanced by thermal conduction of the plates.

Metal plates are generally made of steel because of its ferromagnetic property that helps to attract the arc and keep it inside the stack. In this type of arc chute, the arc is initially guided inside the plates by means of diverging arc runners, which are simply a pair of specially designed arc horns. Subsequently, the arc moves deeper into the chute due to the electromagnetic forces produced by the current loop.

The voltage distribution of the TRV has a decisive influence on the breaking capability of the arrangement of the metal plates. In the beginning of the TRV, the voltage distribution is established by the resistances of the post- arc sections. After some time, it is the stray capacitance between the plates that determines the grade of voltage disparity along the sections. Therefore, an improvement towards a desired linear voltage distribution along the series of metal plates can be achieved by appropriate design of the stray capacitances between the plates.

A schematic representation of this type of arc chute (known as a *deion chamber*) is shown in Figure 6.6. The principle of arc-extinction applied in deion chambers is generally used in LV circuit-breakers, and still today it is the most economical and practical technology for applications up to 1000 V. With large-size chutes, even circuit-breaker applications for distribution systems can be covered.

An *arc chute with insulated plates* is an alternative approach that relies on elongation and cooling the arc by means of plates made of insulating material. The insulating arcing plates are made of a variety of ceramic materials, such as zirconium oxide or aluminium oxide. This type of arc chute is used in circuit-breakers intended for applications at medium voltage up to 24 kV and for interrupting short-circuit currents up to 50 kA. The arc cooling and extinction is accomplished by a combination of several processes. First, the arc is elongated as it is forced to travel upwards and through a tortuous path that is dictated by the geometry of the insulating plates and their slits. Simultaneously, the arc is constricted as it travels through the slots in the arc plates and as the arc fills the narrow space between the plates. Finally, when the arc gets in direct contact with the walls of the insulating plates, the arc is cooled by these walls. The arc behaves as a flexible and stretchable conductor of which both the length and voltage increases as the arc retracts deeper into the slots. Additional lengthening of the arc can be achieved by a proper design of the insulating plates. This is illustrated in Figure 6.8 where the arc path is stretched from CC initially to BB and then to AA.

Figure 6.9 illustrates another principle of arc elongation using several parallel insulated plates.

Motion of the arc is forced by the action of the magnetic field produced by a coil that is usually embedded between the external supporting plates of the arc chute. The coil is not part of the main path of the current to be interrupted when the circuit-breaker is in the closed position. When the circuit-breaker begins to open, the currents transfers from the main contacts to the arcing contacts where, upon their separation, the arc is initiated. As the contact separation continues, the arc is forced towards the arc runners to which the coil is connected. Then the coil becomes part of the circuit and produces an additional magnetic field, which forces the arc to move deeper inside the arc chute. It is important to have a phase lag between the magnetic

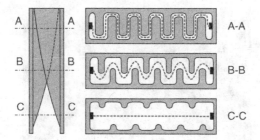

**Figure 6.8** Cross-section of an arc chute with an insulated plate shaped for arc elongation.

flux and the current being interrupted so that, at current zero, there is still a force being exerted on the extinguishing arc. The heating of the arc plates results in the emission of a large amount of vapours that must be exhausted through the opening at the top of the arc chute.

In spite of the fact that the air magnetic technology is one of the oldest techniques of current interruption, its application offers advantages over other media by being free of overvoltage problems that can be caused by the switching (on or off) of inductive loads.

Another advantage of air magnetic circuit-breakers stems from their capability to interrupt short-circuit currents with abnormally delayed (missing) current zero, which can occur under some fault conditions (see Section 10.1.2). Because of the high arc resistance, they decrease the natural $X/R$ ratio of the circuit and are thus able to influence the waveform of the fault current to such an extent that they can advance the occurrence of a current zero. Thus, they are capable of interrupting short-circuit currents with more than 100% asymmetry, which is common in applications related to the protection of large generators where the decay of the DC component under short-circuit conditions may lead to a delayed current zero.

However, as the dielectric strength of the contact gap restores rather slowly after interruption of the current, air-blast circuit-breakers should not be the preferred choice for interrupting capacitive currents. Another disadvantage is the fact that they are prone to environmental

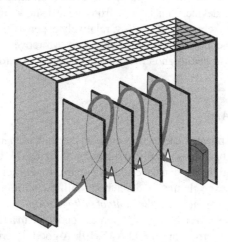

**Figure 6.9** Principle of arc elongation using several parallel insulated plates.

**Figure 6.10**    Test of a DC circuit-breaker interrupting 20 kA in a 1.2 kV circuit.

pollution. Due to the exhaust of hot gases at the arc exit, in practice a large volume of open space is required above the device. This is clear from the test shown in Figure 6.10.

Air magnetic circuit-breakers are bulky, noisy, relatively expensive, and require more maintenance than the modern vacuum and $SF_6$ circuit-breakers. Moreover, materials that nowadays are considered as harmful to health and environment, such as asbestos, may have been used for the arc chutes.

## 6.2.4    Arcs in Open Air

Arcs that appear in faulted power systems may conduct currents of many kiloamperes. The arc voltage can vary considerably, but is usually of the order of 10 to 20 V cm$^{-1}$ for arc currents up to approximately 50 kA and the arc can have a length up to 2 m. For long arcs, the voltage drop in the arc column largely exceeds the anode-cathode-fall, the sum of the *anode-fall voltage* and *cathode-fall voltage*, also known as the *contact voltage drop*.

The values of arc voltages, given by various sources, tend to be in a range of 10 to 14 V cm$^{-1}$ in a wide range of current up from 50 to 80 kA [5–10]. A good description of the associated phenomena is given by Peelo [11].

In the case of a three-phase short-circuit, for example, in a busbar system, there will often be only two arcs, each of which will conduct a phase short-circuit current. Short-circuit fault arcs will attempt to move as a result of the force created by the magnetic field of the current loop. The arc will travel in the direction away from the current source. In order to limit damage from travelling arcs, the busbars are often insulated (coated) or split up into sections with intermediate walls with bushings.

Open arcs in air can emerge between EHV- and/or UHV-transmission lines and earth as *secondary arcs*, with currents of several hundreds of amperes [12–15].

The secondary arcs follow the primary fault-current arcs in single-phase tripping and reclosure schemes (see also Section 4.2.5); they are of interest particularly in terms of their extinction mechanism. The voltage that drives the secondary-arc current originates from inductive and capacitive coupling with the sound phases and with parallel lines in the case of multi-circuit towers. This voltage cannot be considered a hard source but tends to produce a constant current, even as the arc evolves in length [16, 17]. Numerous system tests have shown that secondary-arc currents are low in magnitude, generally less than 100 A [18–21] and can be symmetrical or asymmetrical [22]. The extinction times of secondary arcs are usually less than 1 s, often low enough to permit reclosure times between 0.5 and 1.0 s, even 0.33 s has been reported. This is demonstrated in a study of secondary arcs (with length up to 9.3 m and voltage gradients rising up to 68 V cm$^{-1}$ at extinction) following primary fault currents of 8 kA [23].

Modelling of free-burning arcs in air is described in several sources [24, 25].

Also in air-break disconnectors, switching of low capacitive or inductive currents up to several amperes with arcs in atmospheric pressure is common practice (see Section 10.3.2.3). Spontaneous extinction of arcs in air at current above several amperes requires a lot of time, as the arc length has to increase beyond a critical length under the influence of aerodynamic (wind) and convective (thermal) flow forces.

Extended arc duration in open air is unacceptable, even at low current.

## 6.2.5   Current Interruption by Compressed Air

The efficiency of current interruption by arc-extinction in air can be significantly increased by cooling the arc through a blast of compressed air. These types of circuit-breakers are called *air-blast circuit-breakers* (or *compressed-air circuit-breakers*).

In air-blast circuit-breakers, the interrupting process is initiated by establishing the arc between a pair of contacts and – simultaneously with the initiation of the arc – by opening a pneumatic valve, which produces a blast of high-pressure air against or along the arc. The intensive air flow provides efficient cooling of the arc column.

Depending on the direction of the air flow, there are three basic types of blast orientations:

- *axial-blast* type, in which air-blast is directed along the arc path (as shown in Figure 6.11a).
- *cross-blast* type, in which air blast is directed perpendicular to the arc path; and
- *radial-blast* type, in which the air blast is directed radially (as shown in Figure 6.11b).

The axial type is generally in use for higher-voltage applications, whereas the cross-blast principle has been applied only at medium voltage for very high interrupting currents.

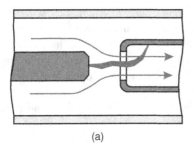 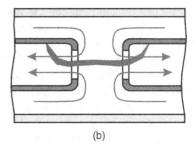

(a)                                                              (b)

**Figure 6.11** Two types of conducting nozzles in air-blast circuit-breakers: (a) single-flow nozzle; (b) double-flow nozzle.

To cool the arc effectively the gas flow in air-blast circuit-breakers must be properly directed towards the arc column. Effective control of the air flow is achieved by using either insulating or conducting nozzles. Both types of nozzle can be designed as either a single-flow (Figure 6.11a) or a double-flow nozzle (Figure 6.11b). With the single-flow design, the air consumption is substantially reduced, whereas the double-flow nozzle provides better interrupting performance.

There are many other factors that influence the performance of an air-blast circuit-breaker. From all those factors, the operating pressure of the compressed air and the nozzle diameter are of crucial importance for the intensity of the gas flow and the efficiency of arc cooling. It is also important that the pneumatic air plant supplies sufficient volume of air to ensure that the arc extinction is still achieved, even in the case when the current is not interrupted at the first or second current zero.

Significant improvements in the breaking capability of a circuit-breaker can be achieved by connecting a number of interrupters in series. It is necessary to ensure that, during the interruption process, each of the interrupters operates under exactly the same conditions. This requirement applies to the aerodynamic, mechanical and electrical conditions. From an aerodynamic point of view, identical flow conditions must be maintained for each interrupter. Electrically, the transient recovery voltages should be divided equally across each interrupter. Due to the unbalanced inherent capacitance, which exists across the contact gap itself and between the line and earthed parts of the circuit-breaker, it is common practice to use grading capacitors or resistors to balance the voltage distribution. The number of interrupters used in series may be determined either by interruption demands or by voltage-level requirements. The maximum withstand voltage per interrupter is limited by the internal air pressure. Usually this voltage is the criterion for the number of interrupters to be used per pole.

Air-blast technology was introduced in the 1930s. It was the preferred technology for extra-high voltages until the advent of the $SF_6$ technology.

## 6.3  Oil as an Interrupting Medium

### 6.3.1  Introduction

In the 1890s, the idea was launched that the arc could be extinguished in oil. At first sight, it does not seem feasible due to the flammability of the medium. However, oil – as fuel and a source of heat – requires ignition and oxygen. There is no oxygen in oil used as a medium for arc extinction since it is a mixture of different hydrocarbons. During arcing, oil evaporates and its vapours dissociate, but the arc cannot ignite the oil.

The general advantages of oil are its good dielectric characteristics (see Figure 6.3), as well as its very good thermal conductivity (see Figure 6.2, the curve $H_2$), since an arc in oil basically exists in compressed hydrogen that is released from oil by the arc. Essentially, the oil circuit-breaker is a hydrogen interrupter. The arc that develops between the contacts, when immersed in oil, causes this oil to decompose, releasing thus a substantial amount of hydrogen gas. Hydrogen features the highest thermal conductivity and the lowest viscosity of all gases. The thermal conductivity is higher during the period of gas dissociation, which results in a more rapid cooling and deionization of the arc. Together with the pressure rise and the resulting directed flow of the dissociation products, arc extinction is achieved. The dielectric strength of hydrogen is 5 to 10 times higher than that of air. These features can largely explain its excellent arc extinction and dielectric characteristics.

However, as a result of oil decomposition by the arc, particularly after breaking high fault currents, the oil carbonizes. The oil also absorbs oxygen and moisture when in contact with air. The dielectric characteristics of oil very much depend on the contents of both free carbon (resulting from the oil decomposition) and moisture absorbed from the air. This is why special attention needs to be paid to the purity and quality of oil in service in order to preserve the switching and insulating performance of an oil circuit-breaker. Therefore, it requires frequent maintenance.

Modelling an arc in oil is a complex and difficult task, since the model has to cover three states: liquid oil, gas, and arc plasma. The pressure within oil circuit-breakers can reach a value of several MPa. The large tank surface of oil circuit-breakers has to withstand strong forces if the arc is not successfully interrupted. This constitutes a potential fire hazard.

Another disadvantage is the large amount of gas generated that must be vented from the circuit-breaker, often taking oil with it.

The coefficient of thermal expansion of oil is very high. Oil expands more than 7% for each step of 100 K of temperature rise. Such changes in temperature are common in outdoor installations. This is another reason why the tank containing a large oil volume should not be fully closed.

## 6.3.2   Current Interruption in Bulk-Oil Circuit-Breakers

*Bulk-oil circuit-breakers* use a large volume of oil in the tank where the contacts are situated. The design of this type of circuit-breaker is rather simple. In the most basic version, there are no interruption chambers. The arc is divided into two partial arcs, and is drawn by means of a cross-arm. All poles are within one enclosure and so are the arcs of the three phases. The arcs are merely confined between the walls of a large oil tank and the extinguishing process is accomplished by elongation of the arcs, by high pressure of the vaporized oil, and by the natural turbulence that is created by the heated oil. To attain a successful interruption, it is necessary to develop a long arc with long arcing time that is difficult to control. This bulk-oil circuit-breaker concept is illustrated in Figure 6.12.

When an arc is drawn in oil, the vaporized oil forms a gas bubble that completely surrounds the arc and the bubble in turn is surrounded by oil. The bubble contains a high percentage of

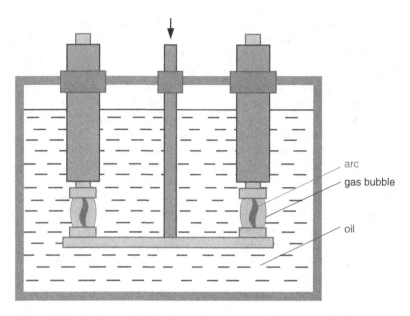

**Figure 6.12**   Outline of a bulk-oil circuit-breaker without interruption chambers.

hydrogen 40 to 45%, as well as several different hydrocarbons: acetylene 10 to 12%, methane 4 to 6%, and others. Besides that, around 40% of the total volume of the bubble consists of oil vapours. The remaining volume consists of small portions of other gases.

It is obvious that the arc exists in a gaseous medium, which in oil circuit-breakers consists mostly of hydrogen. Therefore, the theory of interruption developed for gas circuit-breakers is also applicable to oil circuit-breakers.

The circumstance that positively affects arc extinction in oil is that gases in the bubble heated by the arc increase in pressure and tend to expand, while the inertia of the oil and the tank walls prevent the expansion. Therefore, the pressure in the gas bubble rises, reaching several megapascal. The pressure rise contributes to a more rapid cooling and deionization of the arc. In addition, it enhances the dielectric strength of the remaining arc column after the final current zero that prevents re-ignitions and restrikes.

The bulk-oil circuit-breakers without interruption chambers (circuit-breakers with an open arc in oil) were used for voltages of up to 15 kV and breaking currents of not more than 200 to 300 A. In addition to the low breaking capacity, a particular problem is the open arc, which is difficult to control and to stabilize, especially in the case of longer arcs. Apart from the mentioned ranges, the use of these circuit-breakers becomes inconvenient, their dimensions are huge, and their price rises rapidly with their rating. In addition, fire hazard is significant due to the presence of large volumes of oil. Vaporized oil or oil in small droplets, sprayed from a leak in the over-pressurized circuit-breaker tank, can easily explode when it is ignited by a spark.

These disadvantages were overcome by the introduction of interruption chambers immersed in oil. Their use started in the 1930s. The basic operating principle of the simplest interruption chamber in oil circuit-breakers is shown in Figure 6.13. Contacts are placed in an isolating

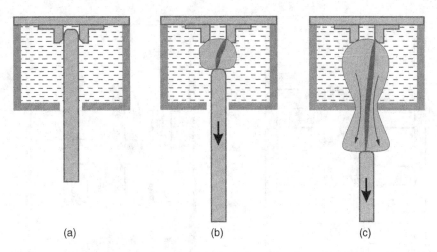

(a)                              (b)                              (c)

**Figure 6.13**   Outline of an interruption chamber in oil circuit-breakers: (a) closed position; (b) closed arc; (c) arc cooled by hydrogen blast.

interruption chamber filled with oil (a). After contact separation, the arc remains inside the chamber where the pressure rises (b). When the moving contact leaves the chamber, a strong gas and oil flow extinguishes the arc at current zero (c). The interruption chamber geometry improves the cooling of the arc considerably.

Later, the design of the interruption chambers was significantly improved by pumping mechanisms creating a cross flow of oil that provides additional cooling to the arc (see Figure 6.14).

The development of bulk-oil circuit-breakers was limited to voltage ratings up to 330 kV. The limitations were mainly because of the use of a very large amount of oil (> 50 000 litre for a 330 kV circuit-breaker), the necessity of very high opening and closing speed, and the use of a large and powerful operating mechanism. The interrupting capability of these designs

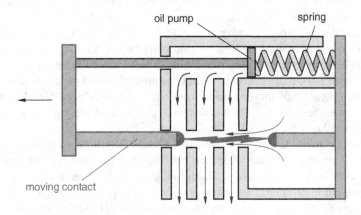

**Figure 6.14**   Cross-section of an interruption chamber with a cross flow of oil in an oil circuit-breaker.

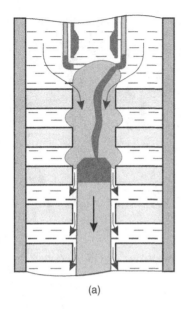 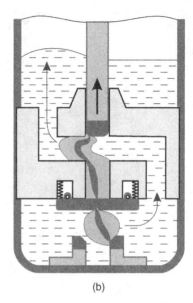

(a)                                            (b)

**Figure 6.15**  Schematic illustration of axial blast (a) and cross blast (b) interruption chamber design in minimum-oil circuit-breakers.

corresponds approximately to that of minimum-oil circuit-breakers with a current-dependent arc-extinction characteristic.

Bulk-oil circuit-breakers have been very popular in the USA and they are still widely used. Minimum-oil circuit-breakers conquered the market in Europe.

### 6.3.3  Current Interruption in Minimum-Oil Circuit-Breakers

Further development of oil circuit-breakers was directed at a new circuit-breaker design, which uses small volumes of oil. This type of circuit-breaker is known as a *minimum-oil circuit-breaker*. The main difference between the minimum-oil and the bulk-oil circuit-breakers is that the minimum-oil circuit-breaker uses oil only for its interrupting function while a solid insulating material is used for its dielectric purposes, in contrast with bulk-oil circuit-breakers where oil serves for both purposes. Unlike bulk-oil circuit-breakers, the contacts and interruption chamber are placed inside a hollow insulator (instead of in a bulky metal tank) at live potential. These insulators are made from reinforced fibre glass for MV applications and porcelain for higher voltages. By this design, the volume of oil required is minimal and therefore greatly reduces the fire risk.

The heart of the minimum-oil circuit-breakers is the interruption chamber. Interruption chambers can be conveniently classified into *axial-blast* and *cross-blast* interrupters. Many types of both chambers exist. Figure 6.15a shows an example of an axial-blast interruption chamber design. The interrupting process takes place in the following manner. When the arcing contact moves inside the series of vents, the arc vaporizes the oil and the gases that are formed (mainly hydrogen) increase the pressure and force the arc into the vents. The arc plasma is frequently renewed, since the ionized gas is being removed from the arcing zone via the vents. Ultimately, when the pressure inside the interruption chamber becomes sufficiently high, and the length of the arc is also sufficient, the arc is extinguished at power-frequency current zero. The chamber should be made of a qualified resin reinforced with fibre glass that can withstand a high pressure. Peak pressures measured in arcing chambers are generally less than 7 MPa [26]. A higher pressure increases the dielectric strength of the contact gap and thereby speeds up the current interruption.

The interruption behaviour of the minimum-oil circuit-breakers depends on the current and on the arc energy. At high current, the generated pressure is high and may even reach levels that could result in the destruction of the interruption chamber. However, with lower current, the generated pressures are low and the arcing times increase until a certain critical current level. Also, at lower currents, during normal operation, the self-blast effect cannot develop fully and the arc is unstable during a major part of the arcing time. As a consequence, it becomes more difficult to achieve interruption. This is particularly the case for axial-blast interruption chambers.

Figure 6.15b shows a cross-blast interruption chamber design. First, the arc is drawn in the lower part of the chamber where the pressure starts to rise. It forces the oil to move along a zigzag channel. When the moving contact comes out of the lower part of the chamber, metal flaps close and the arc splits into two parts. The cross-flow of the oil gives extra cooling to the arc and forces the arc to move into the vent. An increased voltage is necessary to keep the longer part of the arc inside the vent. Before the arc can escape from the vent it short circuits itself at the entry to the vent. This process continues during the whole arcing period until current zero is reached and the arc extinguishes. The hot gases, emerging from the vent to the outside interruption chamber, are still ionized and it is essential to ensure by correct vent design that no breakdown occurs between the vent and the other part of the circuit-breaker.

The closing speed of the contacts is very important for minimum-oil circuit-breakers. For example, it may be desirable to reach the fully closed position before the peak making current is reached, thus permitting the use of lighter mechanisms. In general, the closing speed should be high enough so that arcing is minimized in order to reduce the generation of high gas pressures and contact burning, both of which have negative effects on the contacts. A problem of absorbing the kinetic energy of this high-speed contact during no-load operations can arise.

Minimum-oil circuit-breakers are widely used in the transmission and distribution networks. Presently, minimum-oil circuit-breakers still do their job in service but they have left the scene of circuit-breaker development. Minimum-oil technology became out of date when vacuum circuit-breaker technology in distribution voltages and $SF_6$ technology in the (sub-)transmission range, became available.

## 6.4   Sulfur Hexafluoride ($SF_6$) as an Interrupting Medium

### 6.4.1   Introduction

The steady growth of electrical power systems in the world, which is visualized by the increase in installed power, rated voltages and transmitted energy, as well as the growing demand for reliability of the installed system components, has had a significant impact on the development of HV circuit-breakers. Circuit-breakers with classical media for extinguishing electric arcs (air and oil) were no longer able to satisfy all these demands in an economically competitive way. A decisive revolution in the development of HV circuit-breakers was caused by the introduction of *sulfur hexafluoride* ($SF_6$) gas.

Sulfur hexafluoride is a synthetic gas and had its first industrial application in 1937 when it was used in the USA as an insulating medium for cables (patent by F.S. Cooper of General Electric). With the advent of the nuclear power industry in the 1950s, $SF_6$ was produced in large quantities and its use extended to circuit-breakers as an interruption medium [27].

$SF_6$ has a unique combination of physical properties: high dielectric strength (about 3 times that of air at atmospheric pressure), high thermal interruption capabilities (about 10 times that of air), and high heat-transfer performance (about twice that of air). For that reason, $SF_6$ has been successfully used by the electricity industry in HV equipment for electric energy transmission and distribution. There are four major types of electrical equipment that use $SF_6$ for insulation and/or interruption purposes: HV circuit-breakers, gas-insulated substations (GIS) and, on a much smaller scale, gas-insulated transformers and *gas-insulated transmission lines* (GIL).

Other non-electrical industrial applications include: metallurgy (aluminium production, magnesium casting), electronics (semiconductor production, production of flat panel screens), scientific equipment (nuclear fuel cycle, high performance radar and meteorological measurements), civil engineering (insulating windows), other industries (tyres, sport shoes), medical and military applications.

## 6.4.2   Physical Properties

Pure $SF_6$ is a colourless, odourless, tasteless, non-toxic, inflammable, non-explosive, chemically and biologically inert and thermally stable gas. Although the gas is non-toxic, it will not support life, and equipment containing $SF_6$ must not be handled without adequate ventilation, especially because it is heavier than air. Its compatibility with materials used in electrical apparatus is similar to that of nitrogen up to temperatures of about 180 °C.

$SF_6$ gas has a much higher density than air. Under conditions of insufficient mixing with air, $SF_6$ has a tendency to accumulate at ground surface level and low-lying areas (e.g. trenches) may contain high concentrations of the gas and the necessary precautions must be taken to avoid asphyxiation. The mixing with air by convection and diffusion is slow, but once it has mixed, it does not separate again.

With a molecular weight of 146.05, $SF_6$ is about 5 times heavier than air and is one of the heaviest known gases. At room temperature and pressure (20 °C and 0.1 MPa absolute) it is gaseous and has a density of 6.07 kg m$^{-3}$. Since its critical point is at 45.54 °C (critical temperature) and 3.759 MPa (critical pressure), it can be liquefied by compression. It is normally available as a liquefied gas in standard gas bottles. The high molecular weight of $SF_6$ gas causes its high density as well as its high thermal capacity.

Under normal conditions, $SF_6$ is chemically inert and insoluble in water. Its reactivity is amongst the lowest of all substances. Under normal conditions, it affects no substance with which it comes in contact.

These characteristics are a result of the symmetrical structure of the $SF_6$ molecule. Six atoms of fluorine are gathered around a centrally situated atom of sulfur forming an octahedron (see Figure 6.16).

The chemical bond between fluorine and sulfur is known as one of the most stable existing atomic bonds. Six of them grant the molecule very high thermal and chemical stability, giving

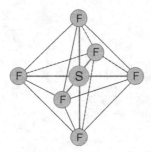

**Figure 6.16**  Structure of a $SF_6$ molecule.

$SF_6$ gas the attributes of an inert gas. The high symmetry of polyatomic $SF_6$ molecules results in small intermolecular forces and, in this way, it prevents $SF_6$ from liquefying at low temperatures at the pressures used in HV equipment[3], that is, 0.1 to 0.9 MPa. Figure 6.17 shows the dependence of pressure on temperature at a constant $SF_6$ density. Since, in a sealed gas zone, $SF_6$ pressure changes with temperature, it is not advisable to use a manometer (pressure measuring instrument) for monitoring gas-leakage. As the gas density remains constant, measurement of density is generally used for this purpose.

The gas properties, being mainly responsible for the high arc-interruption capability of $SF_6$, are:

- high thermal conductivity;
- high dielectric strength;
- fast thermal recovery; and
- fast dielectric recovery.

These $SF_6$ properties explain a rapid transition from the conducting arc plasma to the insulating state of the gap as well as the capability of the gas to withstand the rise in the recovery voltage.

The thermal conductivity of arc plasma generally comprises three components. The first is a consequence of the regular kinetic transfer of energy (by collisions between the particles). The second is a result of the diffusion of energy caused by dissociation and ionization. The third mechanism includes a transfer of energy from atom to atom by emission and absorption of radiation.

Figure 6.2 shows the dependence of the thermal conductivity of $SF_6$, nitrogen, and hydrogen plasma on temperature. It can be seen that the curves have peaks at various temperatures. Very pronounced peaks are a consequence of the dissociation of gas, attributed to polyatomic molecules, and ionization. The cause of the peak is the dissociation of particles, which diffuse to the cold regions carrying the dissociation energy, while undissociated particles diffuse from

---

[3] The rated filling pressure of an $SF_6$ circuit-breaker and its rated temperature range are related. $SF_6$ liquefaction may not happen inside a declared rated temperature range because a decrease in density causes an inadmissible loss of operational capability.

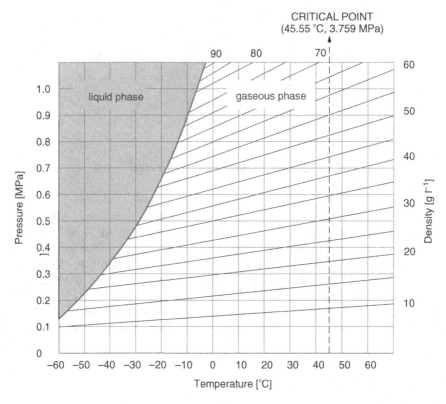

**Figure 6.17**   Relationship between pressure, temperature, and density of SF$_6$ [2].

cold to hot regions carrying no dissociation energy. Peaks occur for the first ionization as well as for the second and higher degree of ionization.

Because of its low dissociation temperature and high dissociation energy, SF$_6$ is an excellent arc-extinction medium. In comparison to nitrogen, the temperature range is much lower and narrower. This is the reason for the very efficient cooling of the arc in SF$_6$ near current zero, which causes a decrease in the arc radius. Figure 6.18 shows a representation of arc temperature as a function of the radius for arcs near current zero in nitrogen and in SF$_6$. Within the temperature ranges, when the thermal conductivity of the gas is very high, the temperature gradient is rather small since heat easily flows radially outwards.

Where the thermal conductivity is low, as it is near the edge of the arc where gas temperatures are low, the temperature gradients are high. The radius of a SF$_6$ arc is much smaller than that of a nitrogen arc. If the hot central core is removed from both SF$_6$ and nitrogen arcs (by rapid cooling at current zero) with the arbitrary minimum conducting temperature level shown, the arc in nitrogen would continue to conduct while the cooler arc in SF$_6$ would be extinguished. The ultimate consequence is a significant reduction in the thermal time constant of the SF$_6$ arc (below 0.5 μs, which is 100 times smaller than the thermal time constant of arcs in air). The small time constant is essential for the successful interruption of short-line faults.

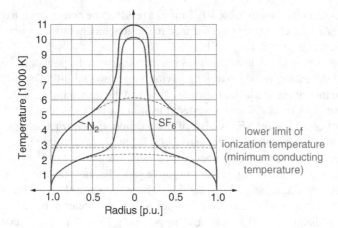

**Figure 6.18**   Arc temperature as a function of radius for nitrogen and SF$_6$ [2].

When an electric arc cools in SF$_6$, it remains conductive until a relatively low temperature, thus minimizing current chopping before current zero, and thereby avoiding the appearance of high overvoltages.

After current zero, SF$_6$ gas establishes its dielectric strength much faster than air (see Figure 6.19). The reason for this is the strong *electronegative character* (*electron affinity*) of SF$_6$ and its molecular fragments SF$_4$, SF$_2$, F$_2$, and F.

The electronegativity of SF$_6$ is based mainly on two mechanisms: resonance capture (SF$_6$ + e$^-$ → SF$_6^-$) and dissociative attachment of electrons (SF$_6$ + e$^-$ → SF$_5^-$ + F). Detailed data of breakdown voltages of SF$_6$ compared with air and N$_2$ can be found in [29].

The probability of a molecule capturing electrons reveals sharp peaks, so-called "resonances", which occur at electron energies that are specific for each gaseous molecule. Resonance capture of electrons at lower energies is called "*non-dissociative attachment*". Very low energy resonant peaks occur in SF$_6$ [30]. Already at 0.01 eV, there is a resonance inside the

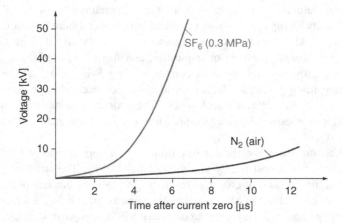

**Figure 6.19**   Recovery of dielectric strength in SF$_6$ and air after current zero [28].

attachment cross-section of about $3.2 \times 10^{-18}$ m$^2$ as a result of the non-dissociative attachment, leading to SF$_6^-$. At 0.35 eV, there is a resonance in the attachment cross-section as a result of the dissociative attachment, leading to SF$_5^-$ + F.

This strong interaction of electrons, having high mobility, with the polyatomic SF$_6$ molecule causes a rapid deceleration to negative ions having very low mobility. Due to the very low mobility of these ions, there is a lack of current carriers in the remaining arc column following a period of arcing. This makes SF$_6$ an excellent medium for arc extinction.

The capture of electrons by SF$_6$ is also of crucial importance for the rapid recovery of dielectric strength of the gas after current zero and is primarily responsible for its high electric breakdown strength. It reduces the number of free electrons and makes the gas dielectrically more stable. SF$_6$ is peculiar amongst electronegative gases because the predominant negative ions formed by electron attachment, SF$_6^-$ and SF$_5^-$, have anomalously high threshold energies for collisional detachment. The threshold energies for electron release by (SF$_6^-$ + SF$_6$) and (SF$_5^-$ + SF$_6$) collisions are 90 and 87 eV, respectively [31]. The collisional-detachment thresholds are much greater than the electron affinities for these chemical species, that is, 1.06 eV for SF$_6$ and 2.7 to 3.7 eV for SF$_5$. Therefore, a breakdown in SF$_6$ is only possible at relatively high field strengths.

Thanks to the above properties of SF$_6$ gas, SF$_6$ technology has taken an absolutely dominant role in the last few decades in high-voltage switching, and SF$_6$-based circuit-breakers are presently superior in their performance to alternative gas technologies (air-blast, minimum-oil).

### 6.4.3  SF$_6$ Decomposition Products

At temperatures above 750 K, SF$_6$ begins to dissociate into its constituent parts. Already below 3000 K, all SF$_6$ molecules will be dissociated into a large number of radicals (lower sulfur-fluorides: S$_2$F$_2$, SF$_5$, SF$_4$, SF$_3$, SF$_2$, SF, S$_4$, S$_2$, F$_2$) and above 4000 K, practically all these molecular fragments are completely broken down to their parent atoms, sulfur and fluorine (see Figure 6.20).

During the interruption process, the arc core reaches temperatures of the order of 10 000 K. At such high temperatures, the gas, which is inert at temperatures below 750 K, becomes a mixture of chemically highly reactive fragments and ions. This mixture has some undesirable properties [33]. It quickly reacts with other substances, such as vaporized contact metal, water vapour, air, the vessel wall, plastics, or impurities, and forms decomposition by-products. Some of these are highly toxic and corrosive compounds (e.g. S$_2$F$_{10}$, SO$_2$F$_2$, SOF$_2$, SO$_2$, CF$_4$, SF$_4$, and HF). Regarding potential effects on health, the existence of these by-products is of importance only if they enter the local atmosphere. Their concentrations inside switchgear are of no direct significance, especially since the presence of absorbents, such as molecular sieves, will purify the gas.

After arc extinction, when the temperature drops below approximately 1250 K, the gas will begin to recombine almost totally and reform mainly SF$_6$. The small amounts of gas that do not recombine, because of chemical reactions, increase the amount of decomposition products and the impurity level of SF$_6$. The formed impurity quantities are directly related to the electric-discharge energy. Therefore, the degree of SF$_6$ decomposition in circuit-breakers after switching is higher than in other types of SF$_6$-filled equipment.

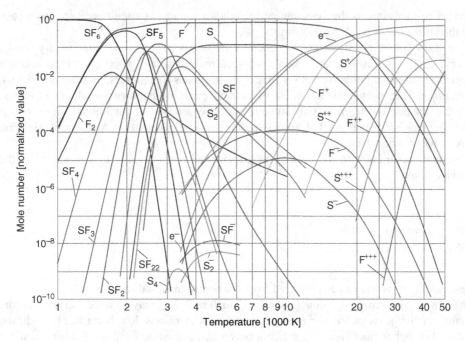

**Figure 6.20** Abundance of dissociation products of $SF_6$ versus temperature [32].

During or shortly after a discharge (in the sub-second range), besides lower sulfur-fluorides (i.e. compounds with less than 6 fluorine atoms), different metal fluorides can be formed, the most important being $CuF_2$, $AlF_3$, $WF_6$, and $CF_4$. These products, generally known as primary decomposition products, are usually present in the form of fine, non-conductive, dust-like deposits, which may appear on the bottom of the circuit-breaker enclosure and on the surfaces of insulators. During normal operation, they have no detrimental effect on their dielectric performance. In the case of copper fluoride, it appears as a milky white powder that acquires some blue tinges when exposed to the atmosphere due to a reaction which yields a dehydrated salt.

Some of the decomposition products are chemically stable. Others are very unstable, particularly in the presence of water. If exposed to moisture, the above-mentioned decomposition products hydrolyse and form secondary decomposition products, as illustrated in the following reactions:

- $F + H_2O \rightarrow HF + OH$
- $SF_4 + H_2O \rightarrow SOF_2 + 2HF$
- $SOF_2 + H_2O \rightarrow SO_2 + 2HF$
- $SOF_4 + H_2O \rightarrow SO_2F_2 + 2HF$
- $CuF_2 + H_2O \rightarrow CuO + 2HF$.

These reactions imply the formation of significant quantities of hydrogen fluoride (HF). Hydrogen fluoride is an extremely strong acid. It is, therefore, very important to take appropriate measures in the selection of materials used inside circuit-breaker interrupters and the utilization of protective coatings.

The most commonly used metals generally do not deteriorate and remain very stable. However, phenolic resins, glass, glass reinforced materials, and porcelain can be severely affected. Hydrogen fluoride vigorously attacks any materials containing silicon dioxide ($SiO_2$), for example, glass and porcelain. The use of these materials in equipment for which $SF_6$ is used for arc-extinction is, therefore, only suitable under certain special conditions. Some type of insulating materials, such as PTFE (polytetrafluorethylene[4]), bisphenol A, and cycloaliphatic epoxies will not be affected.

The main contributor to the toxicity of arc-affected $SF_6$ is $SOF_2$. It is produced in the presence of oxygen:

- $S + O + 2F \rightarrow SOF_2$
- $SF_4 + O \rightarrow SOF_2 + 2F$
- $SF_3 + O \rightarrow SOF_2 + F$

These reactions involve oxygen released from the contact materials during arcing.

Additional very toxic by-products of arcing, such as $S_2F_{10}$, may be formed. However, the quantity of $S_2F_{10}$ formed under arcing conditions is extremely low because $SF_5$ radicals, produced at high temperatures, form $S_2F_{10}$ only when cooled very rapidly – a condition not likely to apply in the arc.

In order to avoid formation of $SOF_2$, HF, and some other by-products of arcing, $SF_6$ needs to be exceptionally pure at the first filling of the circuit-breaker.

The lower sulfur fluorides and many of the other by-products can be effectively neutralized by molecular sieves, by a (50:50)% mixture of sodium hydroxide (NaOH) and calcium oxide (CaO), or by activated *alumina* (dried aluminium oxide $Al_2O_3$). They are very effective, and they adsorb the acidic and gaseous products practically irreversibly. The recommended amount to be used is approximately equal to 10% of the weight of the gas. Removal of the acidic and gaseous contaminants is accomplished by circulating the gas through filters containing the above-described materials. These filters are usually attached to the circuit-breaker itself or they may be installed in specially designed devices for handling $SF_6$.

At the same time, the filters dry the $SF_6$ gas and maintain a low dew point[5] for gas filling.

Figure 6.21 shows the dew point as a function of the gas moisture content.

As already indicated, pure $SF_6$ is chemically inert. Therefore, it cannot cause corrosion. However, in the presence of moisture, the primary and secondary decomposition products of sulfur hexafluoride form corrosive electrolytes, which may cause damage and operational failure in electrical equipment.

If the formation of decomposition products cannot be avoided by the use of appropriate construction methods, corrosion can be largely eliminated by careful exclusion of non-suitable materials.

---

[4] PTFE is best known by the brand name *Teflon*.

[5] The dew point is the temperature below which the water vapour in a volume of humid air at a given constant pressure will condense into liquid water at the same rate at which it evaporates.

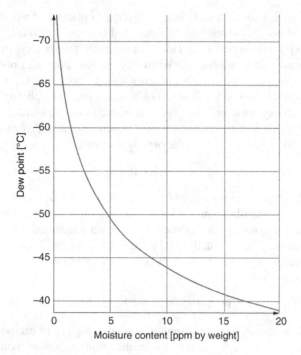

**Figure 6.21**    Dew point as a function of the gas moisture content [2].

## 6.4.4    Environmental Effects of SF$_6$

### 6.4.4.1    Introduction

Atmospheric pollution due to emissions of manufactured gases and materials cause two major global effects. One is *stratospheric ozone depletion*, creating a hole in the ozone layer, and the other is *global warming of the atmosphere* that is, increase in the average temperature of the earth's surface or the *greenhouse effect*.

In addition, ecotoxicology should be considered, that is, the toxic effect of materials and gases released having effects on the environment and life.

In this respect, a number of studies have been published regarding the safe and sustainable use of SF$_6$ [34–36].

### 6.4.4.2    Ozone Depletion

*Ozone depletion* describes two distinct but related observations: a slow, steady decline of about 4% per decade in the total volume of ozone in the earth's stratosphere (ozone layer) since the late 1970s, and a much larger, but seasonal, decrease in stratospheric ozone over the earth's polar regions during the same period. The latter phenomenon is commonly referred to as the *hole in the ozone layer*.

Since the ozone layer prevents most harmful ultraviolet (UV) light from passing through the earth's atmosphere, it is recognized by the international community that a variety of biological consequences, such as increases in skin cancer, damage to plants and reduction of plankton populations in the oceans, may result from the increased UV exposure due to ozone depletion.

Three forms of oxygen (isotopes) exist: atomic oxygen (O), normal oxygen ($O_2$), and ozone ($O_3$). Ozone is formed in the stratosphere when oxygen gas molecules photo-dissociate to atomic oxygen. The atomic oxygen then combines with $O_2$ to create $O_3$. Ozone molecules absorb UV light between 310 and 200 nm, which is followed by splitting of ozone into a molecule of $O_2$ and an oxygen atom. The oxygen atom then joins up with an oxygen molecule to regenerate ozone. This is a continuing process which terminates when an oxygen atom recombines with an ozone molecule to make two $O_2$ molecules:

$$O + O_3 \rightarrow 2O_2$$

Ozone can also be destroyed by a number of free radical catalysts, the most important of which are halogen elements, especially atomic chlorine (Cl) and bromine (Br). The main source of Cl atoms in the stratosphere is photo-dissociation of *chlorinated fluorocarbon compounds* (CFC), commonly called *freons*, usually used as aerosol spray propellants, foaming agents, refrigerants, and solvents. Once in the stratosphere, the Cl atom is liberated from the parent compounds by the action of UV light:

$$CFCl_3 + photon \rightarrow CFCl_2 + Cl$$

The Cl atom can then destroy ozone molecules through a variety of catalytic cycles. In the simplest one, a Cl atom reacts with an ozone molecule, forming chlorine monoxide (ClO) and leaving a normal oxygen molecule. The chlorine monoxide can react with a second molecule of ozone to yield another chlorine atom and two molecules of oxygen:

$$Cl + O_3 \rightarrow ClO + O_2$$
$$ClO + O_3 \rightarrow Cl + 2O_2$$

Once a free Cl atom appears, it can immediately react again with $O_3$, thus forming a repetitive, catalytic cycle for each individual atom, passing many times through the reactions. A single chlorine atom is able to react with 100 000 ozone molecules before it is neutralized by other reactions. This fact plus the amount of chlorine released into the atmosphere by CFCs yearly demonstrates how dangerous CFCs are to the environment.

In the case of $SF_6$, the only halogen constituent is the fluorine. Laboratory studies have shown that fluorine atoms participate in analogous catalytic cycles. However, in the earth's stratosphere, due to the high chemical affinity of fluorine for hydrogen, fluorine atoms react rapidly with water and methane and form hydrogen fluoride (HF). Therefore, the above catalytic reaction scheme is practically impossible with fluorine atoms.

Taking account of the fact that the concentration of $SF_6$ is more than 1000 times lower than that of CFCs and that virtually no free fluorine is formed from $SF_6$ under the circumstances described above, it is clear that $SF_6$ does not contribute to the destruction of stratospheric ozone and ozone depletion.

### 6.4.4.3  Greenhouse Effect

The average global temperature of the earth results from a balance between the heating effects of solar radiation and the cooling associated with the infrared radiation from the earth. Some of the infrared radiation is reflected back to the surface of the earth and therefore does not escape from the atmosphere. This is known as the *greenhouse effect*. It results in an increase in the average temperature of the earth's surface compared with the temperature that would have prevailed if no greenhouse effect had existed. An imbalance in the earth's normal greenhouse effect arises when the emissions of *greenhouse gases* contribute to an enhanced greenhouse effect, which shifts the balance between incoming and outgoing radiation so that more radiation is retained, thus causing changes in the climate system.

*Greenhouse gases* are atmospheric gases that absorb a portion of the infrared radiation emitted by the earth and return it to earth by emitting it back. Potent greenhouse gases have strong infrared absorption in the wavelength range from 7 to 13 μm [37]. They occur both naturally in the environment (e.g. water vapour, carbon dioxide, methane and nitrous oxide) and as gases that may be released (e.g. $SF_6$, *fully fluorinated compounds* FFCs, combustion products, nitrous, and sulfur oxides). Today, the natural phenomenon is by far predominant, and human activities give only a small contribution to the total greenhouse effect. However, as the man-made contribution is growing, this is currently a concern to many. According to several studies, the actual trend, if it is not reversed, will cause a significant increase in the average temperature on the planet and the global climate will change.

$SF_6$ is an efficient absorber of infrared radiation, in particular for wavelengths around 10.5 μm. Additionally, unlike most other naturally occurring greenhouse gases (e.g. $CO_2$, and $CH_4$), $SF_6$ is largely immune to chemical and photolytic degradation. Therefore, its contribution to global warming is expected to be cumulative and in fact permanent. The atmospheric lifetime of $SF_6$[6] is very long, estimated at around 3200 years. The strong infrared absorption of $SF_6$ and its long lifetime in the environment are the reasons for its extremely high *global warming potential* (GWP)[7], which for a 100 year time horizon is estimated to be ~22 800 times larger than the GWP of $CO_2$ [38], the predominant contributor to the greenhouse effect by volume.

The high GWP of $SF_6$ gas became widely known in 1995 [39], and consequently, this gas was put on the list of greenhouse gases in the Kyoto Protocol [40] of 1997, an international agreement to control the emission of man-made greenhouse gases. The following greenhouse gases are specified in this document:

- carbon dioxide ($CO_2$);
- methane ($CH_4$);
- nitrous oxide ($N_2O$);
- hydrofluorocarbons (HFCs);
- perfluorocarbons (PFCs); and
- sulfur hexafluoride ($SF_6$).

The latter three substances are fluorinated greenhouse gases or F-gases.

---

[6] The atmospheric lifetime of $SF_6$ is the time taken for a given quantity of $SF_6$ released into the atmosphere to be reduced via natural processes to ~37% of the original quantity.

[7] *Global warming potential* (GWP) is a measure of how much a given mass of greenhouse gas is estimated to contribute to global warming. It is a relative scale that compares the gas to carbon dioxide, whose GWP is by definition equal to one. A GWP is calculated over a specific time interval that must be stated whenever a GWP is quoted. Commonly, a time horizon of 100 years is used.

$SF_6$ is explicitly listed as the gas with the highest GWP. This obliges all participating countries to achieve $SF_6$ emission reduction. Most national governments have signed and ratified the Kyoto Protocol aimed at reducing greenhouse gas emissions.

The European Parliament and Council have initiated regulation with regard to certain fluorinated greenhouse gases [41]. All European producers, exporters, and importers of F-gases are obliged to report the quantities of all imported or exported amounts of these gases to countries outside the community in a prescribed format [42].

Furthermore, additional European regulation [43] defines the certification of the acquisition of knowledge for carrying out activities in connection with recovery of certain F-gases from HV switchgear. This means that maintenance and recovery work on $SF_6$-switchgear is allowed by certified staff only to ensure minimum $SF_6$ emissions.

In many countries where $SF_6$ gas is used in switchgear, measures are now taken to limit the emission of $SF_6$. Examples are the mandatory programme of the Environmental Protection Agency (EPA) in the USA, the F-gas regulation in Europe and the $SF_6$ reduction programmes in Japan [44].

In the European F-gas regulation (2007) it is required that all larger systems containing $SF_6$ gas should be inspected on a regular basis and emissions should be prevented as much as possible during maintenance, refilling, and dismantling. Although there is currently an exception for hermetically sealed switchgear containing less than 6 kg $SF_6$, there might be additional measures in the future for this kind of application due to an increasing pressure from Non-Governmental Organizations (NGOs) and political parties to limit the emissions of non-carbon greenhouse gases.

Australia was the first country in the world to actually impose a levy on the import of $SF_6$ gas, from mid-2012 to mid-2014 [45]. As a result, importers of $SF_6$ (in bulk or contained in products and equipment) were required to pay an *equivalent carbon price* related to the global warming potential of $SF_6$. An equivalent carbon price of AUS $ 23 per ton of $CO_2$ equivalent has to be paid. For $SF_6$, due to its very high $CO_2$ equivalent of 23 900 (as taken by the Australian government), this implied an amount of 23 900 × 23 = AUS $ 549 700 ≈ € 440 000 ≈ US $ 577 000 per ton (2012).

The *Western Climate Initiative* (WCI) [46] was created in 2007 for the western hemisphere. Participation in the WCI reflects the strong commitment of each partner jurisdiction to take cooperative actions to address climate change and implement a joint strategy to reduce greenhouse-gas emissions. The members now include the state of California and the Canadian provinces of British Columbia, Ontario, Québec, and Manitoba. Methods to determine emissions are also defined. For example, external parties are now mandated to check the calculations and confirm the declared emissions.

In Japan, voluntary agreement and self-commitment actions led to a reduction of the $SF_6$ emission rate during operation of HV equipment from 1.1% in 1998 to less than 0.2% in 2011.

Regional average emission rates presently vary between far less than 1% to more than 10% of the stored gas. Leakage losses from HV GIS are estimated to be in the range 0.5 to 1% and handling losses around 1 to 2% per year. For distribution-voltage equipment of the sealed-for-life type, the total losses during operation are still much lower [47].

The high GWP of $SF_6$ triggered various efforts to reduce its emissions from $SF_6$ insulated transmission and distribution equipment. These efforts involve improved sealing of equipment, improved gas handling procedures, systematic gas reuse, voluntary emission-reduction programmes in the electrical industry, and governmental regulatory actions.

Unlike for electrical applications, the $SF_6$ gas used for non-electrical applications is usually not recoverable. Most of this gas is immediately released into the atmosphere because of the "open application" of the gas. There is a clear decreasing trend in the $SF_6$ emissions from the electrical industry since the time when the high GWP of $SF_6$ became known in 1995. Since 1996, the electrical industry considerably reduced emission with a further falling trend for the future, although new $SF_6$ equipment continues to be installed.

The concentrations of different gases relevant to the environment are regularly monitored by several scientific bodies. In 2009, the concentration of $SF_6$ in the atmosphere was reported [48] as being about 6.4 pptv[8]. This concentration is very low but it is constantly increasing at a rate of approximately 8.7% per year [37].

If the concentration continues to increase at this rate, the concentration could be about 50 pptv within less than 30 years. At these concentrations, the expected global warming attributable to $SF_6$ is still very small in comparison with the expected increase in the global temperature attributable to $CO_2$. Nevertheless, the long atmospheric lifetime of $SF_6$ is a potential hazard. It is, therefore, essential that all types of release of $SF_6$ into the atmosphere be reduced to an absolute minimum. When handling $SF_6$ it is, therefore, necessary to adopt procedures to keep the gas in a closed cycle, avoiding any release to the environment.

The contribution of $SF_6$ to the man-made greenhouse effect is in the range of 0.1% [49]. Good handling practices and procedures, such as those defined in IEC standards [50–53] contribute to ensuring that this very small impact is effectively maintained over a long time. These procedures should be regarded as minimum requirements to ensure the safety of personnel working with $SF_6$ and to minimize the $SF_6$ emissions. The safety concept is very well known and established and the focus is nowadays on the environmental compatibility.

Today, specialized high-quality equipment, measuring devices, and $SF_6$ handling systems for gas purification and recycling are commercially available, capable of restoring contaminated gas to its as-new condition [54]. Further improvements of such equipment can be expected in the light of the growing demand for minimizing $SF_6$ gas emissions.

Technical grade $SF_6$ – according to IEC 60376 [52] – is widely commercially available. If the gas does not comply with the specification, it is disposed of according to local or international regulations on waste management.

The purification efficiency of used $SF_6$ is good when the gas contains less than 4000 ppmv of acidity, expressed in terms of HF. When the acidity is higher, the gas is too corrosive to be passed through the purification process. Also, when used $SF_6$ contains more than 7500 parts per million in volume of air, the efficiency of the process is rather poor and the gas must be returned to the manufacturer for reprocessing.

There is controversy about how much $SF_6$ gas is leaking into the atmosphere. In spite of the very low value of annual leakage losses, it is clear that one can expect a major share of the total amount of $SF_6$ gas ever produced to be in the atmosphere one day. Even if the requirements in terms of sealing the gas zones containing $SF_6$ are more rigorous, this has no long-term significance because there is no possibility of destroying $SF_6$ gas molecules when they leak

---

[8] pptv = parts per trillion by volume (i.e. $10^{-12}$ mol of $SF_6$ per mol of dry air).

from the gas zone. The only portion of the entire world production of $SF_6$ which will never be released into the atmosphere is the quantity of the gas which can be dissociated through environmentally acceptable processes.

Generally, measures for the reduction of $SF_6$ emission and its contribution to global warming can be both short term and long term.

The best short term measures include:

- preventing the release of $SF_6$ into the environment by improvement of sealing systems and $SF_6$ handling procedures;
- systematic recycling of used $SF_6$ and destroying contaminated gas;
- voluntary emission-reduction programmes;
- controlled use of the gas in order to reduce its application strictly to those areas where $SF_6$ has obvious superior characteristics, such as in HV circuit-breakers; and
- designing equipment requiring a smaller quantity of $SF_6$.

Clearly, the most effective long-term measure is not to use $SF_6$ at all. Therefore, the ideal long-term solution is the use of other gases or gas mixtures in place of $SF_6$, which are environmentally acceptable [37]. Until such new technology is found, however, $SF_6$ technology will remain essential.

### 6.4.4.4   Ecotoxicology and Potential Effects on Health

$SF_6$ is not toxic and has no reported potential to be acute or chronically ecotoxic [53]. As its solubility in water is very low, it presents no danger to surface and ground water or the soil. A biological accumulation in the nutrition cycle does not occur. Therefore, $SF_6$ does not harm the ecosystem.

It has been recognized that $SF_6$ is:

- not carcinogenic (not causing cancer);
- not mutagenic (not causing damage to the genetic constitution); and
- not nitrifying (no enrichment in the food chain).

However, $SF_6$ by-products can cause irritation of the skin, eyes, and respiratory tract. In high concentrations, they can cause pulmonary oedema, which may cause respiratory failure in the case of sufficient exposure time.

During normal service, $SF_6$ remains inside the equipment and the by-products formed are neutralized by molecular sieves as well as by natural recombination processes. $SF_6$ can become present in the atmosphere from leakage or if a gas-filled compartment fails to keep the gas inside, for example in the unlikely case of an internal fault arc burning through the enclosure. When evaluating health risk, it is necessary to differentiate clearly between leakage and internal-fault situations leading to a sudden release of $SF_6$.

In the case of leakage, it is necessary to consider the effects of long-term exposure to the gaseous by-products. The by-products, which are formed by arcing and by low-energy discharges, and are released due to leakage from $SF_6$-filled equipment, reach negligible concentrations in the workplace atmosphere. Under worst-case conditions, the concentrations found for both MV and HV cases are four orders of magnitude lower than the threshold limit value [53] (TLV)[9].

In the case of a sudden release of $SF_6$ due to an internal fault, mandatory evacuation and ventilation procedures imply only a momentary exposure. Significant concentrations of by-products can occur in an equipment room. However, the calculated concentrations do not exceed a defined limit for short-term exposure [37]. In this case, account should be taken of all possible sources of toxic emanations and this requires detailed knowledge of all products formed. In this respect, a full treatment should consider contributions from metal vapour, burnt plastics, cable insulation, paint, and so on, as well as those that are attributable to $SF_6$.

It has also been stressed that in any internal-fault situation, corrosive and/or toxic fumes are produced whether or not $SF_6$ is present. In cases where these fumes enter the ambient atmosphere of the switchgear room, it has been shown that non-$SF_6$-related arc products are likely to be the dominant contributors to overall toxicity. This further strengthens the view that the use of $SF_6$ in electric-power equipment does not significantly add to the risks associated with an internal fault.

$SF_6$ containing by-products have an unpleasant, pungent smell ("rotten eggs" due to $H_2S$) that in itself gives an irritating effect. Because of this characteristic, even small quantities of gaseous by-products may give rise to unmistakable warning indications within seconds, before any risk of poisoning.

It can be concluded that as long as safety procedures are followed, there is minimal risk specifically associated with the use of $SF_6$ in electric-power equipment. The most important safety aspect is prevention of asphyxiation.

## 6.4.5   $SF_6$ Substitutes

The extremely high GWP of $SF_6$ (explained in Section 6.4.4.3) mandates that users actively pursue means to minimize releases into the environment, one of which is the use of other gases or gas mixtures instead of $SF_6$.

$SF_6$ substitute gaseous dielectrics are more difficult to find than it seems at first sight, because the large number of basic and practical requirements that a gas has to satisfy need extensive studies and many tests to be performed to allow confident use. Efforts have been made throughout several decades to find a better gaseous dielectric than $SF_6$ [55]. It was found that there are gases superior in dielectric strength to pure $SF_6$, but they are not acceptable for various reasons, such as their environmental impact, toxicity or flammability.

For example, the gas must have a high dielectric strength, which requires the gas to be electronegative. However, strongly electronegative gases are usually toxic, chemically reactive, and environmentally harmful, with low vapour pressure, and implying decomposition products from gas discharges that are extensive and unknown. Non-electronegative gases that are benign and environmentally ideal, such as nitrogen, normally have a low dielectric strength. For example, $N_2$ has a dielectric strength about three times less than $SF_6$ and lacks the fundamental properties necessary for use in circuit-breakers. Nevertheless, such environmentally friendly

---

[9] The *threshold limit value* (TLV) of a chemical substance is a level to which a worker can be supposedly exposed day after day during a working lifetime without adverse health effects.

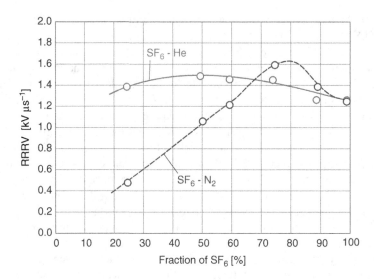

**Figure 6.22**   RRRV withstand capability as a function of $SF_6$–$N_2$ and $SF_6$–He mixture ratio for a total pressure of 0.6 MPa [56].

gases might be used as a main component in mixtures with electronegative gases (including $SF_6$) at higher or comparatively lower pressures and at partial concentrations of a few per cent.

The most desirable $SF_6$ substitute would be a gas that could be put in all existing $SF_6$ equipment, requiring little or no change in hardware, practice, operation, procedures, or ratings.

There are three gaseous dielectric media identified as a potentially acceptable replacement for $SF_6$ regarding both HV insulation and current interruption:

- High-pressure pure $N_2$ for HV insulation applications. Consideration of the use of such environmentally friendly gases, where $SF_6$ is not absolutely required, should be investigated and promoted.
- $SF_6$/$N_2$ mixtures (see Section 6.5):
  - mixtures of low concentrations (below 15%) of $SF_6$ in $N_2$ for HV insulation; and
  - mixtures of nearly equal quantities of $SF_6$ and $N_2$ for both electrical insulation and current interruption.
- $SF_6$/He mixtures for current interruption. A mixture of $SF_6$ and helium has shown promising performance when used in gas-insulated circuit-breakers. Helium complements $SF_6$ in its cooling capability. The dielectric performance of $SF_6$/He mixtures for a total pressure of 0.6 MPa seems to be ~10% higher than pure $SF_6$ for virtually all mixture compositions from 75 to 25% $SF_6$ (see Figure 6.22). Therefore, the mixture should be investigated further.

Sufficient data are presently available to demonstrate the potential of the gaseous media mentioned above, but not to sufficiently prove their performance. An analysis of the data indicates that there is no gas immediately available for use as an $SF_6$ substitute in existing electrical utility equipment. For gas-insulated transmission lines and gas-insulated transformers, the limitation is primarily due to the need for re-certification and possible re-rating of equipment that is already in use. For gas insulated circuit-breakers, there are still significant open questions concerning the performance of gases other than pure $SF_6$.

However, various gas mixtures do show promise for use in new equipment, particularly if the equipment is specifically designed for use with such a gas mixture.

Several studies have been performed on $CO_2$ [57–60] and $CF_3I$ [61, 62]. Since 2012 live-tank circuit-breakers operating with $CO_2$ as insulation and interruption medium are on the market for 72 to 84 kV and are field-tested at 145 kV rated voltage [63].

## 6.5   $SF_6$ – $N_2$ Mixtures

There are two reasons why mixtures of $SF_6$ and $N_2$ are studied as arc-extinction media [64–66].

The first is related to the effort to improve dielectric properties and performance of HV circuit-breakers in extremely cold climates (temperatures below –40 °C) where $SF_6$ used under pressure in circuit-breakers may liquefy and thus lose part of its current-interruption properties. With a temperature decrease at constant volume, pressure also drops, while the gas density remains constant. At a certain temperature, when liquefaction of $SF_6$ starts, both the pressure and the gas density suddenly start to drop. As a result, the dielectric and interruption properties of the circuit-breaker also suddenly deteriorate. To avoid liquefaction of $SF_6$, circuit-breakers for low-temperature applications must be filled with an adequate pressure at 20 °C. Since condensation temperatures of component gases are determined by their partial pressure, not by the total pressure, a corresponding reduction of the dielectric strength can be compensated by adding a gas that liquefies at much lower temperatures than $SF_6$ – in practice usually nitrogen. It was found that for such applications, $SF_6/N_2$ mixtures with 50% $SF_6$ are efficient arc-interrupting media. Besides $SF_6/N_2$, other mixtures are in use, including $SF_6/CF_4$ and $SF_6/He$.

The second reason comes from the need to prevent the release of $SF_6$ into the atmosphere as a response to the concerns over the possible impact of $SF_6$ on global warming, explained in more detail in Section 6.4.4. Obviously, the use of $SF_6/N_2$ mixtures is not a long-term solution to the problem of eliminating the potential contribution of $SF_6$ to global warming. Clearly the most effective way to do this is not to use $SF_6$ at all, which is an option that is neither technically nor economically feasible at present. The use of gas mixtures containing $SF_6$ can only help to reduce the rate at which $SF_6$ enters the environment in the short to mid-term and thereby gain more time for innovations.

A number of studies [67–69] have shown that there is a significant positive synergy between the two components in $SF_6/N_2$ mixtures whereby the addition of small amounts of $SF_6$ to $N_2$ dramatically increases the dielectric strength of the mixture over the $N_2$ value. The increase in dielectric strength of the mixture tends towards saturation as the amount of $SF_6$ is increased above 50%.

Electrical breakdown in an $SF_6/N_2$ mixture shows less susceptibility to non-uniform fields than pure $SF_6$. The presence of $N_2$ in $SF_6$ tends to reduce the undesirable influence of contact surface roughness and asperities. In the presence of free conducting particles, $SF_6/N_2$ mixtures perform well compared with pure $SF_6$. This is why it is considered that $SF_6/N_2$ mixtures may be as good as, if not better, than pure $SF_6$ as an insulating gas at absolute pressures above 1 MPa. Therefore, $SF_6/N_2$ mixtures have great industrial potential for gas-insulated transmission lines [70, 71]. Nevertheless, there is reluctance in the industry to use gas mixtures in power-system applications, primarily because $SF_6$ adequately serves present needs. There are also probable disadvantages because of the increased complexity in handling, recycling, and purifying the mixtures.

The excellent behaviour of $SF_6/N_2$ mixtures with small portions of $SF_6$ with regard to dielectric performance is not so easily achieved when it comes to the arc-extinction properties. Here, the arc-extinction principle plays an important role, and there can be no general common conclusion in terms of the effect on the breaking performance.

Despite all these measures, a reduction of breaking and switching capability of $SF_6$ circuit-breakers, depending on the design of the circuit-breaker, is unavoidable. The effect of using the SF6/N2 gas mixture is that the interrupting capability is reduced by approximately one step, that is, a 40 kA circuit-breaker with pure $SF_6$ gas would be derated to 31.5 kA with the above gas mixture. The capacitive current switching capability would be derated by one voltage rating: a 245 kV pure $SF_6$ breaker would derate to 170 kV when filled with a gas mixture [72].

# References

[1]  ASEA Pamphlet Gas Insulated Switchgear, NT 42-102 E (1982).
[2]  Solvay Fluor und Derivate GmbH Brochure: Sulphur Hexafluoride.
[3]  Slepian, J. (1929) Theory of Deion circuit-breakers. *Trans. Am. Inst. Electr. Eng*, **48**(2), 523–527.
[4]  Pugliese, H. and von Kannewurff, M. (2013) Discovering DC: a primer on DC circuit breakers, their advantages, and design. *IEEE Ind. Appl. Magazine*, **19**(5), pp. 22–28.
[5]  Browne, T.E. (1955) The electric arc as a circuit element. *J.Electrochem. Soc.*, **102**(1), 27–37.
[6]  Ackermann, P. (1928) A study of transmission line power-arcs. *The Engineering Journal*, **XI**(5).
[7]  Tretjak, G.T., Kaplan, V.V. and Kender, E.I. (1935) Free Burning Long Power Arcs in Air, CIGRE Report No. 324.
[8]  Eaton, J.R., Peck, J.K., and Dunham, J.M. (1931) Experimental studies of arcing faults on a 75 kV transmission system. *Trans. AIEE*, **50**(4), 1469–1478.
[9]  Strom, A.P. (1946) Long 60-cycle arcs in air. *AIEE Trans.*, **65**, 113–118.
[10] Kohyama, H., Kamei, K., Yoshida, D. *et al.* (2013) Study of Interrupting Duties of Delayed Zero Crossing Current in Generator Main Circuit, CIGRE A3/B2 Symposium, paper 421, Auckland.
[11] Peelo, D.F. (2004) Current interruption using high voltage air-break disconnectors. Ph.D. Thesis, Eindhoven University, ISBN 90-386-1533-7 (available through http://alexandria.tue.nl/extra2/200410772.pdf).
[12] Megahed, A.I., Jabr, H.M., Abouelenin, F.M. and Elbakry, M.A. (2003) Arc Characteristics and a Single-Pole Auto-Reclosure Scheme for Alexandria HV Transmission System. Int. Conf. on Pow. Syst. Transients, New Orleans.
[13] He, B., Lin, X. and Xu, J. (2008) The Analysis of Secondary Arc Extinction Characteristics on UHV Transmission Lines. Int. Conf. on High Voltage Eng. and Appl., Chongqing, China, November 9-13.
[14] CIGRE Working Group A3.22 (2008) Technical Requirements for Substation Equipment Exceeding 800 kV, Field experience and technical specifications of Substation equipment up to 1 200 kV, CIGRE Technical Brochure 362.
[15] CIGRE Working Group A3.22 (2011) Background of Technical Specifications for Substations exceeding 800 kV AC, CIGRE Technical Brochure 456.
[16] Monseth, I.T. and Robinson, P.H. (1935) *Relay Systems*, McGraw-Hill Book Company, New York.
[17] Maikapar, A.S. (1960) Extinction of an open electric arc. *Elektrichestvo*, **4**, 64–69.

[18] Edwards, L., Chadwick, J.W., Reich, H.A. and Smith, L.E. (1971) Single-pole switching on TVA's Paradise-Davidson 500 kV line: design concepts and staged fault tests. *IEEE Trans. Power Ap. Syst.*, **PAS-90**(6), 2436–2450.

[19] Hasibar, R.M., Legate, A.C., Brunke, J. and Peterson, W.G. (1981) The application of high-speed grounding switching for single-pole reclosing on 500 kV power systems. *IEEE Trans. Power Ap. Syst.*, **PAS-100**(4), 1512–1515.

[20] Kappenman, J.G., Sweezy, G.A., Koschik, V. and Mustaphi, K.K. (1982) Staged fault tests with single phase reclosing on the winnipeg-twin cities 500 kV interconnection. *IEEE Trans. Power Ap. Syst.*, **PAS-101**(3), 662–673.

[21] Fakheri, A.J., Shuter, T.C., Schneider, J.M. and Shih, J.H. (1983) Single phase switching tests on the AEP 765 kV system – extinction time for large secondary arc currents. *IEEE Trans. Power App. Syst.*, **PAS-102**(8), 2775–2783.

[22] Hasibar, R.M. and Taylor, C.W. (1983) Discussion of: Fakheri, A.J., Shuter, T.C., Schneider, J.M., Shih, J.H. Single phase switching tests on the AEP , 765 kV system – extinction time for large secondary arc currents. *IEEE Trans. Power Ap. Syst.*, **PAS-102**(8), 2775–2783.

[23] Anjo, K., Terase, H. and Kawaguchi, Y. (1968) Self-extinction of arcs created in long air gaps. *Elec. Eng. Jpn.*, **88**(4), 83–93.

[24] Terzija, V. and Kochlin, H.-J. (2004) On the modeling of long arc in still air and arc resistance calculation. *IEEE Trans. Power Deliver.*, **19**(3), 1012–1017.

[25] Kizilcay, M. and Pniok, T. (1991) Digital simulation of fault arcs in power systems. *Eur. Trans. Electr. Power*, **1**(1), 55–60.

[26] Flurscheim, C.F. (ed.) (1982) *Power Circuit Breaker Theory and Design*, Peter Peregrinus Ltd, IEE Power Engineering Series 1, revised edition, ISBN 0-901233-62-X.

[27] Dufournet, D. (2009) Circuit breakers go high voltage. *IEEE Power Eng. Mag.*, 34–40

[28] Boggs, S.A. and Schramm, H.-H. (1990) Current interruption and switching in sulphur hexafluoride. *IEEE Elec. Insul. Mag.*, **6**(1), 12–17.

[29] CIGRE Study Committee 15 (1974) Breakdown of gases in uniform fields. Paschen curves for nitrogen, air and sulfur hexafluoride. *Electra*, **32**, 61–82.

[30] Raju, G.G. (2006) *Gaseous Electronics – Theory and Practice, The Book*, CRP Press, Taylor & Francis Group, ISBN 0-8493-3763-1.

[31] Christophorou, L.G. and Brunt, R.J. (1995) SF6/N2 mixtures basic and HV insulation properties. *IEEE Trans. Diel. Electr. Insul.*, **2**(5).

[32] Ragaller, K. (1977) *Current Interruption in High-Voltage Networks*, Plenum Press, ISBN 0-306-40007-3.

[33] CIGRE Working Group B3.25 (2014) SF6 Analysis for AIS, GIS and MTS, CIGRE Technical Brochure 567.

[34] CIGRE Taskforce B3.02.01 (2003) $SF_6$ Recycling Guide (Revision 2003), CIGRE Technical Brochure 234.

[35] CIGRE Taskforce B3.02.01 (2005) Guide for the Preparation of Customised Practical $SF_6$ Handling Instructions, CIGRE Technical Brochure 276.

[36] CIGRE Working Group B3.18 (2010) $SF_6$ Tightness Guide, CIGRE Technical Brochure 430.

[37] Christophorou, L.G., Olthoff, J.K. and Green, D.S. (1997) Gases for electrical Insulation and arc interruption: Possible Present and Future Alternatives to Pure $SF_6$. NIST Technical Note 1425.

[38] Intergovernmental Panel on Climate Change (2007) 4th Assessment Report Working Group 1 Ch. 2.10.2: Climate change.

[39] CIGRE Working Group 23.02 (2001) $SF_6$ in the Electric Industry, Status 2000.

[40] United Nations, Kyoto Protocol to the United Nations Framework Convention on Climate Change (1998) http://unfccc.int/resource/docs/convkp/kpeng.pdf, accessed 4 April 2014.

[41] Regulation (EC) No. 842/2006 of the European Parliament and of the Council of 17 May 2006 on certain fluorinated greenhouse gases. Official Journal of the European Union, L161/1 (2006).

[42] Commission Regulation No. 1493/2007 of 17 December 2007 establishing, pursuant to European Regulation (EC) No. 842/2006 of the European Parliament and Council, the format of reports submitted by producers, importers, and exporters of certain fluorinated greenhouse gases. Official Journal of the European Union, L332/7 (2007).

[43] Commission Regulation No. 305/2008 of 2 April 2008 establishing, pursuant to European Regulation (EC) No. 842/2006 of the European Parliament and Council, minimum requirements and the conditions for mutual

recognition for the certification of personnel recovering certain fluorinated greenhouse gases from high-voltage switchgear. Official Journal of the European Union, L92/17 (2008).

[44] Fushimi, Y., Ichikawa, Y., Oue, T., Yokota, T. *et al.* (2004) Activities for Huge $SF_6$ Emission Reduction in Japan. CIGRE Conference, Paper B3-213.

[45] www.cleanenergyfuture.gov.au/cleanenergy-future/our-plan/ and www.environment.gov.au/atmosphere/ozone/ sgg document calculating the equivalent carbon price on synthetic greenhouse gases, accessed 4 April 2014.

[46] WCI, Inc., USA (Western Climate Initiative),(2014) http://www.wci-inc.org/ accessed 4 April 2014.

[47] Biasse, J.-M., Otegui, E. and Tilwitz-von Keiser, B. (2010) Benefits of proper SF6 handling to reduce SF6 emissions for sustainable Electricity Transmission and Distribution. China Int. Conf. on Elec. Distr. (CICED).

[48] US National Oceanic and Atmospheric Administration, (2009) NOAA Earth System Research Laboratory GMD Carbon Cycle - Interactive Atmospheric Data Visualisation, (http://www.esrl.noaa.gov/gmd/ccgg/iadv), accessed 4 April 2014 .

[49] Ecofys Emission Scenario Initiative on Sulphur Hexafluoride for Electric Industry (ESI-SF6), Update on global SF6 emissions trends from electrical equipment, ed. 1.1.07 (2010).

[50] IEC 61634, 1995 (1995) High-voltage switchgear and controlgear - Use and handling of sulphur hexafluoride $SF_6$ in high-voltage switchgear and controlgear.

[51] IEC 60480, 2nd edn, 2004-10 (2004) Guidelines for the checking and treatment of sulphur hexafluoride $SF_6$ taken from electrical equipment and specification for its re-use.

[52] IEC 60376, Ed. 2.0, 2005-06 (2005) Specification of technical grade sulphur hexafluoride $SF_6$ for use in electrical equipment.

[53] IEC TR 62271-303 Ed. 1.0, 2008-07-23 (2008) High-voltage switchgear and controlgear – Part 303: Use and handling of sulphur hexafluoride $SF_6$.

[54] Alexander, B., Robbie, D., Marenghi, M. and Kiener, M. (2012) SF6 and a world first. *ABB Rev.* No. **1**, pp. 22–25.

[55] Okubo, H. and Beroual, A. (2011) Recent trend and future perspectives in electrical insulation techniques in relation to sulphur hexafluoride $SF_6$ substitutes for high-voltage electric power equipment. *IEEE Electr. Insul. Mag.*, March/April, **27**(2), 34–42.

[56] Grant, D.M., Perkins, J.F., Campbell, L.C. *et al.* (1976) Comparative Interruption Studies of Gas-Blasted Arcs in SF6/N2 and SF6/He Mixtures. Proc. 4th Intern. Conf. on Gas Disch. And their Appl., IEE Conf. Publ. No. 143, pp. 48–51.

[57] Suzuki, K., Nishiwaki, S., Kawano, H. *et al.* (2008) Characteristic of Large Current Making Switch to Verify Making Performance of High Voltage Switchgear. Int. Conf. on Gasdisch. and their Appl.

[58] Colombo, A., Barberis, F., Berti, R. *et al.* (2007) CO2 and its Mixtures as an alternative to SF6 in MV circuit-breakers. CIGRE SC A3 Int. Techn. Coll., Rio de Janeiro.

[59] Udagawa, K., Koshizuka, T., Uchii, T. *et al.* (2011) CO2 Circuit Breaker Arc Model for EMTP Simulation of SLF Interrupting Performance. Int. Conf. on Pow. Syst. Transients, Delft, Paper 64.

[60] Wada, J., Ueta, G. and Okabe, S. (2013) Evaluation of breakdown characteristics of $CO_2$ gas for non-standard lightning impulse waveforms: breakdown characteristics in the presence of bias voltages under non-uniform electric field. *IEEE Trans. Dielect. Electr. Insul.*, **20**(1), 112–121.

[61] Kasuya, H., Kawamura, Y., Mizoguchi, H. *et al.* (2010) Interruption capability and decomposed gas density of CF3I as a substitute for SF6 gas. *IEEE Trans. Dielect. Electr. Insul.*, **17**(4), 1196–1203.

[62] Kamarudin, M., Albano, M., Coventry, P. *et al.* (2010) A Survey on the Potential of CF3I gas as an Alternative for $SF_6$ in High-Voltage Applications. UPEC Conference.

[63] ABB Brochure High-voltage $CO_2$ circuit-breaker type LTA, Publication 1HSM 9543-21-06en (2012).

[64] Kynast, E. (2005) Investigations Concerning Discussed Alternatives to SF6 in HV Equipment for Insulating and Arc-Extinguishing Properties. CIGRE SC A3/B3 Colloquium, paper 306.

[65] CIGRE Working Group 23.01 (2000) Guide for $SF_6$ gas mixtures, CIGRE Technical Brochure 163.

[66] Boeck, W., Blackburn, T.R., Cookson, A.H. *et al.* (2004) (Task Force D1.03.10 on behalf of CIGRE Working Group D1.03), $N_2/SF_6$ Mixtures for Gas-Insulated Systems. CIGRE Conference, Paper D1-201.

[67] Christophorou, L.G. and Brunt, R.J. (1995) SF6/N2 mixtures basic and HV insulation properties. *IEEE Trans. Dielect. Electr. Insul.*, **2**(5), 952–1003.

[68] Rokunohe, T., Yagahshi, Y., Endo, F. and Oomori, T. (2006) Fundamental insulation characteristics of air, N2, CO2, N2/O2, and SF6/N2 mixed gases. *Electr. Eng. Jpn*, **155**(3), 619–625.

[69] Diessner, A., Finkel, M., Grund, A. and Kynast, E. (1999) Dielectric Properties of N2/SF6 mixtures for use in GIS or GIL. Int. Symp. on High Voltage, Paper No. 3.67.S 1 8.

[70] Koch, H. (2011) *Gas Insulated Transmission Lines (GIL)*, John Wiley & Sons, ISBN 978-0-470-66533-6.

[71] CIGRE Working Group B3/B1.09 (2008) Application of Long High Capacity Application of Gas-Insulated Lines in Structures, CIGRE Technical Brochure 351.

[72] Peelo, D.F., Bowden, G., Sawada, J.H. *et al.* (2006) High voltage Circuit Breaker and Disconnector Application in Extreme Cold Climates, CIGRE Session Paper A3-301.

# 7

# Gas Circuit-Breakers

## 7.1   Oil Circuit-Breakers

From a historical perspective, oil circuit-breakers were the first circuit-breakers designed for high-power applications. The interruption of current in oil circuit-breakers is performed in oil by way of the dissociation product $H_2$ gas (see Section 6.3), and therefore the oil circuit-breaker is classed as a gas circuit-breaker. At the beginning of the twentieth century, their breaking capacity was sufficient to meet the demands of the power systems at that time. Oil circuit-breakers are still used in the electrical power system worldwide, but development stopped long ago.

Oil circuit-breakers are classified according to the volume of oil used: *bulk-oil* and *minimum-oil circuit-breakers*.

Bulk-oil circuit-breakers (see Section 6.3.2 for the interruption principle) were of simple design: a set of contacts immersed in oil and not equipped with interruption chambers. One of the first plain-break circuit-breakers in the USA was designed and built by J.N. Kelman [1] in 1901. It was capable of interrupting a short-circuit current of 300 to 400 A at 40 kV supply voltage. Kelman's circuit-breaker (Figure 7.1) was made up of two open wooden barrels filled with a combination of water and oil. The two breaks, connected in series, were operated by a common handle. This circuit-breaker was in service for less than a year, from April 1902 to March 1903, and after a number of current interruptions at short-time intervals, blazing oil was spewed over on the surrounding woodwork, starting a fire. This led to the start of serious research in circuit-breaker technology.

In the 1930s, interruption chambers were fitted to the existing plain-break oil circuit-breakers but no significant modification was made to the oil tank itself. The bulk-oil circuit-breakers were widely applied in the USA.

For voltages of up to 72.5 kV, the three phases were enclosed in one tank of oil (Figure 7.2a). For higher voltages, oil circuit-breakers were designed with three independent metal tanks, as illustrated in Figure 7.2b).

Bulk-oil circuit-breakers require large tanks and need a large volume of oil to provide the necessary insulating distance between the live parts of the circuit-breaker and the earthed tank. So, for example, a 145 kV circuit-breaker required approximately 12 000 litre of oil, whereas

*Switching in Electrical Transmission and Distribution Systems*, First Edition.
René Smeets, Lou van der Sluis, Mirsad Kapetanović, David F. Peelo, and Anton Janssen.
© 2015 John Wiley & Sons, Ltd. Published 2015 by John Wiley & Sons, Ltd.

**Figure 7.1**   Kelman's oil circuit-breaker built in 1901 (40 kV, 300 A) [1].

in the case of a 245 kV circuit-breaker, the volume could increase to over 50 000 litre. A special foundation was necessary to give support to the weight and resist the forces during operation of the breaker.

Bulk-oil circuit-breakers were manufactured till the mid-1990s and are still in service, sometimes having eight breaks in series [2].

Compared with the present technologies, the bulk-oil circuit-breakers had a rather low mechanical and electrical endurance. They required frequent maintenance to remove free

(a)                                                              (b)

**Figure 7.2**   (a) Single-tank oil circuit-breaker; (b) multiple-tank oil circuit-breaker.

carbon particles resulting from the oil decomposition caused by the arc burning in oil. Otherwise, a progressive deterioration of the insulating properties of the oil could occur with a significant risk of explosion and a potential fire hazard to the substation.

In order to reduce the price of circuit-breakers because of the high cost of oil, new circuit-breakers with a small volume of oil, so-called *minimum-oil circuit-breakers*, were developed. These circuit-breakers became more popular in Europe than in the USA, see Section 6.3.3 for their interruption principle.

The main difference between bulk-oil and minimum-oil circuit-breakers lies in the fact that minimum-oil breakers use oil for arc-extinction only, while a solid insulating material is applied for insulation. In the case of bulk-oil circuit-breakers oil serves both purposes.

Minimum-oil HV circuit-breakers fall into the class of *live-tank circuit-breakers*. By insulating the live parts from earth via solid supporting insulators, the volume of oil is greatly reduced. The insulators are made of reinforced fibreglass for MV applications and of porcelain for the higher voltage-ratings.

For a long time, until the 1970s, minimum-oil circuit-breakers dominated in the field of breaker technology, in particular for voltages up to 145 kV where only one interrupter or break per pole was sufficient. That ensured simplicity, low weight, low price and relatively simple, but still rather frequent, maintenance.

Modular concepts of live-tank minimum-oil circuit-breakers with two breaks at 170 kV (for example Figure 7.3) and four breaks per pole at 420 kV were developed for the first 380 kV network in the world in Sweden (1952). There were designs made for multi-break minimum-oil circuit-breakers for 750 kV networks, but they lost out to competition from

**Figure 7.3**   Live-tank 170 kV minimum-oil circuit-breaker with two breaks per pole.

air-blast circuit-breakers, first applied in 525 kV, and later on in 735 kV, and 765 kV networks, respectively in the USSR (1960), in Canada (1965) and in the USA (1969).

## 7.2   Air Circuit-Breakers

The basic principle of current interruption in air at atmospheric pressure is trivial: the contacts are opened and the arc is elongated.

The interrupting capabilities of this approach were rather limited. In order to increase the interrupting capabilities of these simple air-break interrupting devices, either *deion chambers* can be added or air can be blown across the contacts. Current interruption in atmospheric air is still in use today, almost exclusively in LV circuit-breakers (see Section 6.2.2). For medium voltage, thanks to deion chambers with ceramic arcing plates, circuit breaking in atmospheric air is feasible, providing that an external magnetic field is applied. This principle works well, in spite of the relatively low dielectric withstand strength of air at atmospheric pressure.

With insulating deion chambers, the cooling of the arc and its final extinction is achieved by arc elongation as the arc is forced to move into the narrow space between the insulating plates. In addition, the arc gets cooled when it makes contact with the insulating plates. The motion of the arc is caused by the magnetic field. The "magnetic blast" is produced by a coil that is usually embedded into the external supporting plates. An example of this type of circuit-breaker is shown in Figure 7.4.

Since compressed air has a higher dielectric strength and better thermal properties than air at atmospheric pressure, circuit-breakers with higher ratings can be designed by applying the air-blast principle of interruption. The air-blast principle is based on a blast of compressed air

**Figure 7.4**   Air circuit-breaker, rated 15 kV with magnetic blast principle.

**Figure 7.5**   Air-blast circuit-breaker with 14 interrupters per pole for 765 kV from 1968 (ASEA).

directed at the arc, preferably along its length, as an axial blast. This technology was, for more than fifty years, the technology for extra-high voltages until the advent of $SF_6$ circuit-breakers.

The development of the air-blast extinction principle started in Europe in the 1920s; further progress was achieved in the 1930s, and the air-blast circuit-breakers were widely installed in the 1950s, having an interrupting capability of 63 kA that even increased to 90 kA in the 1970s. Many of these circuit-breakers are still in service today, in particular in North America, for the higher voltage levels of 500 and 800 kV.

Air-blast circuit-breakers have the advantage of high interrupting capability in combination with short interruption time. The interrupters, however, have a relatively small dielectric withstand capability. This limitation comes from the opening speed of the contacts. The opening speed could be effectively increased by choosing multi-break designs. Therefore, as can be seen in Figure 7.5, an air-blast circuit-breaker for rated voltages above 420 kV initially needed 10 or even 12 interrupters in series per pole. More advanced designs followed requiring only six interrupters in series at 500 kV.

The main difficulty encountered with this type of design is to ensure that during the interruption process each of the interrupters operates under the same conditions, both aerodynamically and electrically. From an aerodynamic point of view, the same gas flow must be maintained for each interrupter. To avoid pressure drops that may affect the gas flow, the use of individual blast valves for each interrupter becomes necessary. Electrically, the transient recovery voltages must be distributed equally across each set of contacts. To improve the voltage distribution across the gaps, it is common practice to use grading capacitors or resistors.

Another difficulty that arises is the rather high air pressure, as high as 3 MPa, necessary for arc extinction. Therefore, air-blast circuit-breakers need powerful compressors and are very noisy when they operate, especially when the arc is blasted into the surrounding air, as is the case for the *free-jet* type circuit-breakers shown in Figure 7.6.

**Figure 7.6**   Interrupters of a free-jet type air-blast circuit-breaker.

Air-blast circuit-breakers are very fast, reliable, and uniquely suited for interruption of very high currents. Therefore, they are still used almost exclusively as master circuit-breakers in high-power laboratories (see Section 14.2.2).

In existing applications, certain designs of MV and HV air-blast circuit-breakers have proved to be very robust and reliable. Even after several decades of operation a significant extension of lifetime can be assumed for those designs through a program of refurbishment and re-testing [3].

## 7.3   SF$_6$ Circuit-Breakers

### 7.3.1   Introduction

The outstanding dielectrical properties of SF$_6$ gas were already known some forty years before the first industrial application of SF$_6$ for current interruption, in 1953 [4,5]. Those were high-voltage, 15 to 161 kV load-break switches with a breaking capacity of 600 A. The first high-voltage SF$_6$ circuit-breaker was developed in 1956 by Westinghouse (see Figure 7.7).

A few years thereafter, in 1959, Westinghouse produced the first high-voltage SF$_6$ circuit-breaker with a high short-circuit-current capability. This dead-tank circuit-breaker could interrupt 41.8 kA at 138 kV and 37.6 kA at 230 kV. Such a performance was already significant, but the three interrupters per pole and a high SF$_6$ pressure of 1.35 MPa necessary for the arc blast needed to be reduced. The double-pressure design prevailed only in the US market, from the early 1960s until the mid-1970s. They got competition from the single-pressure or *puffer circuit-breakers*.

Initially puffer circuit-breakers were limited to applications with lower interrupting ratings but, after 1965, puffer circuit-breakers with high interrupting capability came to market.

During the 1970s and 1980s, the rated voltage of puffer circuit-breakers increased to 800 kV, with a remarkable simplicity in design and a drastic reduction of both the number of interrupters per pole and the operating energy; an example is illustrated in Figure 7.8. This resulted in increased reliability of the circuit-breakers. Reliability data are given in Section 12.1.

**Figure 7.7**    $SF_6$ circuit-breaker with rated voltage 115 kV from 1956 (Westinghouse).

**Figure 7.8**    Dead-tank $SF_6$ circuit-breaker rated 72.5 kV (reproduced with permission from Siemens AG).

**Figure 7.9**   Two-break GIS circuit-breaker in Japan with rated voltage 1000 kV.

In 1980, the first single-break 245 kV, 40 kA puffer circuit-breaker was developed and type tested. The progress made in the 1990s with single-break 420 and 550 kV circuit-breakers, and 1000 and 1100 kV breakers with two interrupters per pole, in Japan [6], Figure 7.9 and also later for the 1100 kV projects in China [7,8], underlines the dominance of $SF_6$ circuit-breakers over the whole range of power-transmission voltages [9].

The improvement in the design of puffer interrupters made it possible to apply operating mechanisms with considerably lower energy. An impression of the progress made is visualized in Figure 7.10. The drive energy cannot be brought below a certain level.

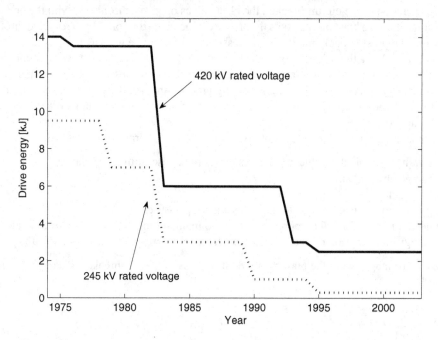

**Figure 7.10**   Reduction of drive energy per pole for a 245 kV and a 420 kV circuit-breaker in past years [2].

Since the 1980s, a new interruption principle, the *self-blast technology*, was introduced. Self-blast technology uses the energy of the arc itself to (partly) produce the pressure necessary to blast the arc and to obtain current interruption.

In parallel with the development of puffer circuit-breakers, the *rotating-arc technology* was studied. In contrast with the classical interruption process, where arc cooling is done by moving the gas, in rotating-arc interrupters, the arc is set in motion by the magnetic field of the current that is to be interrupted. Low operating-energy requirements motivated this direction of research but, so far, the performance of this interrupting principle has not been as effective as other technologies.

Summarizing, there are four types of $SF_6$ circuit-breakers:

- double-pressure SF6 breakers;
- puffer type breakers, operating on single pressure at steady-state;
- self-blast breakers, also operating on single pressure at steady-state; and
- breakers using the rotating-arc technology.

## 7.3.2   Double-Pressure $SF_6$ Circuit-Breakers

Double-pressure $SF_6$ circuit-breakers are in fact designed as air-blast circuit-breakers with axial blast, with the air being replaced by $SF_6$. Double-pressure $SF_6$ circuit-breakers are mostly of the dead-tank type, with the interrupters mounted inside a tank, similar to the tanks used for oil circuit-breakers and filled with low-pressure $SF_6$ gas. A second, high-pressure reservoir, separated from the low-pressure tank by a blast valve contains compressed $SF_6$ gas. During operation of the circuit-breaker, the blast valves open synchronously with the opening of the contacts and a blast of $SF_6$ cools the arc. After each interruption, $SF_6$ is recycled through filters and compressed back into the high-pressure reservoir to be used again.

Since, at 1.6 MPa, the $SF_6$ gas liquefies at approximately 10 °C, heaters have to be mounted at the high-pressure reservoir to avoid liquefaction of the high-pressure $SF_6$ gas. Apart from the extra energy consumption in the gas-heating process, the heating elements also introduce an extra failure possibility for the circuit-breaker; when the heating does not work, the circuit-breaker cannot operate.

The most important disadvantages of double-pressure $SF_6$ circuit-breakers are:

- their high mechanical complexity that results from the use of the blast valves;
- their large physical size;
- the large amount of $SF_6$ gas needed;
- the propensity for leakage due to higher operating pressures;
- the need for an additional compressor system to maintain the high pressure of $SF_6$; and
- the energy consumption of heaters that are necessary to prevent liquefaction of $SF_6$.

These disadvantages were the reason why double-pressure circuit-breakers disappeared from the market.

## 7.3.3   Puffer-Type $SF_6$ Circuit-Breakers

A common characteristic of all puffer-type circuit-breakers is the compression of $SF_6$ gas during the opening operation inside the interrupting chamber. The gas is compressed by the

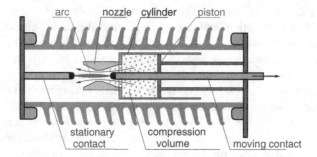

**Figure 7.11**   Operating principle of an SF$_6$ puffer circuit-breaker.

motion of the cylinder against a fixed piston. In most of the puffer designs, the moving contact plays the role of a compression cylinder (Figure 7.11). The increased pressure forces a gas flow through the nozzle and along the arc plasma channel. The gas flow stops when the moving contact is at the end of its stroke and reaches the final open position. It is important to provide efficient blasting throughout the whole range from minimum to maximum arcing time. This can be achieved by properly dimensioning the compression stroke.

Puffer interrupters have a current-dependent pressure-build-up characteristic. At a no-load operation, without an arc, the maximum pressure inside the compression volume is typically twice the filling pressure (see curve (a) in Figure 7.12). However, during the high-current interval the arc, burning between the arcing contacts, blocks the gas flow through the nozzle;

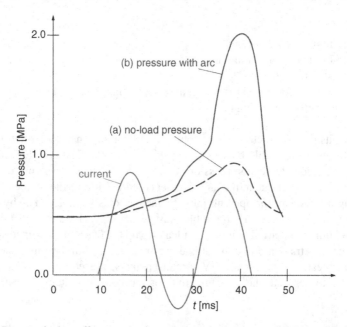

**Figure 7.12**   Pressure in the puffer compression volume at no-load operation (a); and during interruption of an asymmetrical short-circuit current of 40 kA (b) [10].

this is called *nozzle clogging*. The arc inside the nozzle reacts to the gas flow as a valve with a time variable cross-section, and causes an additional pressure rise of the gas in the compression volume. In a situation when the nozzle is temporarily clogged or its diameter is effectively reduced by the arc during high instantaneous values of current, a part of the arc energy in the nozzle cannot be efficiently released. Therefore, the energy content of the gas in the compression volume increases. In this way, the arc energy contributes to the pressure rise. The maximum pressure inside the compression volume can be several times the maximum no-load pressure, curve (b) in Figure 7.12. Thus, the resulting pressure during current interruption becomes higher. When the current decreases towards zero, the arc diameter also decreases, leaving more and more free space for the gas to flow. A full gas flow is active at current zero, giving the maximum cooling power when it is most needed.

Arc clogging causes ablation of the nozzle material. This material evaporates and gives an additional gas mass inside the compression volume. This effect can notably increase the gas pressure, the density and the gas-mass flow during the high-current interruption; additionally, it may influence the contact-travel characteristic of a puffer interrupter [11].

The high pressure in the puffer requires a high operating force from the breaker mechanism in order to prevent slowing down, stopping, or even reversal of the movement of the circuit-breaker contacts, because it counteracts the force of the operating mechanism. Thus, the blast energy required for opening is almost entirely supplied by the operating mechanism. The higher the current required to be interrupted, the higher the force that is needed. Therefore, puffer circuit-breakers require powerful operating mechanisms with a large energy output. The energy required for closing is far less than that for opening.

The way in which the accumulated energy of the compressed gas is used for forced arc cooling varies. Some basic nozzle design principles for $SF_6$ puffers can be distinguished (see Figure 7.13):

- single-blast with insulating nozzle;
- partial double-blast with insulating nozzle;
- full double-blast with insulating nozzle;
- full double-blast with combined insulating and conducting nozzle; and
- full double-blast with conducting nozzle.

Each design has its own characteristics for optimal utilization. The arc-extinction capability of an $SF_6$ puffer with full double-blast is higher than the single-blast type for the same blast pressure. However, since a larger compression volume is needed to achieve a particular pressure in a double-blast interrupter due to the larger outlet (Figure 7.13c d, and e), the output energy of the operating mechanism for opening must be increased as well. This is why often either single-blast, (Figure 7.13a), or partial double-blast technology, (Figure 7.13b), is used. The compromise solution with partial double-blast leads to up to 20% higher extinction capability, requiring neither an extra compression volume nor an increase in the opening energy.

Puffer circuit-breakers, in fact all HV circuit-breakers, are equipped with two contact systems in order to separate the two functions of the contact system, as can be seen in Figure 7.14a:

- conduction of the normal-current, assigned to the *main contacts*; and
- ability to withstand the electric arc, assigned to the *arcing contacts*.

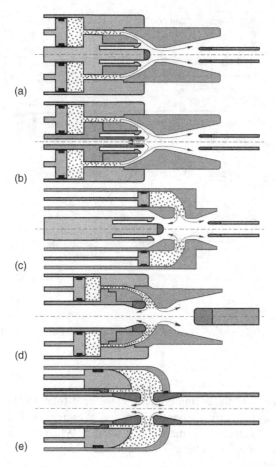

**Figure 7.13** Nozzle design principles for SF$_6$ puffers: (a) single-blast with insulating nozzle; (b) partial double-blast with insulating nozzle; (c) full double-blast with insulating nozzle; (d) full double-blast with combined insulating and conducting nozzle; (e) full double-blast with conducting nozzle [12].

Current flows through the main contacts outside the nozzle. This position of the main contacts gives a better cooling possibility and makes it easier to achieve larger cross-sections of the conductors, to facilitate higher values for the normal current.

Only for the lower normal-current ratings, are both main and arcing contacts placed inside the nozzle, as in Figure 7.14b.

In the closed position, the current flows from the upper current terminal to the main stationary contact, the main moving contact, the contact rod, passing the tulip contact, towards the lower current terminal.

The compression cylinder, main moving contact, nozzle, and moving arcing contacts move together with the contact rod which is connected to the operating mechanism by an isolating rod mounted inside the supporting insulator. During opening operation, the main contacts separate first. The current commutates to the arcing contacts and the pressure starts to build up inside the compression volume.

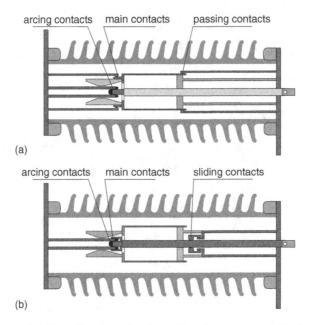

**Figure 7.14** Contacts for $SF_6$ puffer circuit-breakers: (a) main contacts outside the nozzle; (b) arcing contacts inside the nozzle.

During closing operation, the arcing contacts close first, followed by the main contacts. Arcing contacts conduct the current from the initiation of current flow at a pre-strike and the following pre-arc, until the instant the main contacts touch. When the contacts reach the fully closed position, the compression volume is refilled with cold gas and the breaker is ready for the next opening operation.

In the upper part of the interrupter is a filter with an adsorbent, a molecular sieve of activated alumina, to remove moisture and most of the $SF_6$ decomposition by-products after arcing, in particular hydrogen fluorides (see Section 6.4.3). The process goes rapidly and is effective, the level of corrosion, due to $SF_6$ decomposition products, is very small.

An interrupter also has a safety rupture disc, which bursts at a predefined pressure.

A great advantage of the puffer-type circuit-breakers is the simplicity of the interrupters, providing reliability and excellent mechanical endurance of the total device. Also contributing to the reliability at the highest voltages is the reduced number of interrupters per pole. The excellent properties of $SF_6$ made it possible to develop puffer-type circuit-breakers for voltages above 800 kV with only two interrupters per pole (see Figure 7.15). Several manufacturers have developed puffer-type circuit-breakers for rated voltages of 550 kV, with a single interrupter per pole.

The advantages of puffer-type circuit-breakers are exploited best in metal-enclosed $SF_6$ circuit-breakers for dead-tank and GIS applications because they provide possibilities to reduce the size of the switchgear considerably.

Puffer-type circuit-breakers have a relatively long stroke and require high operating forces that can only be provided by rather complex and powerful operating mechanisms with a high

**Figure 7.15**  Pole of an 800 kV SF$_6$ puffer circuit-breaker with four interrupters per pole next to a 123 kV three-pole operated circuit-breaker with one interrupter per pole (ABB).

energy output. The operating energy required by the puffer circuit-breaker is much higher than the opening energy for an oil circuit-breaker. A simple and reliable spring-operating mechanism can only be applied for short-circuit currents below 40 kA. Above 40 kA, pneumatic or hydraulic mechanisms are needed and that makes these breakers more costly and, compared with breakers equipped with a spring mechanism, less reliable.

## 7.3.4   Self-Blast SF$_6$ Circuit-Breakers

*Self-blast breakers are single-pressure circuit-breakers but with a different technology.* They differ from puffer circuit-breakers in the way the gas pressure for cooling is being built up. In a puffer circuit-breaker, the energy to compress the SF$_6$ gas comes from the operating mechanism, whereas a self-blast circuit-breaker (for a major part) uses the thermal energy released by the arc to heat the gas and to increase its pressure. The name self-blast was introduced by a manufacturer who brought the first self-blast breaker to market. Different names are in use by other manufacturers: self-compression, auto-expansion, arc-assisted, thermal-assisted, auto-puffer, and so on, all describe the same principle.

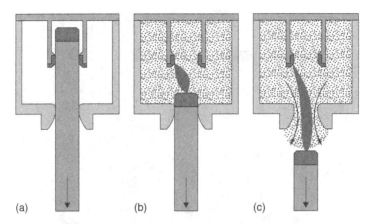

(a)                              (b)                              (c)

**Figure 7.16**  Basic principle of self-blast technology in SF$_6$ circuit-breakers: (a) closed position; (b) closed-in arc; (c) arc blasting and extinction.

In every gas circuit-breaker, the thermal energy of the arc partially contributes to an increase in the gas pressure in the interrupter.

Service experience has shown that circuit-breaker failures due to insufficient breaking capacity are very rare. The majority of failures on site are of mechanical origin (see Section 12.1.3) and switchgear manufacturers have concentrated their engineering efforts on developing simple and reliable operating mechanisms. To make this possible, the fundamental issue becomes reducing the retarding forces on the operating mechanism during an opening operation. This way of thinking guided the breaker development in the direction of self-blast interruption. This has led to a reduction in drive energy as shown in Figure 7.10. The self-blast interruption principally represents a milestone in the development process of HV circuit-breakers.

The interruption chamber based on the self-blast principle as described above is very simple, as depicted in Figure 7.16. The contacts are mounted in an isolating interruption chamber, see Figure 7.16a. After contact separation, the arc burns in a closed volume for some time during which the thermal energy released by the arc accumulates, resulting in a substantial pressure rise inside the interruption chamber, see Figure 7.16b. When the moving arcing contact comes out of the nozzle, a strong gas flow cools the arc (Figure 7.16c).

This scenario, however, cannot work for the complete range of currents that the circuit-breaker must interrupt. Interrupting small currents may be a problem, since in this case the arc energy is not sufficient to generate a high enough pressure for efficient arc blast. Therefore, over the past 20 years, circuit-breakers have been developed that combine the self-blast and the puffer principle.

A relatively low blast pressure is able to interrupt current up to approximately 30% of the rated short-circuit-breaking current, and a corresponding part of energy is required from the operating mechanism. For higher currents, the arc itself provides the energy needed to generate sufficient pressure for efficient cooling of the arc.

There are other ways to relieve the effect of the arc-generated pressure on the operating mechanism, such as in the self-blast double-volume interrupters equipped with suitable over-pressure valves. These valves prevent the slowing down and/or stopping of the movement of

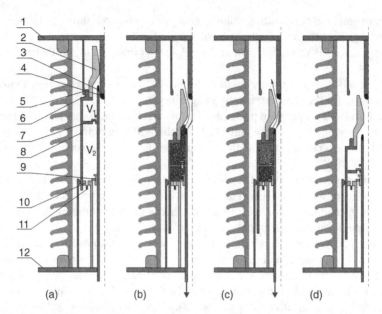

**Figure 7.17** SF$_6$ self-blast double-volume interrupter with suitable overpressure valves: (a) closed position; (b) interrupting small current; (c) interrupting high current; (d) open position. V$_1$ – thermal volume; V$_2$ – puffer volume; 1 – current upper terminal; 2 – nozzle; 3 – moving arcing contact; 4 – stationary arcing contact; 5 – stationary main contact; 6 – moving main contact; 7 – overpressure valve; 8 – puffer cylinder; 9 – refill valve; 10 – piston; 11 – overpressure valve; 12 – current lower terminal.

the contact system. An example of a self-blast interrupter using this principle is shown in Figure 7.17.

When interrupting relatively small currents, up to a few kiloamperes, the interrupter works similarly to the classical puffer breaker. SF$_6$ gas is compressed in the *puffer volume* V$_2$ and flows through the fixed *thermal volume* V$_1$ and the nozzle throat axially along the arc, Figure 7.17b. There is not sufficient gas pressure developed to close the overpressure valve (7) and the fixed thermal volume and puffer volume form one puffer volume.

When interrupting higher short-circuit currents, thermal energy released in the arc column accumulates in the thermal volume V$_1$, where the pressure increases due to the rise in temperature and due to the compression of the gas between the puffer cylinder and the stationary piston. The gas pressure in volume V$_1$ continues to increase until it is high enough to force the overpressure valve (7) to close. All SF$_6$ gas required for interruption is now trapped in the fixed thermal volume V$_1$ where further increase in pressure is solely due to heat from the arc (Figure 7.17c). At about the same time, the gas pressure in the puffer volume reaches a value sufficient to open the overpressure valve (7). Since the gas in the puffer volume is released by the overpressure valve, there is no additional energy from the operating mechanism necessary to overcome the compression of SF$_6$ gas and to maintain the contact speed necessary for recovery after current interruption.

From the moment when the nozzle throat opens up, mechanically and/or by reduction of the arc diameter at lower instantaneous current values, the pressure difference between the thermal volume and the surrounding volume generates a gas flow through the nozzle, along the arc. It extinguishes the arc at current zero.

On closing, the refill valve opens, and the $SF_6$ gas is drawn into the thermal and puffer volumes.

This type of self-blast interrupter requires 50 to 70% less operating energy than classical puffer interrupters for the same interrupting capability.

Figure 7.18 shows the operating principle of another HV circuit-breaker interrupter design, also applying a combination of the self-blast and the puffer principle [13]. This interrupter is a classical puffer type enhanced with an auxiliary piston on the support of the stationary contact.

During opening, just after contact separation, the so-called *closed-in arc* appears for a few milliseconds, causing a pressure rise in the volume between the compression and the auxiliary pistons. For the interruption of higher currents, the thermal energy of the arc becomes dominant and causes a significant increase in pressure of the closed-in $SF_6$ gas.

The pressure increase under the auxiliary piston generates additional force acting in the same direction as the driving force, and assists the operating mechanism in performing its opening operation. The pressure rise in the suction volume may partly or even fully compensate the action of the pressure rise in the puffer volume. In this way the interrupter also utilizes the self-compression principle. After the instant when the auxiliary piston exits the nozzle, a strong gas stream cools the arc, driven by the accumulated arc energy in the compression volume and opening-spring energy. From this moment, till the end of the contact travel, there is no difference between this interrupter and the classical puffer. At current zero, the arc disappears and the current is interrupted.

When interrupting small currents, the flow of $SF_6$ gas through the nozzle is driven by a combination of the puffer effect of the compression piston and the suction effect of the auxiliary piston.

The volume as a whole, consisting of the suction and the puffer volumes, is closed and reduced in size in a few milliseconds after contact separation. This means that the volume will have a somewhat higher pressure, and this lowers the probability of restrike in the case of interruption of capacitive currents.

From the point of view of mechanical simplicity, this interrupter is practically as simple as the classical puffer. Combining self-blast, self-compensated and puffer interrupters with a low-energy spring-operating mechanism allows a very reliable operating mechanism.

All designs mentioned above belong to the family of *single-motion* interrupters. This is the simplest design, with only one moving set of arcing and main contacts.

## 7.3.5   Double-Motion Principle

For the highest voltage ratings, interrupters must have a longer contact gap and a high opening speed. From the total amount of energy necessary to interrupt the current, the parts needed for fast motion of the contacts increase rapidly with the contact speed since the kinetic energy of moving bodies is proportional to the product of mass and the square of the speed.

There is thus again a demand for high-energy operating mechanisms for circuit-breakers at the highest voltages. The single-motion interrupters with a combination of the self-blast, the self-compensated and the puffer principle for arc extinction can be further optimized by

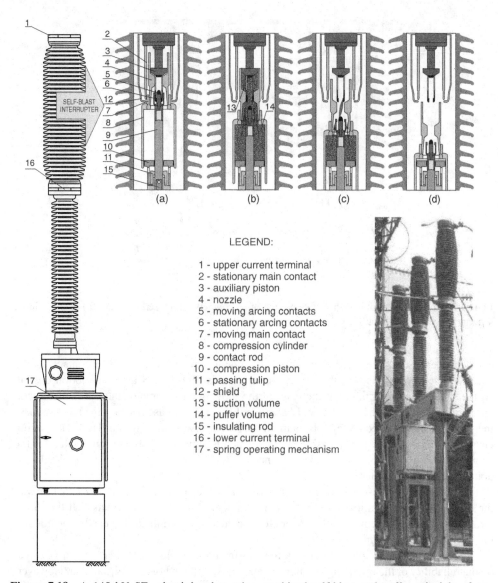

**Figure 7.18**  A 145 kV SF$_6$ circuit-breaker using combined self-blast and puffer principle of arc extinction with a spring-operating mechanism: (a) closed position; (b) closed-in arc; (c) process of arc-extinction; (d) open position.

making use of the so-called *double-motion principle* that consists of moving the two arcing contacts in opposite directions.

Figure 7.19 shows the interrupter of a circuit-breaker with the double-motion principle. The movable upper contact system is connected to the nozzle by means of a linkage. This makes it possible to move the lower contact system and the upper contact system in opposite directions. In this way, the speed requirement for the operating mechanism is reduced, since the effective

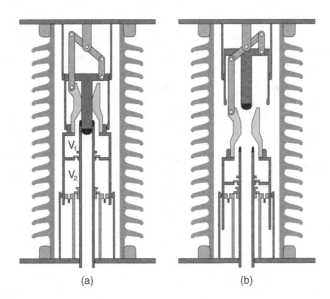

(a)                                          (b)

**Figure 7.19** SF$_6$ interrupter with double motion of its contacts: (a) closed position; (b) open position. $V_1$: thermal volume; $V_2$: puffer volume.

contact speed is the speed of the movable upper contacts relative to the lower contacts. If the speed of each contact is half the relative speed, it is still 100% of a single-motion design. At the same time, assuming that the mass of the moving parts remains the same, the kinetic energy is reduced by a factor of four. In practice this is not the case because the moving mass increases for higher rated voltages because of the longer insulating distance. The total opening energy also includes the compression energy, which is approximately the same value for the single-motion and double-motion design since the self-blast principle can be combined with double-motion techniques.

There are also double-motion designs that also move the upper shield to optimize in this way the electric field distribution in order to achieve a better dielectric performance. If the shield moves at a speed different from the upper arcing contact speed, this design can be regarded as the *triple-motion principle*.

The reduction of the operating energy reduces the dynamic loads and has a positive effect on the reliability of the operating mechanism and, thus, of the whole circuit-breaker. However, at the same time, the simplicity of single-motion interrupters is lost and consequently, the reliability of such interrupters might be lower.

## 7.3.6   Double-Speed Principle

An alternative to the double-motion principle is an interrupter with a double contact speed, as is depicted in Figure 7.20.

The basic idea of the double-speed contact system is to divide the mass of the moving contact parts into two, an upper and a lower part, and temporarily a part of the kinetic energy is transferred from the lower to the upper mass.

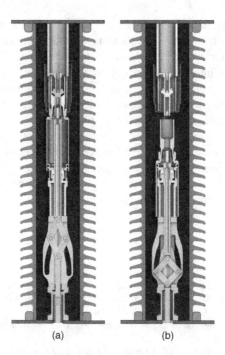

            (a)                              (b)

**Figure 7.20**   $SF_6$ interrupter with double-speed contact stroke: (a) closed position; (b) open position.

   A couple of shaped guides are used for steady drive of the moving contact and the direction
of its motion is in a straight line. At the same time, the shape of the guides controls the speed of
the moving contact. The first part of the guides is rectilinear and parallel, providing identical
movement on both sides. The curvilinear part of the guides begins approximately 12 mm
before the arcing-contact separation. Then, an acceleration of the upper masses begins, as well
as a retardation of the lower masses of the kinetic chain. The increased speed is maintained
only for 10 ms after contact separation. After that, the guides become rectilinear and parallel,
equalizing both speeds again. The remaining contact travel is similar to the case without the
double-speed mechanism.
   Either double speed or double motion of contacts together with combined self-blast, self-
compensation and puffer principle, allow for a significant reduction of energy needed for the
opening operation. In comparison with the first generation of puffer circuit-breakers, such a
combination enables a reduction in the required operating energy by almost an order of magnitude. This illustrates the progress HV $SF_6$ breaker development has made since the 1970s.
   Furthermore, the self-blast principle also gives a reduction in the amount of $SF_6$ gas used
for arc extinction. Thus the total volume of the interrupter housing can be reduced by up to
40% [2]. This is of both economic and environmental importance.

### 7.3.7   $SF_6$ Circuit-Breakers with Magnetic Arc Rotation

In nearly all $SF_6$ circuit-breakers the arc is cooled by forcing $SF_6$ gas to flow along the arc.
In $SF_6$ circuit-breakers with magnetic arc rotation, the arc is set in motion and rotates through

the motionless SF$_6$ gas. The motion is realized by the Lorentz force in a magnetic field that is generated by the current. The effect is the same as in puffer technology: an efficient cooling and successful current interruption can also be achieved, provided that the magnetic field created by the current to be interrupted is strong enough.

A great advantage of magnetic arc rotation lies in the fact that this method requires a short total contact-gap length, low operating forces and, therefore, very low energy of the operating mechanism. SF$_6$ circuit-breakers with magnetic arc rotation can be designed with less powerful and, therefore, less expensive mechanisms and are more compact than puffer circuit-breakers. A further advantage of arc rotation is that contact wear is reduced since the arc root is forced to move. Until now, however, arc rotation has not been as effective as other arc extinction principles for current interruption in SF$_6$ gas for high voltage. In most cases, puffer assistance is needed to improve the performance at the lower current levels.

# References

[1] Wilkins, R. and Cretin, E.A. (1930) *High Voltage Oil Circuit Breakers*, McGraw Hill.
[2] Dufournet, D. (2009) Circuit breakers go high voltage. *IEEE Power Energy Mag.*, 34–40.
[3] Janssen, A.L.J., van Riet, M., Smeets, R.P.P. *et al.* (2014) Life Extension of Well-Performing Air-Blast HV and MV Circuit-Breakers. CIGRE Conference, paper A3-205.
[4] Lingal, H.J. (1953) An investigation on the arc quenching behaviour of sulfur hexafluoride. *AIEE Trans.*, **772**, 242–246.
[5] Methwally, I.A. (2006) New Technological Trends in High-Voltage Power Circuit Breakers. IEEE Potentials, pp. 30–38.
[6] Yamagata, Y., Toyoda, M., Hemmi, R. *et al.* (2007) Development of 1 100 kV Gas Circuit Breakers - Background, Specifications, and Duties. IEC/CIGRE UHV Symposium, Beijing, Paper 2-4-3.
[7] Lin, J., Gu, N., Wang, X. *et al.* (2007) The Transient Characteristics of 1 100 kV Circuit Breakers. IEC/CIGRE UHV Symposium, Beijing, Paper 2-4-4.
[8] Riechert, U. and Holaus, W. (2012) Ultra high-voltage gas-insulated switchgear - a technology milestone. *Eur. Trans. Electr. Power*, **22**(1), 60–82.
[9] Brunke, J. (2003) Circuit-breakers: Past, present and future. *Electra* **208**, 14–20.
[10] ABB Publication, Live Tank Circuit Breakers - Application Guide, Edition 1.1 (2009-06).
[11] Gajić, Z. (1998) Experience with Puffer Interrupter Having Full Self-compensation of Resulting Gas Pressure Force Generated by the Electrical Arc. CIGRE Conference, Rep. 13-103, Paris.
[12] Berneryd, S. (1982) Design and Testing Principles for SF$_6$ Circuit Breakers, ASEA HV Apparatus, LKU 12-80 Rev. 1.
[13] Kapetanović, M. (1996) SF$_6$ Circuit breakers for outdoor installation 72.5 to 170 kV, 40 kA, Type SFEL. 11th CEPSI Conference, Kuala Lumpur.

# 8

# Current Interruption in Vacuum

## 8.1 Introduction

Since the first patent from 1890 [1] and the first practical experiments in the 1920s, switching in vacuum has been considered as an effective way to control the power flow in electrical supply networks.

Already in 1926, experiments were carried out to interrupt 101 A at 16 kV by a vacuum switch, shown in Figure 8.1. The results were reported as follows [2]: "*A noticeable feature of the vacuum-switch tests is that every oscillographic record shows that the arc produced at the opening of the switch is extinguished at the end of the first half-cycle after the separation of the contacts. Only the very best oil-switch operations give these results.*"

Figure 8.2 shows the first published oscillogram of current interruption of 101.2 A at a voltage of 16.1 kV.

The basic principle is that vacuum gives the best-known insulating medium in the steady state since vacuum provides no means to support conduction [3]. Current interruption and dielectric recovery after interruption is achieved due to the natural diffusion of arc residues. This is unlike other technologies (e.g. gas circuit-breakers), where the design and performance are closely related to gas flow by externally energized mechanical means.

Switching in vacuum is fundamentally different from switching in the "classical" extinguishing media $SF_6$, air, and oil, since the switching behaviour is entirely determined by the metal vapour that is released from the contacts by the vacuum arc [4]. Therefore, a closer look into vacuum, vacuum arcs, and vacuum-interrupter contact material is necessary.

## 8.2 Vacuum as an Interruption Environment

The word vacuum comes from the Latin word "vacuus" which means "empty", but in practice, no volume of space can ever be perfectly empty. The classical notion of a perfect vacuum with gaseous pressure of exactly zero is only a philosophical concept and has never been observed.

In everyday use, the term *vacuum* is used to describe a volume of space with diluted gas, such that its gaseous pressure is less than atmospheric pressure. This concept of vacuum was revived when technological means permitted the partial removal of air from a vessel. Reduced

*Switching in Electrical Transmission and Distribution Systems*, First Edition.
René Smeets, Lou van der Sluis, Mirsad Kapetanović, David F. Peelo, and Anton Janssen.
© 2015 John Wiley & Sons, Ltd. Published 2015 by John Wiley & Sons, Ltd.

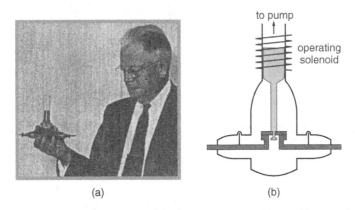

(a)                                                      (b)

**Figure 8.1**   Schematic drawing of the first vacuum switch with its inventor Dr. Sorensen in 1962.

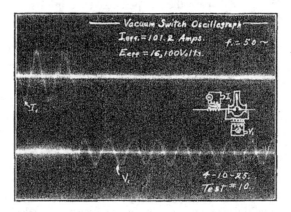

FIG.   5—OSCILLOGRAM SHOWING CURRENT OPENED BY VACUUM
SWITCH AND VOLTAGE ACROSS SWITCH AT OPENING

$I_{eff}$ = 101.2 amperes    $E_{eff}$ = 16,100 volts    $f$ = 50 cycles

**Figure 8.2**   The first published oscillogram of current interruption in vacuum. (Reproduced with permission of IEEE [2]).

pressure in the volume creates a *partial vacuum* of different levels. The residual gas pressure is a primary indicator of the *quality of the vacuum,* which is divided into ranges with somewhat arbitrary boundaries [4]:

low vacuum:            $10^5$ to $10^2$ Pa[1];
medium vacuum:         $10^2$ to $10^{-1}$ Pa;
high vacuum:           $10^{-1}$ to $10^{-5}$ Pa;
ultra-high vacuum:     $<10^{-5}$ Pa.

---

[1] The SI unit for pressure is the pascal [N m$^{-2}$], abbreviated as Pa. The bar is the technical unit for pressure. 1 Pa = $10^{-2}$ mbar.

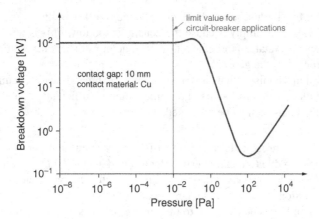

**Figure 8.3** Paschen curve (schematic).

Vacuum is primarily measured by its *absolute pressure*, lower pressures indicate higher vacuum quality. At absolute pressures that differ by more than two orders of magnitude from the atmospheric pressure, this difference in pressures remains virtually constant. Therefore, a complete characterization requires further parameters. One of the most important parameters is the mean free path of residual gases. This is the average distance that molecules will travel between collisions with each other. As the gas density decreases, the mean free path increases and, when the mean free path is longer than the confining enclosure, *high vacuum* is reached. Only high vacuum can be used as an arc-extinction medium. The reason for this is the sudden drop of dielectric strength of vacuum above the pressure of $10^{-2}$ Pa, as shown in Figure 8.3, the so-called *Paschen curve* [5] giving the breakdown voltage of a two-contact system as a function of contact distance times pressure.

The mean free path of air molecules at atmospheric pressure is very short, roughly 70 nm, so the molecules are in a constant state of collision.

As pressure is reduced, a transition point is reached where the mean free path is equal to or greater than the dimensions of the confining enclosure. Under this condition, the molecules will collide more frequently with the walls of the enclosure than with each other. In this region of dimensions, the gas is said to be in a state of *molecular flow* and the continuum assumptions of fluid mechanics do not apply. The transition point is specified by a dimensionless parameter called the Knudsen number[2]. For a cylindrical tube, the Knudsen number is defined as the ratio of the mean free path of gas molecules to its radius. When the ratio is < 0.01, the gas flow is viscous. If the ratio is > 1, the flow is molecular. The range between these two limits is called the transition range.

In the transition range, there is a drop in the dielectric strength compared with the gas in a state of viscous flow. This occurs as a consequence of the increased free path of the charged particles. Forces exerted on the charged particles by an electric field increase their kinetic energy, influencing the development of an initial avalanche into a final dielectric breakdown. This is known as the *Townsend avalanche effect*, belonging to gases in a state of *viscous flow*.

---

[2] The Knudsen number is defined as the ratio of the molecular mean free path length to a certain representative physical length scale.

At low pressure, the Townsend mechanism no longer works. In high vacuum, there is no possibility of an avalanche simply because the density is too low. This is why the breakdown mechanism in gases in the state of molecular flow is completely different, resulting in extremely high dielectric strength of the gap.

There are several mechanisms that can cause a dielectric breakdown in vacuum [6]. Usually they are linked to the creation of localized plasma, which is sufficiently dense for an electron avalanche. Localized plasma may be produced:

- on the cathode side through the explosion of a microscopic emitter site caused by the intensive overheating due to locally very high current density (*Joule effect*);
- on the anode side which is bombarded by beams of highly energetic electrons (which also results in the emission of X-rays under certain conditions);
- through a local rise in gas density due to the release of gases absorbed in or on the surface of the contact material; and
- when loosely attached charged metallic particles, present on the surface of the vacuum interrupter walls, are released, either by a shock or by the effect of electrostatic forces, and impact on the contact that attracts them.

All these effects have to do with the contact surface and cause the breakdown strength to be surface dependent, rather than volume determined as in gases, fluids and solids. Since vacuum-interrupter contacts have to switch as well as to insulate (in contrast with $SF_6$ circuit-breakers that have clearly defined main contacts for insulation and arcing contacts for switching), the surface condition of vacuum contacts is undefined.

This has the following, unfavourable, influence on the dielectric withstand capability of vacuum gaps:

- the dielectric withstand capability in vacuum increases approximately proportional to the square root of the distance between two contacts (see Section 9.10.1);
- the dielectric withstand capability of a vacuum interrupter has stochastic behaviour with a larger scatter than in high-density media;
- a delayed breakdown can occur for a recovery voltage at a lower voltage level value than was withstood before, see Section 4.2.4;
- mechanical operations and electric arcing modify the contact surface condition and generate particles;
- the dielectric withstand capability of a vacuum switching device evolves over time and is not constant;
- the voltage withstand level reached after conditioning cannot be considered as permanently acquired.

Therefore, vacuum is not the best defined and constant insulating medium when the reliability of dielectric withstand is essential, for example for disconnector application.

A breakdown mechanism involving detached particles also explains the experimentally proven relation between the breakdown voltage and the mechanical properties of the contact material (see Figure 8.4a). Under electrical stresses, mechanically weaker materials can release a substantial amount of metallic micro-particles that may initiate breakdown.

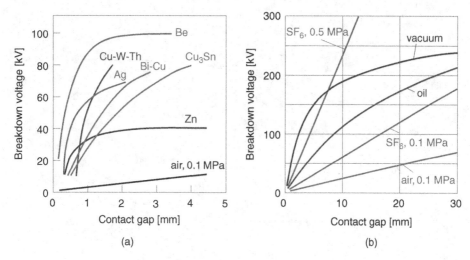

**Figure 8.4** (a) Breakdown characteristics of various contact materials in vacuum [7]; (b) Breakdown voltage of vacuum compared with air, oil, and $SF_6$ [8].

The withstand voltage of a contact gap in high vacuum can be considerably increased by polishing the contacts or applying a high voltage across the gap during a certain time. When proper parameters are applied, these *conditioning* processes modify the contact surfaces in a favourable way. However, in switching devices conditioning can only by used prior to the practical installation of switchgear, not during its service. This is because the switching arc constantly changes the contact micro-geometry and structure.

Breakdown in high vacuum does not follow Paschen's law. The dielectric strength of high vacuum for a contact distance up to several centimetres is higher than the dielectric strength of any other medium used today for arc-extinction (see Figure 8.4b). Vacuum is clearly dominant up to a voltage of about 150 kV. However, for larger distances, the curve no longer proceeds linearly and the withstand voltage cannot be significantly increased by enlarging the contact distance. This is why the application of vacuum circuit-breakers at transmission voltages meets a challenge, mainly because of their dielectrical behaviour (see Section 9.10).

The recovery time of the dielectric strength after arcing in vacuum is extremely short. This is a consequence of a very fast and effective process of diffusion of the plasma residue and metal vapour towards contacts and metal screens on which they condense, quickly removing charge carriers and neutrals from the switching gap. The recovery strength greatly exceeds that of hydrogen, nitrogen, and $SF_6$ at atmospheric pressure. As an example, Figure 8.5 illustrates the high recovery speed of dielectric strength in vacuum after breaking a current of 1600 A with a contact stroke of 6.25 mm.

## 8.3   Vacuum Arcs

### 8.3.1   Introduction

Electric arcs in gases can be divided into high-pressure arcs and low-pressure arcs. In terms of switching technology, the term vacuum means (ultra-) high vacuum, typically $10^{-6}$ to $10^{-4}$ Pa,

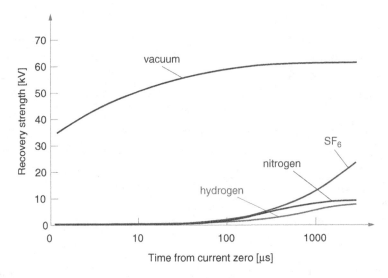

**Figure 8.5**  Speed of dielectric recovery of vacuum compared with hydrogen, nitrogen, and SF$_6$ after breaking a current of 1600 A (6.25 mm gap, pressure of gases 0.1 MPa) [9].

in which the mean free path is considerably longer than the dimensions of the interruption chamber. For this reason, the *vacuum arc* is different from both the high-pressure and the low-pressure arcs. The differences in physical processes that distinguish the vacuum arc from the arc in gases are so significant that the vacuum arc is treated in a separate chapter.

At first sight, the name vacuum arc is a contradiction in terms, since an arc contains positive ions and electrons in its column, while they do not, by definition, exist in absolute vacuum. That is why the vacuum arc can justifiably be defined as an arc that exists only in the metal vapour released from the contacts by the arcing process itself. After arc extinction, the density of the metal vapour drops substantially to zero, leaving a high vacuum in the interrupter. Because of the high-vacuum background, the particles in the former arc path diffuse very rapidly from the inter-contact region and contribute to the fast dielectric recovery that every circuit-breaking device needs.

Hence, in contrast to gaseous media (such as air, evaporated oil, or SF$_6$) for arc extinction, vacuum circuit-breakers do not contain material that would sustain plasma, except for the material emitted from the contact surfaces of the cathode and anode. Therefore, the physical processes related to the vacuum arc can be considered as processes determined by the surfaces of the metal contacts, rather than by the insulating medium in the volume between the contacts.

In vacuum arcs, an intense and physically only partially understood interaction between the solid cathode and the arc plasma occurs at the cathode spots. These are the interfacial regions of the solid contact and the arc plasma, having a very small size (few tens of micrometres) and high current density (order of $10^{12}$ A m$^{-2}$ for steady-state arc spots [10]). The cathode material emission in the spot's area supports the arc. Thus, the cathode spots are responsible for current continuity and for cathode erosion.

At current below around 10 kA, the anode is a passive collector of arc current. At currents above several kiloamperes, the vacuum arc starts to interact with the anode surface as well. This can lead to significant energy input into the anode, causing local melting of the metal and creating *anode spots* (see Section 9.1). Anode spots are much larger in size than cathode spots, and are an essential challenge for vacuum interruption. At current zero, anode spots, unlike cathode spots, due to their large size, have not yet solidified and continue to emit metal vapour into the gap thus slowing down recovery. Technical solutions have been developed to avoid large-scale anode melting, even for the highest currents [11].

## 8.3.2 Cathode- and Anode Sheath

A thin transition zone of about 10 nm above the surface of both contacts is commonly named the *sheath*.

In the cathode-sheath domain (see Figure 8.6), the emitted electrons are accelerated under the voltage difference between the cathode surface and the sheath edge. Outside the sheath, the electron energy is sufficient to ionize the particles. The majority of the newly created ions move towards the cathode surface, accelerated by the potential drop in the sheath. Typically, in a vacuum arc around 10% of the total ion current is directed against the electric field towards the anode. The ion velocity is reduced by collisional interaction between the ions bombarding the cathode surface and the vaporized atoms.

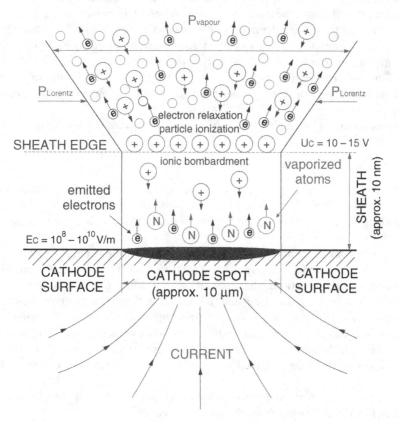

**Figure 8.6** Idealized geometry and the physical processes of the cathode spot and the discharge cone in front of it.

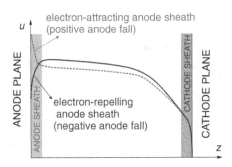

**Figure 8.7**   Voltage distribution in the intercontact gap along a vacuum arc.

Creation of the cathode sheath occurs instantaneously by ionization of vapour in the strong electric field close to the cathode, leading to a fast increase in the ion density and the electric field strength in this particular region.

The anode sheath can be both electron-repelling and electron-attracting (see Figure 8.7). The anode sheath is electron-repelling (negative) when the random thermal motion of the plasma electrons, adjacent to the anode, leads to a flux which is incident to the anode carrying a current exceeding the current demands of the external power supply. The situation may be reversed, and an electron-attracting (positive) anode sheath is formed when the arc carries a sufficiently high current in the plasma of sufficiently low density. Unlike the negative anode sheath, however, the positive anode sheath is not stable and its instability can lead to a large and rapid growth of the sheath potential, externally observable as spikes on the arc voltage.

Vacuum arcs occur in two main modes: *diffuse* and *constricted* (Figure 8.8).

### 8.3.3   The Diffuse Vacuum Arc

The diffuse mode is a unique characteristic of a vacuum arc, whereas the constricted mode is similar to the appearance of an arc in gases. The possibility of the vacuum arc to exist in a diffuse mode is the most significant and fundamental difference between the vacuum arc and the arc in gases.

The diffuse vacuum arc is characterized by the existence of a number of fast moving emission centres (cathode spots), independently existing apart from each other and each carrying 30 to 100 A. The emitted particles (ions, electrons, and vapour atoms) generally have no interaction with those emitted by a neighbouring spot. In this sense, the diffuse vacuum arc consists of a large number of individually active arcs in parallel operation. In spite of carrying a parallel current, these sub-arcs tend to repel each other and the diffuse vacuum arc tends to occupy the entire available cathode surface. They also react to external magnetic fields in an opposite way as Ampère's law dictates: the so-called *retrograde motion*, not yet satisfactorily understood physically.

Above a certain current value, the individual arc channels start to interact in a contractive way, leading to a single-plasma channel: the constricted arc.

Low-current vacuum arcs (below a few kiloamperes) exist in the diffuse mode, whereas high-current vacuum arcs are characterized by the constricted mode, in the absence of external

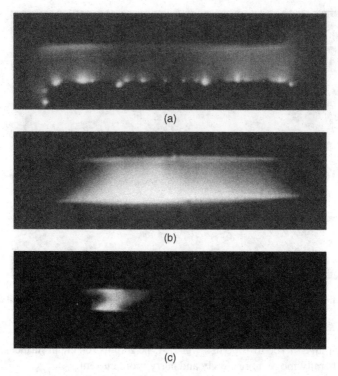

**Figure 8.8** High-speed photographs (exposure time 13 μs, cathode at bottom): (a) diffuse vacuum arc prior to current zero (2 kA); (b) diffuse high-current (60 kA) vacuum arc with axial magnetic field; (c) constricted vacuum arc (40 kA) with radial magnetic field [12].

magnetic fields. The current, at which the transition occurs varies with the shape, size and materials of the contacts, and with the rate of change of arc current. Diffuse vacuum arcs can more easily interrupt current than constricted arcs.

In all vacuum interrupters, the arc is initiated by contact separation under current load. At the instant the contacts separate, in particular at the last contact bridge that galvanically connects, the current density becomes extreme and a molten metal bridge appears between the contacts. As the contacts continue to separate, this bridge is heated up by the arc current and becomes unstable. Its rupture results in the appearance of a metal-vapour arc originating from the liquid-bridge explosion. This arc will adopt the diffuse or constricted mode, possibly even evolve from one to the other, and will be maintained until current zero.

By using modern optical methods, the typical diameter of the cathode spot was found to be a few tens of micrometres [13] (see an example in Figure 8.9).

Cathode spots have a lifetime in the microsecond and sub-microsecond range and they apparently move randomly across the cathode surface, constantly creating new emission sites.

Observations with a relatively low time resolution already show the movement of cathode spots. Surface contaminants strongly affect spot motion as well as current density. Thus, fast moving spots carrying current of 1 to 10 A, the so-called type 1 spots, of a few micrometres in diameter, have been observed on oxide-coated metallic surfaces. These spots move with

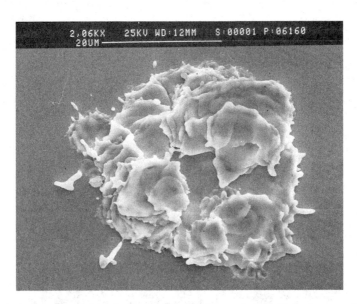

**Figure 8.9**   Cathode-spot crater on the surface of a cathode created by 43 A of current during 100 ns.

velocities in the range of 100 to 1000 m s$^{-1}$ [14]. Type 2 spots (on clean surfaces as in vacuum interrupters) generally move more slowly and carry more current.

At present, the physical processes in the cathode spot are far from being revealed. Two cathode spot models have been proposed, depending on the cathode material emission process: cathodic evaporation and explosive generation of the material.

The first model is based on evaporation of the cathode material from the cathode spot. The high spot temperature is attributed to Joule heating by the high-density current ($>10^{12}$ A m$^{-2}$) [10] and the deposition of energy by ions streaming to the cathode from the ionization region in front of it. The model also assumes the appearance of a positive sheath in front of the spot that produces a high electric field (of the order of $10^{10}$ V m$^{-1}$) which lowers the work function and significantly amplifies electron emission from the cathode.

According to the second model, a cathode spot operates because of a repeated sequence of explosive and short-lived emission events. Explosive generation of plasma is based on intensive Joule heating of a micro-protrusion at the cathode that functions as an emission centre. It is assumed that the cathode-material emission is in the form of plasma generated by the explosion of micro-protrusions. The emission centre has a finite lifetime and cannot provide a stationary cathode spot. The concept of explosion emission explains plasma generation in non-stationary cathode spots.

Regardless of the mechanism of the cathode-material emission, at the macroscopic level the cathode spot is the production point of a low density, quasi-neutral plasma. It is made up of electrons and ions that are typically double charged [15]. One of the characteristics of this plasma is the high speed of the ions, which have energy higher than the arc voltage would suggest. These ions, which emanate from the cathode spot, can reach the anode and create an ionic current in the opposite direction to the main electron current. Because of their high speed, the ions created by the cathode spots have a short transit time through the contact gap

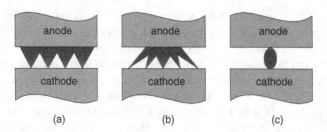

**Figure 8.10**  Basic modes of a vacuum arc: (a) diffuse mode; (b) arc constriction at anode; (c) constricted arc.

(typically < 1 μs). Thus the plasma created by a cathode spot disappears very rapidly when the last single spot stops functioning at current zero.

The peripheral area of the cathode spot is an intense source of neutral metal vapour, which is ionized directly in front of the cathode and spreads in a diffuse manner from the cathode spot towards the anode. It is roughly conical in shape, the apex of the cone being at the cathode spot, as shown in Figure 8.10. The cone shape is a result of the opposite effects of the electromagnetic forces, which tend to concentrate the current lines of the arc (*pinch forces*) and the pressure of the metal vapour, which tends to expand the plasma.

The very efficient extinction of an arc in vacuum, often at the first current zero, is explained by its diffuse mode and shape, due to which the anode remains cold. After the change of polarity, it cannot easily emit electrons and ions needed as charge carriers for the arc current to flow. Therefore, at the instant of the first current zero, the arc is finally extinguished. The vacuum regains its insulating capability almost immediately and is able to withstand the transient recovery voltage.

In the diffuse mode of discharge, there is no classical arc column, and the erosion of contacts is extremely low (a few tens of micrograms per coulomb [16]).

Ideally, a stable diffuse vacuum arc persists between contacts until the current reaches its natural zero. In practice, however, the current flow is usually interrupted prior to this moment at actual current values between 2 and 10 A, depending upon the contact material. Failure to carry the current gradually to zero is called *current chopping* (see also Section 4.3.1.2). This phenomenon may result in overvoltages roughly proportional to the value of the chopped current. These arise from the magnetic energy, still stored in the inductances, and may be especially troublesome when switching off inductive loads, such as stalled or starting motors and transformers in a no-load operation.

The absence of an arc column in a low-current diffuse vacuum arc suggests that the nature of the current-chopping phenomenon can be coupled to the dynamics of the cathode spot and arc instability, which is a manifestation of an ion deficiency in the anode sheath. This deficiency may result from a discontinuity in the ionized mass flow from the cathode spot into plasma, due to an unsuccessful explosive-emission event during the repetitive spot's crater initiation. In low-current vacuum arcs a large number of such individual instabilities occur. The probability for an instability increases drastically at lower currents. After instability, in experimental circuits, current remains reduced with respect to the stationary value, implying an increased probability of a next current reduction. If a subsequent instability indeed takes place, the current is reduced even more, leading to an even higher probability of instability [17, 18].

Such a stepwise fall of current is observed very often, especially in the last few microseconds of the arc's existence when the arc current usually reaches zero in a number of steps [19].

The instability phenomena, and thereby chopping current, depend largely on the cathode material. The materials that ensure a low chopping current have low melting temperatures and high vapour pressures and are usually not resistant to erosion. Hence, the requirements for contact materials to have low chopping currents and a high breaking capacity are conflicting. It is thus necessary to make a careful selection of the contact compound in order to find an optimum for the foreseen application. Presently, the contact materials most often used are CuCr for circuit-breaker and AgWC for contactor applications. Application of these contact materials reduces chopping current to an acceptable value, roughly similar to $SF_6$ circuit-breakers.

## 8.3.4    The Constricted Vacuum Arc

The diffuse model of a vacuum arc is shown schematically in Figure 8.10a. In the diffuse mode, the only active contact is the cathode, while the anode is a passive collector of the cathode plasma jet. In this discharge mode, the anode is not luminous. As the vacuum-arc current increases above a few kiloamperes, the anode begins to play an active role in the vacuum arc. First, some constriction of the inter-contact plasma occurs on the anode side (Figure 8.10b) being compressed by its own magnetic field. As the arc current is further increased, the following changes are taking place:

1.  The arc voltage increases and develops a fairly large noise component;
2.  the anode potential drop becomes positive, and
3.  the density of anode ions increases.

Bright spots (one or more) appear on the anode surface, associated with local anode melting. This intermediate current mode of discharge is termed the *foot-point mode* [20] . In this mode, the erosion of anode material may exceed that of the cathode, yet the overall net loss of material from the anode is low.

With a further increase of arc current, an anode spot at a temperature near the boiling temperature of the anode material appears, and the anode becomes very active. Characteristically, this anode spot is very bright and is an abundant source of vapour and ions. At moderately high current, several small anode spots are observed instead of a single large spot. A further increase in the current may lead to fusion of these spots into a single larger one. In the anode-spot mode, anode erosion becomes significant.

Finally, a well-defined arc column bridges the inter-contact gap between the anode spot and the cathode (Figure 8.10c). The cathode spots in this mode tend to assemble at the cathode end of the arc column. This arc mode is called the *constricted arc*.

Transition into the constricted-vacuum-arc mode requires that the current supplied to the anode be larger than the random electron current in the plasma. In order for the arc current to exceed the random current, a positive anode voltage drop must develop. The critical current for transition is strongly influenced by the gap geometry, the cathode material, and the anode material. Large diameter-per-gap ratios give a high critical current. High vapour pressure and more readily ionized cathode materials result in a higher critical current. More refractory properties and higher thermal conductivity of the anode material result also in a higher critical current.

A vacuum arc in constricted mode has an increased arc voltage and arc power, strong anode and cathode erosion, a high temperature and high density of the inter-contact plasma. Moreover, when an arc enters the constricted mode the energy is dissipated via a reduced contact surface, causing local overheating and considerable vaporization. Therefore, if such a constricted arc does not move across the contact surface, the ability to extinguish the arc at the natural current zero is drastically reduced by the presence of relatively hot and vapour-emitting surface parts – also after current zero. The probability of creating new emission sites is enhanced, together with the probability of a reduced dielectric withstand capability due to the high vapour density.

## 8.3.5    Vacuum-Arc Control by Magnetic Field

### 8.3.5.1    General

There are no mechanical ways to cool the vacuum arc, and the only possibility to influence the arc channel is by means of interaction with a magnetic field. Such a magnetic field can be realized through the contact geometry creating the path of the current through the contact system. A very efficient way to improve the high-current breaking capability by direct magnetic interaction with the arc is to adapt the contact geometry.

Two different principles are used in order to avoid vacuum arc constriction when breaking high currents:

- the *radial magnetic field* (RMF) *principle*, also named *transversal magnetic field* (TMF) *principle*, where the constricted arc column is forced by a self-generated radial magnetic field to rotate rapidly across the outer part of the contact surface circumference;
- the *axial magnetic field* (AMF) *principle* where, due to the self-generated axial magnetic field, the vacuum arc stays in a diffuse mode.

Both radial and axial magnetic fields are created by special current paths provided in the structure below the contact surface or by the contacts themselves.

### 8.3.5.2    The Radial Magnetic-Field (RMF) Principle

The constricted arc can be regarded as a conductor through which a current flows parallel to the contact axis. If a radial magnetic field is applied to this conductor, the resulting electromagnetic (Lorentz) force will cause rotation of the arc across the contact surface.

Spiral-type contacts (Figure 8.11) are used to obtain this result. These contacts generate a radial magnetic field that causes an azimuthally directed electromagnetic force to act on the vacuum arc. The contracted arc rotates across the contact surface at a speed that can be as high as 400 m s$^{-1}$ [21]. The speed of the constricted arc is limited because charge carriers have to be produced. Since the vacuum arc in the constricted mode behaves like a high-pressure arc, metal vapour can approach atmospheric pressure. Consequently, the surface temperature of the anode is approximately equal to the boiling temperature of the contact material. An

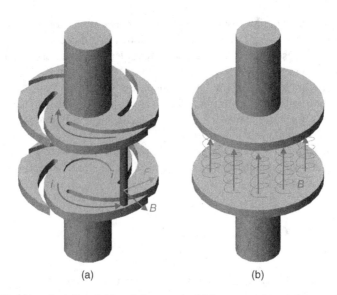

(a)                                         (b)

**Figure 8.11** Working principles of: (a) radial magnetic field contacts; (b) axial magnetic field contacts. (*I* – current, *B* – magnetic flux density, *F* – electromagnetic force) [22].

upper boundary to the speed is estimated from the time the arc needs to heat up the surface of the contact, in order to obtain metal vapour of sufficient density. This high speed ensures that there is less contact erosion and melting and also significantly improves the current interrupting capability.

A certain trade-off of the slot width for the spiral-type contacts is necessary [23]. If the slot width is too large, the arc cannot cross it and it can make the arc stationary and thus overheat it, since the arc is in the constricted mode. If the width is too small, the slot may be filled by fused contact material debris thus modifying the current path. This leads to reduction of the radial magnetic field and immobilization of the arc.

Even being mobile, the rotating vacuum arc remains constricted. Therefore, the energy brought by the arc causes overheating and considerable melting of the contact material. The high pressure at the arcs roots expels the molten contact material in the form of droplets. This process is a means of cooling the contact. The energy is taken away with the expelled material, which then condenses on the surrounding walls. This leads to relatively high contact erosion.

The main advantage of the radial-magnetic-field contact system lies in its simple physical structure. Another advantage of the spiral contacts is that, in the closed position, the current can flow directly via the stem, thereby ensuring lower power losses at normal current.

Radial-magnetic-field contact systems can ensure an improvement of the current-interrupting capability up to 50 kA. For a further increase in the switching capability, from 63 to 80 kA, an axial magnetic field contact system can be used.

Another variant of radial-magnetic-field arc control is to apply contrate or cup-shaped contacts, as shown in Figure 8.12. In this case, because the current has to follow the given path, the radial magnetic field is generated below the contact surface at a certain angle to the

**Figure 8.12** Schematic view of a contrate contact system designed to create a radial magnetic field for arc control [23]. (Reproduced with permission of Schneider Electric).

axis of the contacts. There are no discontinuities (slits) in the contact surface to hamper the motion of the arc.

### 8.3.5.3 The Axial Magnetic-Field (AMF) Principle

The switching capacity of vacuum interrupters can also be increased by using contact systems that generate an axial magnetic field (see Figure 8.11b). When a magnetic field is applied in the direction of the flow of current in the arc, the mobility of charge carriers perpendicular to the flow is considerably reduced. This applies in particular to the electrons, which have a much smaller mass than the ions. The electrons gyrate around the magnetic-field lines so that the arc contraction is shifted towards higher current. The arc maintains its diffuse mode, which ensures that only a small amount of energy reaches the contacts. This is reflected in the arc voltage that is significantly lower in comparison with the arc voltage measured in the case of radial-magnetic-field contacts [24].

Application of an axial magnetic field leads to smooth and stable behaviour of the arc voltage. A very low arc voltage of approximately 60 V at a short-circuit current of 63 kA [25] indicates that the arc is forced to stay in diffuse mode even around the arc current peak.

In contrast with the smooth arc voltage of the axial-magnetic-field contact system, the radial-magnetic-field contact system shows a typical shape of the rotating-arc voltage, representing the high-speed motion of the vacuum arc during the high-current phase.

The arc-state diagrams in Figure 8.13 visualize that a vacuum arc maintains its diffuse mode much longer in an axial than in a radial magnetic field. As an example, dashed lines indicate

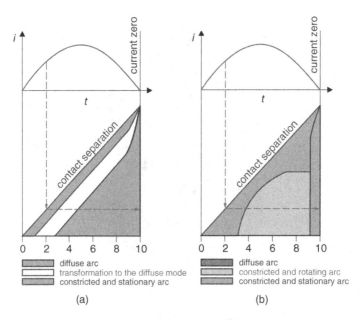

**Figure 8.13**   Arc-state diagram: (a) in axial magnetic field; (b) in radial magnetic field [24].

the duration of the characteristic arc modes when interrupting a high short-circuit current with arcing time of 8 ms.

In the case of an axial-magnetic-field contact system Figure 8.13a, the vacuum arc originates at a single location of the last metallic bridge, and the initial stages of arc contraction are visible just after contact separation. Independent from the current instantaneous value, transition from the constricted- to the diffuse-arc mode practically starts as soon as the contacts separate. The vacuum arc stays in a diffuse mode, in spite of the very high current value, until the first current zero where the current is interrupted.

In the case of radial-magnetic-field contact systems Figure 8.13b, the vacuum arc is first constricted and stationary, then constricted and rotating, and finally, diffuse.

Vacuum-arc behaviour in an axial magnetic field makes AMF contact systems best suited for very high short-circuit currents (> 50 kA). Contact systems that generate an axial magnetic field shift the contraction of the arc towards higher currents; the arc is diffuse and the supply of arc energy, as well as the contact erosion, is reduced drastically. For this short-circuit current range, the more complex AMF contact systems are superior to the conventional RMF contacts and are preferred. Networks with a rated frequency of 16.7 Hz, for example, for railway power supply, are another area for which AMF contacts are of advantage. Due to the long arcing times, vacuum interrupters with AMF contacts are installed in these systems at current levels up to 31.5 kA.

At larger contact distances, however, as applied for instance in vacuum interrupters for higher voltages, the axial field loses its strength.

Regarding arc voltage, the presence of an axial magnetic field has a twofold effect: the anode fall voltage is reduced whereas the voltage along the arc column increases. This effect

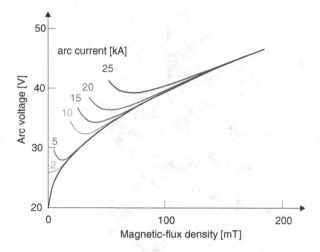

**Figure 8.14**   Dependence of vacuum-arc voltage on magnetic-flux density and arc current [26].

leads to voltage minima at certain combinations of current and magnetic flux density. This is outlined in Figure 8.14.

The main parameters that determine the dimensioning of axial-magnetic-field contacts are the size, and phase relation of the axial magnetic field with current. In the ideal situation, the goal is to have no phase displacement between the high current and the magnetic fields that it generates. However, because of the inherent losses in the contact system, such an ideal case will always be out of reach. An important parameter is the value of the eddy currents produced in the contact by the changing magnetic-flux density. These eddy currents are responsible for both the phase displacement and the reduction of axial magnetic field. The eddy currents become smaller if the surface is divided into two sectors with different directions of the magnetic-flux density. This is because only the areas enclosed by the eddy currents, which passed through in one direction due to the magnetic lines of force, are concerned. Such systems are termed the *bipolar AMF contact systems*, in contrast to to the classical *unipolar AMF contact systems*.

In many axial-magnetic-field contact systems, the axial magnetic field is generated by a coil located behind the contacts (see Figure 8.15).

Another alternative for generation of an axial magnetic field is the application of C-shaped, "horse-shoe" type, iron plates behind the contact. With a proper design, substantial axial field components can be generated in this way. An illustration of this technology is shown in Figure 8.16. This method of axial field generation takes advantage of the fact that, with a proper dimensioning of the iron package, the magnetic field lines tend to close by crossing the contact gap to the opposite package, rather than to close in a circular way to the same package.

Yet another solution is to force the current through a coil outside the interrupter, as illustrated in Figure 8.17.

Since convection is not an option in vacuum, the only practical way of dissipating the generated heat is through conduction via the copper conductors. As a result, the power losses during service reduce the normal current capability (see also Section 9.8).

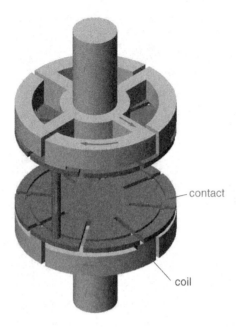

**Figure 8.15**  Generation of axial magnetic field for arc control by coil (segments) directly below the contacts [23]. (Reproduced with permission of Schneider Electric).

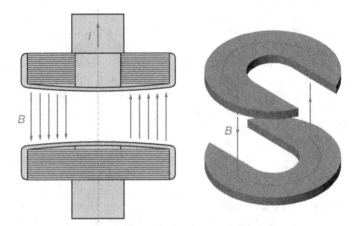

**Figure 8.16**  Generation of axial magnetic field for arc control by iron "horse shoe" packages below the contact surfaces [27]. (Reproduced with permission of Eaton Electric).

**Figure 8.17** Generation of axial magnetic field for arc control by an external winding [23]. (Reproduced with permission of Schneider Electric).

# References

[1] Enholm, O.A. (1890) Device for transforming and controlling electric current, U.S. Patent 441,542.
[2] Sorensen, R. and Mendenhall, M. (1926) Vacuum switching experiments at California Institute of Technology. *Trans. AIEE*, **45**, 1102–1105.
[3] Greenwood, A. (1994) Vacuum Switchgear, IEE, ISBN 0 85296 855 8.
[4] Slade, S. (2008) *The Vacuum Interrupter. Theory, Design and Application*, CRC Press, ISBN 13: 978-0-8493-9091-3.
[5] Kuffel, E. and Zaengl, W.S. (1984) *High-Voltage Engineering. Fundamentals*, Pergamon Press, ISBN 0-08-024212-X.
[6] Picot, P. (2000) Vacuum Switching, Cahier Technique, Schneider Electric, No. 198.
[7] Selzer, A. (1972) Vacuum interruption – a review of the vacuum arc and contact functions. *IEEE Trans. Ind. Appl.*, **8**(6), 707–722.
[8] Müller, A. (2009) *Mittelspannungstechnik, Schaltgeräte und Schaltanlagen*, Siemens AG, Ausgabe 19D2.
[9] Cobine, J.D. (1963) Research and development leading to the high-power vacuum interrupter – a historical review. *IEEE Trans. Power Ap. Syst.*, **PAS-82**, 201–217.
[10] Daalder, J.E. (1978) Cathode Erosion of Metal Vapour Arcs in Vacuum. Ph.D. Thesis, Eindhoven Univ. of Techn. Available at http://alexandria.tue.nl/extra3/proefschrift/PRF3A/7805322.pdf accessed 4 April 2014.
[11] Boxman, R.L., Goldsmith, S. and Greenwood, A. (1997) Twenty-five years of progress in vacuum arc research and utilization. *IEEE Trans. Plasma Sci.*, **25**(6), 1174–1186.
[12] Schramm, H.H. (2003) Introduction to High-Voltage Circuit-Breakers. CIGRE SC A3 Tutorial.
[13] Daalder, J.E. (1981) Cathode spots and vacuum arcs. *Physica C*, **104**(1–2), 91–106.
[14] Rakhovsky, V.I. (1987) State of the art of physical models of vacuum arc cathode spots. *IEEE Trans. Plasma Sci.*, **PS-15** (5), 481–487.
[15] Davis, W.D. and Miller, H.C. (1969) Analysis of the electrode products emitted by dc arcs in a vacuum ambient. *J. Appl. Phys.*, **40**, 2212–2221.
[16] Daalder, J.E. (1976) Components of cathode erosion in vacuum arcs. *J. Phys. D: Appl. Phys.*, **9**(16), 2379.

[17] Smeets, R.P.P. (1989) The origin of current chopping in vacuum arcs. *IEEE Trans. Plasma Sci.*, **17**(2), 303–310.

[18] Smeets, R.P.P. (1987) Low Current Behaviour and Current Chopping of Vacuum Arcs. Ph.D. Thesis, Eindhoven Univ. of Techn. Available at http://alexandria.tue.nl/extra3/proefschrift/PRF5A/8704511.pdf accessed 4 April 2014.

[19] Smeets, R.P.P., Kaneko, E. and Ohshima, I. (1992) Experimental characterization of arc instabilities and their effect on current chopping in low-surge vacuum interrupters. *IEEE Trans. Plasma Sci.*, **20**(4), 439–445.

[20] Miller, H.C. (1983) Discharge modes at the anode of a vacuum arc. *IEEE Trans. Plasma Sci.*, **PS-11**(3), 122–127.

[21] Dullni, E. (1989) Motion of high-current vacuum arcs on spiral–type contacts. *IEEE Trans. Plasma Sci.*, **17**(6), 875–879.

[22] Fink, H., Heimbach, M. and Shang, W. (2000) Vacuum Interrupters with Axial Magnetic Field Contacts. ABB Techn. Rev. No. 1.

[23] Picot, P. (2000) Vacuum Switching, Cahier Technique Schneider Electric No. 198.

[24] Fink, H. (1987) Vakuumschalter zum einsatz in mittelspannungsnetzen, *Vakuum-Technik*, **36**(4), 118–125.

[25] Fink, H., Gentsch, D., Heimbach, M. *et al.* (1998) New Developments of Vacuum Interrupters Based on RMF and AMF Technologies. IEEE 18th Int. Symp. on Disch. and Elec. Insul. in Vac., Eindhoven.

[26] CIGRE Working Group 13.01 (1998) State of the art of circuit-breaker modelling. CIGRE Technical Brochure 135.

[27] Eaton Co., Insulating and Switching Media in Medium Voltage Distribution and Medium Voltage Motor Control. White Paper IA08324006E.

# 9

# Vacuum Circuit-Breakers

## 9.1  General Features of Vacuum Interrupters

In comparison with other types of circuit-breakers, vacuum circuit-breakers are mechanically the most simple. They basically consist of a fixed and a movable contact mounted in a vacuum bottle. When the contacts separate, the arc is supported by ionized metal vapour released from the cathode (the negative contact) that thus provides the arcing medium. This is unlike gas or oil-filled interrupters, where ionized gas is provided by the extinguishing medium between the contacts. When the current approaches zero, the collapse of ionization and the condensation of vapour are very fast, ensuring efficient current interruption, virtually independent of the rate-of-rise of the transient recovery voltage, see Section 8.2.

The first serious demonstration of switching in vacuum took place in 1926 (see Section 8.1), and for many years, switchgear designers were fascinated by the great advantages of vacuum interrupters.

The development of vacuum circuit-breakers [1–3], in spite of the rather simple concept, required far more time and research effort than other interruption devices, due to the lack of supporting technology.

First, there were problems associated with the production of high-degree degassed contact materials, called gas-free contacts. Degassing is required to prevent deterioration of the initial vacuum due to the release of gases that are normally trapped inside the metals. These gases accumulate and affect the vacuum during arcing.

Another problem was the lack of proper technology to weld or braze the external ceramic and glass envelopes to the metallic armatures, that is, ceramic-to-metal seals, necessary to maintain the high vacuum during a life of 20 to 30 years in a hermetically sealed interrupter. An additional weakness of early vacuum interrupters was the occurrence of severe overvoltages due to current chopping when pure copper or the refractory metals tungsten or molybdenum were used as contact material to capture the gas escaping from the contacts during arcing. Another difficulty to overcome was that the highly clean surfaces of the contacts produced strong welds in vacuum, often with normal contact pressures and in a no-load situation.

Besides the mechanical simplicity, vacuum interrupters do not contain gases or liquids, they are not flammable and do not emit flames or hot gases. Due to the absence of inelastic

*Switching in Electrical Transmission and Distribution Systems*, First Edition.
René Smeets, Lou van der Sluis, Mirsad Kapetanović, David F. Peelo, and Anton Janssen.
© 2015 John Wiley & Sons, Ltd. Published 2015 by John Wiley & Sons, Ltd.

collisions between the gas molecules, vacuum has the fastest recovery strength after arc interruption at current zero. This means that there is no avalanche mechanism to trigger the dielectric breakdown, as is the case in gaseous media.

Because of its small contact gap, short arc length, and very low arc voltage, the arc energy released in vacuum is approximately one-tenth of that in $SF_6$, and even less than in oil. The low arc energy keeps the contact erosion to a minimum. Operation of vacuum circuit-breakers requires relatively little mechanical energy and this allows the use of simple spring-driven operating mechanisms that are both reliable and silent.

The advantages offered by vacuum circuit-breakers were the driving force in overcoming technological problems. By the end of the 1950s, after long-lasting efforts to develop a vacuum circuit-breaker, the products started to appear in practice. Advances in plasma physics and developments in contact metallurgy and ceramic welding provided solutions needed for the vacuum interrupter to become reality. Finally, in1962, the General Electric Company brought the first commercial vacuum circuit-breaker onto the market and since then it has been firmly established as a reliable option for current interruption in MV networks.

The basic component of the vacuum circuit-breaker is the vacuum interrupter, also named *vacuum bottle*. The interrupter is a vacuum-tight enclosure made of ceramic and metal parts that are hermetically sealed together. It is completely evacuated to attain a high vacuum. The internal pressure in the bottle is less than $10^{-2}$ Pa. Inside the interrupter, a pair of contacts is mounted. Separation of the contacts is achieved by moving one of them using metal bellows because no gasket would be sufficiently tight to maintain the desired vacuum quality. The arc is sustained by metal vapour from the contacts and elongated during contact separation extinguishing at a current zero and the vapour particles condense on the solid surfaces.

As outlined in Section 8.3.5, the dielectric recovery characteristic of a vacuum interrupter is highly influenced by the creation of *anode spots*, which depend on the type of material and its composition. The anode spots are relatively large "pools" of molten metal created by the current close to its peak. If they are too big and do not solidify at current zero they continue to emit vapour and impair dielectric recovery. Therefore, arc control devices (see Section 8.3.5) in the contact system are designed to prevent anode-spot formation. It is also of importance to minimize energy input and energy density on the contacts during the interruption process. To achieve this, the arc should be kept diffuse in order to minimize the energy input per square millimetre, or it has to be forced into circular motion.

Since there is no mechanical method to control the vacuum arc, the only way to influence the plasma channel is by means of interaction with a magnetic field. Many solutions have been proposed to realize this, but the two most practical methods are:

- interaction between the arc current and the radial magnetic field it generates (see Section 8.3.5.2); or
- application of an axial magnetic field (see Section 8.3.5.3) created by a coil which is either an integral part of the contact system inside the interrupter or mounted outside the interrupter.

The interrupting capability of a vacuum interrupter is also related to the surface area of the contacts. Larger contacts combined with an axial magnetic field have better capability to interrupt high current [4]. The rated normal current is also related to the surface area of the contacts. Thus, the contact area should be large enough to absorb the arc energy without becoming excessively heated, and in closed position it has to provide enough contact points

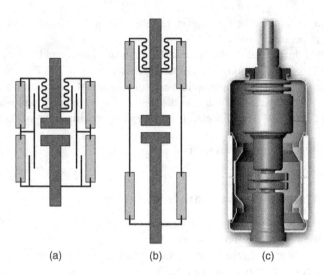

(a)                              (b)                              (c)

**Figure 9.1**  Vacuum interrupter designs, each with the same contact and shield sizes: (a) shorter with larger diameter; (b) longer with reduced diameter; (c) cross-section of a vacuum interrupter – design (a) [5].

of sufficient area to give low enough energy dissipation during passage of the rated normal current.

The interrupting capability of a vacuum interrupter depends on the contact gap, the surface area of the contacts themselves, and the space between the contacts and the vapour shield.

Because of the high dielectric strength of vacuum, a vacuum interrupter can be designed quite small in terms of internal dimensions, however, dielectric strength must also exist outside the interrupter. Therefore, it is largely the external dielectric strength that determines the length of the insulating part of the interrupter. Some interrupters have been specially designed for immersion in $SF_6$, and these could be considerably shorter in length than interrupters for use with dry air external insulation.

Figure 9.1 shows two of the most prominent designs.

In design (a), the contacts are surrounded by a central metal-vapour condensation shield, which serves to protect the inside walls of the insulating ceramics, so that they do not become conductive by condensed metal vapour. The two end shields prevent metal vapour to be reflected back from the end plates to the insulating cylinders. In most designs, the central shield is electrically floating because of the dielectric-stress management [2].

The vapour shields also serve another function: relief of the electrical stress on the joints where the ceramic is attached to metal. These so called triple junctions: insulator, conductor and vacuum can be the source from which discharges originate and lead to a breakdown. Minimizing the stress at triple junctions reduces the probability of breakdown.

Design (a) has a length somewhat greater than its diameter. It has relatively short contact stems, which simplifies the mechanical and thermal design.

Design (b) shows an alternative, in which the diameter of the vacuum interrupter has been reduced at the expense of the length. The central shield becomes part of the envelope and the insulation is now split into two equal parts, one at each end of the interrupter. An additional shield is placed to protect the bellows against molten particles released from the contacts, which may cause puncture of the bellows.

It is important that the central shields have sufficient heat capacity and thermal conductivity to absorb the heat flux without too much temperature rise.

Vacuum technology has been proven to be very suitable for MV applications. Commercial HV interrupters for 72.5 and 145 kV are on the market but have not found wide-spread use yet outside Japan. New HV vacuum circuit-breakers are reaching the market rapidly (see Section 9.10).

Vacuum circuit-breakers are shown in Figure 1.13.


## 9.2   Contact Material for Vacuum Switchgear

The successful operation of vacuum circuit-breakers depends to a considerable extent on the properties and processing of the contact materials.

Contact materials for vacuum application must satisfy a number of different requirements that are sometimes contradictory to each another:

- good dielectric compatibility with vacuum;
- high purity;
- low gas content;
- mechanical strength;
- low contact resistance;
- high electric conductivity;
- high thermal conductivity;
- low tendency to weld and low weld-break forces;
- low current-chopping level;
- low arc voltage;
- favourable thermionic-emission characteristics;
- ample generation of plasma during arcing;
- rapid recovery of dielectric strength after arc-extinction;
- low and uniform erosion due to arcing; and
- adequate gettering[1] effect.

No pure metal has the properties required to make it suitable for use in vacuum interrupters, but composite materials or alloys of two or more metals can potentially satisfy the various requirements.

The vacuum environment offers definite advantages to the contact materials: there is no ambient gas to contaminate the contact surfaces by oxidation or in another way and, therefore, contact materials that are not suitable for applications in gas circuit-breakers can be considered. A clean contact surface also helps to maintain a stable contact resistance throughout the lifetime of a vacuum switching device. The clean contact surfaces, however, can result in severe contact welding.

---

[1] *Gettering* is the chemical capture of active gases with a suitably reactive substance. *Getters* are generally constructed of materials such as zirconium and aluminium, or barium and nickel.

One of the fundamental problems that remains is the contradiction of requirements for a low-current chopping characteristic when interrupting small inductive currents in combination with a high resistance to arc erosion, which essentially limits the breaking capability of vacuum circuit-breakers. The contradictory requirements involve the vapour pressure of the contact material. Namely, the vapour pressure of a good contact material must be:

- high enough in order to sustain the arc as close as possible to the natural current zero of the AC current to reduce the chopping current; but
- not too high, in order to avoid continuation of emission after current zero that might result in re-ignition of the arc.

Contact materials for applications in vacuum can be classified in pure metals and alloys.

### 9.2.1 Pure Metals

Copper (Cu) in its pure form appears to meet most of requirements for use in vacuum interrupters. It has good thermal and electric conductivity, reasonably good dielectric compatibility with vacuum, and is quite good in interrupting high currents. However, it has the major disadvantage of forming very strong welds, which are the result of diffusion between the solid-state lattices when two clean surfaces are pushed together and heated. Refractory metals offer good dielectric strength; their welds are brittle and thus easy to break. However, the main disadvantages of the metals with refractory characteristics are their very efficient thermionic emission, which limits their interrupting capability. They possess low electric conductivity and high chopping levels.

Soft or mechanically weak metals can also be ruled out because they are incapable of withstanding the shock of rapid closing of contacts in the normal operation of vacuum interrupters.

### 9.2.2 Alloys

Since an acceptable compromise contact material cannot be found amongst pure metals, much effort has been put into the development of suitable contact materials and various alloys have been investigated to study their suitability for vacuum-interrupter applications.

*Copper-chromium* (CuCr) contacts have found wide use in vacuum circuit-breaker applications. One example is shown in Figure 9.2. A proportion of approximately 25% by weight of chromium in copper has been shown to be particularly suitable [4]. It is impossible to employ conventional melting techniques in order to manufacture CuCr material with a composition of approximately 50% since chromium and copper are not mutually soluble, even in the liquid state.

There are two powder metallurgy methods that are suitable for manufacturing the CuCr material: (i) densification (sintering and repressing) of CuCr powder mixtures and (ii) infiltration of liquid copper into a porous sintered chromium matrix. It has been verified by tests that the interrupting capability of CuCr contacts produced by the infiltration technique is superior to that produced by the densification technique.

During switching, CuCr composite materials exhibit exceptionally good melting and solidification characteristics, which are the reasons for their superiority as a contact material in a

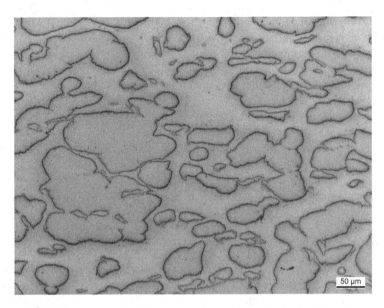

**Figure 9.2**   CuCr 75/25 contact material. (Reproduced with permission of Plansee Powertech AG).

vacuum environment. When subjected to arcing, a CuCr contact surface produces flat shallow pools of melt, which then solidify again to a smooth-surface finish. This results in a number of advantages. Due to their shallow depth, the pools of melt solidify very rapidly, which quickly re-establishes the dielectric strength of the contact gap. The absence of peaks and troughs on the contact faces ensures that the CuCr material maintains a constant high electric strength at very small gaps. This enables the overall size of the interrupters to be reduced. At the same time, the volume of contact material can be minimized, since – when the contact gap is small – a large proportion of the material vaporized in the arc condenses on the contact surfaces and becomes available again.

The high chromium content also reduces the maximum chopping current approximately fivefold – from approximately 16 A for Cu alone to less than 3 A for CuCr.

## 9.3   Reliability of Vacuum Switchgear

Due to difficulties encountered during the production technology that has to assure tightness during decades of operation, it was not until the late 1960s that commercial application of the vacuum-based switching became available on a large scale. Since then, the application of vacuum switchgear for a very wide variety of switching applications in the range up to 40.5 kV has developed rapidly. Nowadays, vacuum switchgear has a dominant position in power-distribution systems [6]. It has a market penetration that is estimated to be in [4]:

- North America and Japan: close to 70%;
- Northern Europe: about 80%;

- Southern Europe: not specified, but there is a general trend away from $SF_6$ to vacuum interrupters;
- China, India, and South-East Asia: tending towards 85%; and
- South America, Africa, Middle East, and Russia: tending towards 80%.

Manufacturers have proven ability to produce highly reliable vacuum switchgear. US data on vacuum circuit-breaker reliability, collected in 1991 to1994, have been reported in [7, 8]. They show a *mean time-to-failure* (MTTF) up to at least several tens of years. Manufacturers currently claim MTTF values of 40 000 vacuum-interrupter-years [9].

## 9.4   Electrical Lifetime

Vacuum switchgear provides very long electrical life because of its low arc voltage and the low erosion of the contacts [10]. Usually, electrical life is limited by contact erosion and deposition of metal vapour on the interrupter's interior ceramics.

Contact erosion rate is proportional to the charge passed through the arc since arcing activity in fact removes mass from the contacts. For the cathode, a number of approximately 10 μg per C of charge is often quoted for a diffuse arc [3]. Slots in the contact, such as those used in a radial magnetic field, usually increase the erosion rate since metal vapour disappears through these slots and is not returned to the contact surface. This is why radial magnetic-field contacts have a shorter electrical life span than axial magnetic-field contact systems [11].

From extrapolations based on measurements, several hundreds of thousands of switching operations under load current are estimated in order to erode 3 mm of contact material [12]. Loss of a layer of 3 mm of contact material is often taken as an end-of-life criterion. The lifetime of a vacuum interrupter is determined by its mechanical lifetime rather than by the electrical lifetime.

With respect to fault-current interruption, there are estimates that vacuum circuit-breakers can perform at least 30 interruptions of the rated short-circuit current. Figure 9.3 shows an arc burn-through of the surface in an axial magnetic-field contact. Figure 9.4 shows excessive erosion on a radial magnetic-field contact. Molten debris has settled between the contact fingers and reduced the effectiveness of the magnetic circuit. In both examples, the end of life had been reached, as demonstrated by short-circuit tests.

## 9.5   Mechanical Lifetime

Vacuum circuit-breakers designed for MV applications make low contact strokes (in the ten-millimetre range), they have low moving masses [13] (in the kilogram range), and they therefore benefit from low operating energy.

The mechanical life of vacuum interrupters is limited by the bellows that enable motion of the moving contact through the vacuum-sealed enclosure. The life-time of the bellows is a function of the number of its contractions, its diameter, the extent of its expansion, and the acceleration of the movement, but it can be designed for several tens of thousands of operations [3].

**Figure 9.3**  End of electrical life characterized by damage to axial magnetic-field contacts.

**Figure 9.4**  End of electrical life, characterized by excessive erosion of radial magnetic-field contacts.

## 9.6 Breaking Capacity

Vacuum switching is recognized for its very rapid dielectric recovery. This is positive for the interruption of currents with a very high value of $di/dt$ and TRVs with very high values of RRRV. This property is exploited in vacuum generator circuit-breakers [14] where high $di/dt$ and RRRV occur in the same switching duty, see Section 10.1, and applications in transformer-limited faults [15] that create high RRRV.

The drawback of this excellent interrupting performance is that vacuum circuit-breakers are also capable of interrupting high-frequency current after a re-ignition or a restrike. Because vacuum circuit-breakers attempt to interrupt the current repeatedly, the recovery voltage increases after every attempt (*multiple re-ignitions*, *multiple restrikes*, see Sections 4.3.3 and 4.2.2). In some cases, very high overvoltages are generated and protection is necessary [16, 17].

Because of this property, vacuum circuit-breakers are sometimes called "hard switching devices". $SF_6$ circuit-breakers, that generally need more recovery time, may fail to interrupt reactive-load currents initially instead of making many attempts before succeeding, but have a lower tendency to generate overvoltages.

A strong point of vacuum interruption is the capability to break currents without external means, such as pressure built-up inevitable in $SF_6$ circuit-breakers. A stationary open gap of vacuum circuit-breakers has a good inherent interruption capability, as demonstrated by the non-sustained disruptive discharges in Section 4.2.4.3. This partly compensates for the lack of reproducibility of dielectric withstand in certain conditions. Stationary open gaps in $SF_6$ circuit-breakers will not interrupt current above a certain level after late or otherwise unexpected breakdown.

Because a vacuum gap is, in principle, always ready to interrupt and no extinction pressure needs to be built-up mechanically, the minimum arc duration, or *minimum arcing time*, is significantly shorter than in $SF_6$ circuit-breakers, that is, several milliseconds in vacuum against 10 to15 ms in $SF_6$ circuit-breakers.

## 9.7 Dielectric Withstand Capability

Dielectric withstand capability is strongly influenced by the contact surface condition and the presence of particles inside the interrupter. Thus, the arcing history is of importance. This is also reflected in the conditioning of the vacuum gap by applying voltage. As a stage in the manufacturing process, all vacuum interrupters are pre-conditioned by a high AC voltage [18]. The intention is to increase the breakdown voltage by removing the surface irregularities and other electric-field electron emitters through breakdowns under controlled conditions. If this procedure is optimized, the initial breakdown voltage can be increased three to four times [18]. When the energy in the conditioning discharge is too high, the energy in the breakdown can create new or more intense field electron emitters that have a net effect of de-conditioning. Therefore, the energy in the HV generator, released at the moment of breakdown, is important [19].

Because breakdown in vacuum is basically determined by the surface condition, the (micro-) geometry of which depends on the arcing and switching history, the dielectric performance of a vacuum gap is less predictable than of an $SF_6$ gap [20, 21], where breakdown is basically governed by the gas. This implies that vacuum gaps have a finite probability of breakdown at relatively low voltage and it makes the design of vacuum interrupters for disconnector applications a challenge [22, 23].

## 9.8 Current Conduction

Vacuum circuit-breakers have butt contacts. This implies that the contact resistance is relatively high, especially after contact surface modification by high-current arcing. This limits the rated normal current that can be handled. In order to reduce the contact resistance and to counteract the "popping" caused by "blow-off" electromagnetic forces that try to open the contact under high current conditions (see Section 2.3), an additional contact closing force must be applied by a set of springs that are energized by the mechanism during the closing operation.

A strong point of vacuum interrupter contacts is their insensitivity to contact oxidation or to contamination by products inside the interrupter's enclosure.

A design challenge is that the transfer of heat, generated at the contact interface, cannot take place by convection, such as in $SF_6$ interrupters. This implies that the heat has to be conducted by the supporting contact stems to the external environment. Radiators are sometimes provided to the contact stems in order to increase the rated normal current.

This limits the rated normal current, especially for high-voltage applications where the interrupters are long because of the requirements of the external dielectric strength.

Even without an arc contacts in vacuum tend to weld upon closing, which is a natural interaction of clean metallic surfaces being pressed together. In the presence of a pre-strike arc, welding can be much more severe due to the thermal energy of the arc. The operating mechanism must be designed to be able to break the weld.

## 9.9 Vacuum Quality

A major requirement set for vacuum interrupters is to provide a high vacuum $10^{-1}$ to $10^{-5}$ Pa over a long time, which is usually 20 to 30 years. This ensures a high breakdown strength for short gaps and half-cycle interruption of power currents throughout the lifetime of a circuit-breaker.

In order to fulfil this requirement, material used for vacuum interrupters should meet the following conditions:

- the materials utilized in the interior of the interrupter must be extremely pure and free of micro-porosity, ruptures, and other defects;
- the contact material has to be totally degassed by heating it in a vacuum furnace to release and remove any gaseous impurities;
- the ceramic-to-metal seals must have an extreme air-tightness; and
- getter material should be mounted inside the vacuum interrupter to capture the free gaseous particals that may remain after assembly.

The experience in MV application shows that loss of vacuum by leakage is an extremely improbable event. Recently, loss of vacuum has been a subject of renewed interest. The primary reason is the fact that many MV devices installed in the field have life times in excess of 30 years. Customers started asking whether they can extend the life of these devices.

In reference [24], the authors conclude that experience with the vacuum interrupters has been excellent in the range of 12 to 38 kV. In rare cases, where vacuum circuit-breakers fail, it is normally related to loss of vacuum. Experience at medium voltage shows that vacuum interrupters can operate in excess of 30 years [9]. However, the same authors report, after a study of over 200 older interrupters, that for interrupters older than 30 years, the mean time-to-failure drops drastically from 40 000 vacuum-interrupter-years to just over 400 vacuum-interrupter-years, due to loss of vacuum [25].

In most practical cases, switching with vacuum circuit-breakers often improves the vacuum quality. The vacuum arc itself exhibits a pumping effect since the metal vapour-plasma jets, emitted from the contacts, trap the particles of the residual gases and embed them in thin film layers deposited on the surfaces of the contacts, shields, and so on. Therefore, operation of the interrupter involving an arc is beneficial in improving the quality of vacuum inside the device.

At the production stage, the pressure is ensured to be below $10^{-4}$ Pa, whereas $10^{-2}$ Pa is the limit that still ensures switching ability. Therefore, the margin of the initial state of the vacuum pressure is rather high and the getter material helps the interrupter to stay within the required (ultra-)high vacuum state.

In the case of vacuum circuit-breakers, continuous supervision of the vacuum quality inside the interrupting unit is possible by the following methods [26]:

The first approach predicts failure by monitoring the actual vacuum inside the vacuum interrupter. This method uses the Penning[2] or Pirani[3] principle and requires a vacuum gauge to be permanently fixed to the vacuum interrupter.

The second approach detects a failure of the vacuum. This can be incorporated in the design of the circuit-breaker, as used in some vacuum contactors, and uses the atmospheric pressure on the bellows as part of a balance of forces when the contactor is closed – that is, if one pole has lost vacuum then the unit cannot close. Continuous monitoring can in theory be achieved via a partial-discharge detector.

Another possibility to give real-time information about the status of the vacuum is a fibre-optic system that conducts light in the case of a normal state of vacuum and blocks the light in the case of loss of vacuum.

In service, it is not common to have any vacuum-quality measurement or detection system for indication of loss of vacuum, because a vacuum-quality monitoring system is in general less reliable than the vacuum interrupter itself. This means that adding these vacuum-measurement or loss-of-vacuum detection systems may reduce the reliability of the vacuum circuit-breaker.

There are several vacuum-integrity monitoring devices on the market that test the voltage withstand capability of the interrupter in open position by measuring voltage withstand.

## 9.10 Vacuum Switchgear for HV Systems

### 9.10.1 Introduction

The great success of vacuum technology in power-distributing systems obviously resulted in exploration of possibilities to develop vacuum switchgear for transmission voltage levels. Another major driver is the search for alternatives for $SF_6$, the ubiquitous high-voltage

---

[2] Pressure by measurement of ionization current under the influence of combined electric and magnetic field.

[3] Method of pressure measurement by monitoring the heat-loss to a low-pressure environment of a heated wire.

switching medium but also a very strong greenhouse gas, see Section 6.4.4.3. CIGRE has summarized the state-of-the art regarding the impact of the application of vacuum switchgear at voltages above 52 kV [27].

There are basically two ways to increase the dielectric strength of the switching gap to the value needed for insulation at transmission levels.

One is to increase the contact distance in a two-contact configuration. However, the breakdown voltage $U_b$ of vacuum gaps is not proportional to the gap length $d$ (as it is in gases), but typically follows the relation:

$$U_b = A \cdot d^\alpha \tag{9.1}$$

where $\alpha$ is a parameter smaller than one and $A$ is a constant. The explanation is that breakdown in vacuum is a surface effect [28], completely governed by the contact-surface condition. In $SF_6$, breakdown is merely a volume effect that scales linearly with the gap length. The breakdown process is then mainly determined by the insulating medium and its pressure rather than by the contact configuration and condition.

The other way is to place two or more gaps in series and, in the case of ideal voltage sharing between the gaps, the necessary withstand voltage level can be achieved with a total contact distance smaller than it would be with a single gap.

These two solutions coexist in the market at voltages above 72.5 kV.

### 9.10.2 Development of HV Vacuum Circuit-Breakers

The first reported commercial development of transmission vacuum circuit-breakers was in the UK in 1968 where 8 vacuum interrupters were connected in series in a circuit-breaker for 132 kV [29]. This breaker has been in service for more than 40 years.

In the mid 1970s, in the USA, a series arrangement of four vacuum interrupters per pole [30] was used as a retrofit kit for bulk-oil circuit-breakers up to a system voltage of 145 kV [31] and further plans were made for up to 14 series interrupters for 800 kV.

Simultaneously with this multi-gap approach, Japanese researchers developed and commercialized single-break vacuum interrupter units up to 145 kV [32].

In 1986, two types of vacuum circuit-breakers were published [33]: a single-break 84 kV vacuum circuit-breaker with a rated breaking current of 25 kA and a prototype vacuum circuit-breaker for the 145 kV voltage level [34].

Commercial single-break vacuum interrupters are today available up to 145 kV, and commercial double-break dead-tank-type vacuum circuit-breakers are developed up to 168 [35] and 204 kV.

Since the beginning of the twenty first century, R&D efforts in HV vacuum interruption have obtained a strong impetus because of concerns on the global warming potential (see Section 6.4.4.3) of the very strong greenhouse gas $SF_6$, see an example in Figure 9.5.

China took a strong lead in research and development of HV vacuum circuit-breakers. Single-break designs up to 252 kV [36] and futuristic conceptual designs of modular EHV circuit-breakers [37] for the 550 kV level and even for 765 kV [38] have been published (see Figure 9.6).

**Figure 9.5** 72/84 kV dead-tank vacuum circuit-breaker with dry air as external insulation. (Reproduced with permission of Meidensha Co.).

## 9.10.3   Actual Application of HV Vacuum Circuit-Breakers

In total, approximately 8300 units of vacuum circuit-breakers with rated voltage $\geq$ 52 kV were delivered to the market by five manufacturers in Japan from the late 1970s to 2010 [40]. Roughly 50% were delivered to power utilities and 50% to industrial users. Cubicle type GIS (C-GIS) represents 50% of the HV vacuum circuit-breakers application, mainly by industrial users.

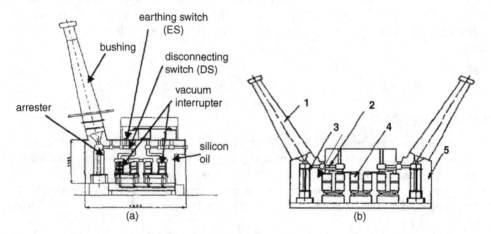

**Figure 9.6** Concepts of a silicone-oil immersed EHV vacuum circuit-breakers. (a) Japanese concept (550 kV) consisting of 4 gaps (reproduced with permission of IEEE) [39]; (b) Chinese concept (750 kV) consisting of 6 gaps. 1 – bushing; 2 – disconnector; 3 – arrester; 4 – vacuum interrupter; 5 – oil. (Reproduced with permission of Xi'an Jiaotong University) [38].

One of the reasons for the rather frequent use of HV vacuum switchgear in Japan is that utilities acknowledge the advantages of less maintenance (compared with $SF_6$ circuit-breakers), the excellent frequent-switching performance, and the suitability for rural distribution systems.

The reliability of HV vacuum circuit-breakers appears to be comparable to $SF_6$ circuit-breakers. A Japanese survey on the failure rate of HV vacuum- and $SF_6$ circuit-breakers installed in 72.5 kV transmission networks was conducted in cooperation with a Japanese utility [27]. Mechanical failure of the operating mechanism was the main failure mode. There were no troubles caused by overvoltages arising from HV vacuum circuit-breakers. The quantity of failures, however, is too low to identify a trend regarding service years.

At present (2013), most research and development efforts devoted to HV vacuum circuit-breakers are concentrated in East Asia. Japanese companies showed the feasibility of mature products 20 years ago, however, they are predominantly applied on their internal market where HV vacuum circuit-breakers take a certain share in special applications.

In China, there are significant developments, and application on a large scale on 72.5 and 126 kV is foreseen in the near future.

Research and development work in Europe has been reported since the mid-1990s [41–43]. Companies in Europe are now bringing HV vacuum circuit-breakers onto the market and have started pilot projects to gain experience in the field [44, 45]. In the modern generation of HV vacuum circuit-breakers, $SF_6$ is also avoided as outside insulation of the vacuum interrupter, and instead, nitrogen or dry-air is used.

US companies, although having an early track record of HV vacuum circuit-breaker development, did not commercialize the HV vacuum circuit-breaker technology. However, products for load switching, notably capacitor banks, with multiple vacuum interrupters in series (up to 9 interrupter units per phase) emerged long ago as HV switches up to rated voltages of 242 kV.

Vacuum technology is seen occasionally as an option in HV disconnectors for increasing their switching capability.

Experimental ('hybrid') designs with $SF_6$ and vacuum interrupters in series have also been reported [46, 47]. The idea is to use the very fast recovery of a vacuum interrupter to withstand the initial TRV (such as appears in short-line fault interruption), whereas an $SF_6$ interrupter with a reduced amount of $SF_6$ should withstand the peak value of the transient recovery voltage.

### 9.10.4  X-ray Emission

X-ray emission is generated in vacuum devices because electrons, accelerated by the electric field in the gap, collide with the metal target contact. In this process, electromagnetic radiation is generated, the energy of which is determined by the voltage across the gap and the intensity of which is determined by the electron current. The biological effect of X-ray radiation on human tissue is expressed as equivalent radiation dose. Its SI unit is the sievert [Sv]; $1 \text{ Sv} = 1 \text{ J kg}^{-1}$.

To benchmark X-ray dose limits, the natural background radiation dose rate of about $0.3 \text{ } \mu\text{Sv h}^{-1}$ can be considered. Investigations show [48] that the X-ray dose rate of interrupters up to rated voltages of 36 kV remains within the limits of $1 \text{ } \mu\text{Sv h}^{-1}$, even during application of power-frequency test voltages much higher than the rated voltage.

It depends on the design and on the contact surface roughness, but the dose rate of X-rays emitted by vacuum interrupters seems to be within the limit of 1 $\mu$Sv h$^{-1}$ up to a rated voltage of 145 kV [49]. From the IEC standard 62271-1 the X-ray emission level shall not exceed 5 $\mu$Sv h$^{-1}$ at 1 m distance at the maximum operating voltage $U_r$.

## 9.10.5   Comparison of HV Vacuum- and HV SF$_6$ Circuit-Breakers

There exists general consensus amongst specialists that applying vacuum technology for HV switchgear is already successful for fault-current interruption [27] . Interruption of very high fault current even in combination with a very high value of rate-of-rise of TRV by vacuum circuit-breakers has been demonstrated and in this context may be even superior to SF$_6$ circuit-breakers. This can be particularly useful in the application of, for example, transformer- and reactor-limited faults (see Sections 3.4 and 3.5).

The main driving force for the development of HV vacuum circuit-breakers is the absence of SF$_6$ gas as well as the reduced maintenance (of the interrupter) and high electrical endurance. In turn, the lack of practical methods to monitor the vacuum quality in service is seen as a disadvantage.

The main challenge for HV vacuum switchgear is in the capacitive- and inductive-load switching duties. Capacitive switching is influenced by the inherently wide variation of vacuum in breakdown statistics, which becomes significant for frequently switching capacitor banks. Test-statistics show that capacitor-bank switching, especially in the case of large inrush currents, at higher voltages is associated with in an increasing occurrence of late breakdown (see Section 4.2.6). A special design of a HV vacuum interrupter may sometimes be advisable for switching of single capacitor banks.

In shunt-reactor (inductive-load) switching, the number of re-ignitions (not the probability of re-ignition) can be large compared with SF$_6$ circuit-breakers (see Section 4.3.3.2). Protective measures when switching small HV reactors, especially when directly connected to the circuit-breaker, are sometimes recommended. Alternatively, designs that are optimized for shunt-reactor switching may be used. The chopping current levels of (HV) vacuum circuit-breakers do not differ essentially from those of SF$_6$ circuit-breakers [50].

In comparison with SF$_6$ circuit-breakers, from a merely technical point of view, vacuum technology has the following strong features [27]:

- vacuum interrupters do not contain greenhouse gas;
- vacuum circuit-breakers (when free of SF$_6$) are disposable at the end-of-life without special issues;
- there is no site contamination in the case of an explosive failure;
- vacuum interrupters are sealed for life (no gas handling infrastructure or interrupter maintenance);
- low operation energy may imply simpler drives and reduced maintenance;
- very fast dielectric recovery after interruption;
- short arcing times make design of a two-cycle circuit-breaker feasible;
- vacuum circuit-breakers can interrupt current after a late restrike;
- restrikes and re-ignitions typically do not damage interrupter internal parts;

- vacuum interrupters function electrically independently of low ambient temperature (there is no liquefaction);
- high number of switching operations even short-circuit interruption (electrical endurance, see Section 12.2.2).

The weaker points of vacuum switchgear for application at the transmission voltage levels are the following:

- costs are probably higher than for $SF_6$ technology for the same ratings (at least to date);
- there is limited HV service experience;
- service life-time is unknown;
- high normal current ratings are difficult to obtain as it is difficult to get heat out of the vacuum interrupter;
- there is no practical way to monitor vacuum quality in service;
- multiple breaks per pole are generally applied above 145 kV, often even at lower voltages;
- the dielectric performance of vacuum interrupters is sensitive to switching history and has significant scatter;
- special designs or protection devices may be necessary in certain reactive switching applications (shunt reactor, capacitor bank switching).

# References

[1] Cobine, J.D. (1963) Research and development leading to the high-power vacuum interrupter – a historical review. *IEEE Trans. Power Ap. Syst.*, **82**, 201–217.
[2] Greenwood, A. (1994) *Vacuum Switchgear*, IEE, ISBN 0 85296 855 8.
[3] Slade, P.G. (2008) *The Vacuum Interrupter. Theory, Design and Application*, CRC Press ISBN 13: 978-0-8493-9091-3.
[4] Garzon, R.D. (1997) *High-Voltage Circuit-Breakers*, Chapter 3, Marcel Dekker, New York, ISBN 0-8247442-76.
[5] Christian, R., Paolo, G. and Kim, H. (2003) The Integrated MV Circuit-Breaker – a New Device Comprising Measuring, Protection and Interruption. 17th Int. Conf. on Electr. Distr., Barcelona.
[6] Falkingham, L.T. (1999) Appendix A to Ph.D. Thesis: Vacuum Interrupter Technology and its Historical Development, available at https://dspace.lib.cranfield.ac.uk/bitstream/1826/838/3/All%20Appendix.pdf accessed 4 April 2014.
[7] Briggs, S.J., Bartos, M.J. and Arno, R.G. (1998) Reliability and availability assessment of electrical and mechanical systems. *IEEE Trans. Ind. Appl.*, **34**(6), 1387–1396.
[8] Hale, P.S. and Arno, R.G. (2000) Survey of Reliability and Availability Information for Power Distribution, Power Generation, and HVAC Components for Commercial, Industrial and Utility Installations. Ind. and Comm. Power Systems Techn. Conf., pp. 31–54.
[9] Renz, R., Gentsch, D., Slade, P. *et al.* (2007) Vacuum Interrupters – Sealed for Life. 19th Int. Conf. on Electr. Distr. (CIRED), 21–24 May, Paper 0156.
[10] Reuber, C., Gritti, P. and Kim, H. (2003) The Integrated MV Circuit-Breaker – A new Device Comprising Measuring, Protection and Interruption. CIRED Conference.
[11] Slade, P. and Smith, R.K. (2006) Electrical Switching Life of Vacuum Circuit-Breaker Interrupters. Proc. of the 52nd IEEE Holm Conf. on Electr. Contacts.
[12] Schlaug, M., Dalmazio, L., Ernst, U. and Godechot, X. (2006) Electrical Life of Vacuum Interrupters. XXIInd Int. Symp. on Disch. and Electr. Insul. in Vac., Matsue.
[13] Dullni, E., Fink, H. and Reuber, C. (1999) A Vacuum Circuit-Breaker with Permanent Magnetic Actuator and Electronic Control. CIRED Conference.
[14] Smeets, R.P.P., te Paske, L.H., Kuivenhoven, S. *et al.* (2009) The Testing of Vacuum Generator Circuit-Breakers. CIRED Conference, Paper No. 393.

[15] Smeets, R.P.P., Hooijmans, J.A.A.N. and Schoonenberg, G. (2007) Test Experiences with New MV TRV Requirements in IEC 62271-100. CIRED Conference, Session I, Paper 0378.

[16] Müller, A. and Sämann, D. (2011) Switching Phenomena in Medium-Voltage Systems – Good Engineering Practice on the Application of Vacuum Circuit-Breakers and Contactors. PCIC Conference, Paper Ro-47.

[17] Schoonenberg, G.C. and Menheere, W.M.M. (1989) Switching Overvoltages in Medium Voltage Networks. CIRED Conference.

[18] Ballat, J., König, D. and Reininghaus, U. (1993) Spark conditioning procedures for vacuum interrupters in circuit-breakers. *IEEE Trans. Electr. Insul.*, **28**, 621–627.

[19] Leusenkamp, M.B.J. (2012) Impulse Voltage Generator Design and the Potential Impact on Vacuum Interrupter De-conditioning. XXVth Int. Symp. on Disch. and Electr. Insul. in Vac., Tomsk.

[20] Betz, T. and König, D. (1998) Influence of Grading Capacitors on the Breaking Capability of Two Vacuum Circuit-Breakers in Series. XVIIIth Int. Symp. on Disch. and Elec. Insul. in Vac., Eindhoven.

[21] Nitta, T., Yamada, N. and Fujiwara, Y. (1974) Area Effect of Electrical Breakdown in Compressed SF6. *IEEE Trans. Power Ap. Syst.*, **PAS-93**(2), 623–629.

[22] Hae, T., Utsumi, T., Sato, T. *et al.* (2013) Features of Cubicle Type Vacuum-Insulated Switchgear (C-VIS). CIRED Conference, paper 1201.

[23] Schellekens, H., Shiori, T., Picot, P. and Mazzucchi, D. (2010) Vacuum Disconnectors. An Application Study. XXIVth Int. Symp. on Disch. and Elec. Insul. in Vac.

[24] Falkingham, L. and Reeves, R. (2009) Vacuum Life Assessment of a Sample of Long Service Vacuum Inter-rupters. CIRED Conference, Paper 0705.

[25] Reeves, R. and Falkingham, L. (2013) An appraisal of the insulation capability of vacuum interrupters after long periods of service. 2nd Int. Conf. on Elec. Pow. Eq. – Switching Techn., paper 1-P2-P-P5, Matsue, Japan

[26] Parashar, R.S. (2011) Pressure Monitoring Techniques of Vacuum Interrupters. CIRED Conference, paper 0234, Frankfurt.

[27] CIGRE Working Group A3.27 (2014) The Impact of the Application of Vacuum Switchgear at Transmission Voltages. CIGRE Technical Brochure.

[28] Latham, R.V. (1981) *High Voltage Vacuum Insulation: The Physical Basis*, Academic Press, Inc., ISBN 0-12-437180-9.

[29] Falkingham, L. and Waldron, M. (2006) Vacuum for HV applications - Perhaps not so new? - Thirty Years Service Experience of 132 kV Vacuum Circuit-Breaker. XIInd Int. Symp. on Disch. and Electr. Insul. in Vac., Matsue.

[30] Shores, R.B. and Philips, V.E. (1975) High-voltage vacuum circuit-breakers. *IEEE Trans. Power Ap. Syst.*, **PAS-94**(5), 1821–1830.

[31] Slade, P., Voshall, R., Wayland, P. *et al.* (1991) The development of a vacuum interrupter retrofit for the upgrading and life extension of 125 kV to 145 kV oil circuit-breakers. *IEEE Trans. Power Deliver.*, **6**, 1124–1131.

[32] Umeya, E. and Yanagisawa, H. (1975) Vacuum Interrupters, Meiden Review, Series 45, pp. 3–11.

[33] Yanabu, S., Satoh, Y., Tamagawa, T. *et al.* (1986) Ten Years Experience in Axial Magnetic Field Type Vacuum Interrupters. IEEE PES, 1986 Winter Meeting, 86 WM 140-8.

[34] Saitoh, H., Ichikawa, H., Nishijima, A. *et al.* (2002) Research and Development on 145 kV, 40 kA One-Break Vacuum Circuit-Breaker. IEEE T&D Conference, pp. 1465–1468.

[35] Matsui, Y., Nagatake, K., Takeshita, K. *et al.* (2006) Development and Technology of High-Voltage VCBs; Brief History and State of Art. XXIInd Int. Symp. on Discharges and Elec. Insul. in Vac., Matsue.

[36] Wang, J., Liu, Z., Xiu, S. *et al.* (2006) Development of High Voltage Vacuum Circuit-Breakers in China. XXIInd Int. Symp. on Disch. and Electr. Insul. in Vac., Matsue.

[37] Homma, M., Sakaki, M., Kaneko, E. and Yanabu, S. (2006) History of vacuum circuit-breakers and recent developments in Japan. *IEEE Trans. Dielect. Electr. Insul.*, **13**(1), 85–92.

[38] Liu, D., Wang, J., Xiu, S. *et al.* (2004) Research on 750 kV Vacuum Circuit-Breaker Composed of Several Vauum Interrupts in Series. XXIst Int. Symp. on Disch. and Elec. Insul. in Vac., Yalta, pp. 315–318.

[39] Okubo, H. and Yanabu, S. (2002) Feasibility Study on Application of High-Voltage and High-Power Vacuum Circuit-Breaker. XXth Int. Symp. on Disch. and Elec. Insul. in Vac., pp. 275–278, Tours.

[40] Ikebe, K., Imagawa, H., Sato, T. *et al.* (2010) Present Status of High-Voltage Vacuum Circuit-Breaker Application and its Technology in Japan. CIGRE Session 2010, Paper A3-303.

[41] Giere, S., Knobloch, H. and Sedlacek, J. (2002) Double and Single-Break Vacuum Interrupters for High-Voltage Application: Experiences on Real High-Voltage Demonstration-Tubes. CIGRE Conference, Paris.

[42] Schellekens, H. and Gaudart, G. (2007) Compact high-voltage vacuum circuit-breaker, a feasibility study. *IEEE Trans. Dielect. Electr. Insul.*, **14**(3), 613–619.

[43] Godechot, X., Ernst, U., Hairour, M. and Jenkins, J. (2008) Vacuum Interrupters in High-Voltage Applications. XXIIIrd Int. Symp. On Disch. and Electr. Insul. in Vac., Bucharest.

[44] Brucher, J., Giere, S., Watier, C. *et al.* (2012) 3AV1FG – 72.5 kV Prototype Vacuum Circuit-Breaker (Case Study with Pilot Customers). CIGRE Conference, Paper A3-101.

[45] Newton, M. and Renton, A. (2013) Transpower's Adoption of non-SF6 Switchgear. CIGRE B3 Symp. Managing Substations in the Power System of the Future, Brisbane.

[46] Smeets, R.P.P., Kertész, V., Dufournet, D. *et al.* (2007) Interaction of a vacuum arc in a hybrid circuit-breaker during high-current interruption. *IEEE Trans. Plasma Sci.*, **35**(4), 933–938.

[47] Cheng, X., Liao, M., Duan, X. and Zou, J. (2010) Study of Breaking Characteristics of High-Voltage Hybrid Circuit-Breaker. XXIVth Int. Symp. on Disch. and Elec. Insul. in Vac., Braunschweig, pp. 449–452.

[48] Renz, R. and Gentsch, D. (2010) Permissible X-Ray Radiation Emitted by Vacuum-Interrupters/ - Devices at Rated Operating Conditions. XXIVth Int. Symp. on Disch. and Elec. Insul. in Vac., Braunschweig, pp. 133–137.

[49] Yan, J., Liu, Z., Zhang, S. *et al.* (2012) X-Ray Radiation of a 126kV Vacuum Interrupter. XXVth Int. Symp. on Disch. and Elec. Insul. in Vac., Tomsk.

[50] Tokoyoda, S., Takeda, T., Kamei, K. *et al.* (2013) Interruption Behaviours with 84/72 kV VCB and GCB. 2nd Int. Conf. on Elec. Pow. Eq. – Switching Techn., paper 2-A1-P-1, Matsue, Japan.

# 10

# Special Switching Situations

There are several switching situations that differ from the duties standardized for switching devices in transmission and distribution systems. The most important ones are treated in this chapter.

## 10.1  Generator-Current Breaking

### 10.1.1  Introduction

Generator circuit-breakers are breakers applied at a rated voltage matching the rated voltage of a generator. They are located between the generator and step-up transformer. When no generator circuit-breaker is applied, an alternative solution is a circuit-breaker at the high-voltage side of the step-up transformer. The advantage of this solution is the less complicated simple high-current connection (*generator busduct*) between generator and transformer. The advantage of having a generator circuit-breaker is the possibility to connect the auxiliary plant to the medium voltage side of the (permanently energized) step-up transformer. This is depicted in Figure 10.1.

The electrical and mechanical performances of a generator circuit-breaker are very different from standard MV distribution switchgear. The only standard available worldwide that covers specifically the requirements for generator circuit-breakers is ANSI/IEEE C37.013 [1]. Apart from the ratings and other relevant characteristics, this standard contains guidelines for the type-testing of generator circuit-breakers.

Regarding the interrupting duties of generator circuit-breakers and general-purpose circuit-breakers, the main points of distinction are:

(a) *Load-current switching*
Load currents for large generation units can be as high as 50 kA, often this makes forced cooling necessary.

Following interruption of the load current, the two circuits at both sides of the generator circuit-breakers oscillate independently, creating a TRV that is a sum of the oscillating source and the load component.

---

*Switching in Electrical Transmission and Distribution Systems*, First Edition.
René Smeets, Lou van der Sluis, Mirsad Kapetanović, David F. Peelo, and Anton Janssen.
© 2015 John Wiley & Sons, Ltd. Published 2015 by John Wiley & Sons, Ltd.

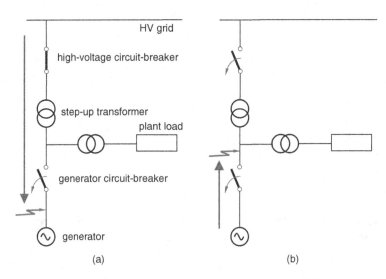

**Figure 10.1**  System-source and generator-source faults.

(b) *System-source faults*

In this situation a major part of the fault current is supplied through the step-up transformer by the system upstream. The total current has the highest value of all other fault situations, because the short-circuit reactance of the transformer and the HV system is usually lower than the (sub-)transient reactances of the generator. Unlike the situation for general purpose HV circuit-breakers, the maximum TRV stress on generator circuit-breakers appears after the maximum short-circuit current stress. A very high RRRV value originates from the small distributed capacitance of the step-up and the auxiliary transformers. In Figure 10.1a, the circuit topology of this fault is outlined in a basic circuit.

For smaller generators, there is a cable connecting the step-up transformer with the generator circuit-breaker. The capacitance of this cable can then reduce the RRRV considerably, depending on the cable length [2]. In certain designs, TRV mitigating capacitors are built inside the circuit-breaker enclosure.

(c) *Generator-source faults*

In this case a major part of the fault current is supplied by the generator (see Figure 10.1b) causing a DC component that can be higher than the symmetrical short-circuit current, resulting in very high asymmetry, possibly with delayed current zeros [3–6]. The AC component, having a constant amplitude in faults in substations and on overhead lines (see Section 2.1), in a generator-source fault has a time-dependent, decreasing, amplitude $A(t)$. This results from the specific transient behaviour of the generator [7].

Reactances of the generator and corresponding time constants are usually quantified by:

$X_d''$  subtransient reactance;

$T_d''$  subtransient time constant;

$X_d'$  transient reactance;

$T_d'$  transient time constant;

$X_d$  synchronous reactance;

$T_a$  armature time constant.

The generator-source fault current $i_{gs}(t)$ is expressed as the sum of a symmetrical, power-frequency current $i_{ac}(t) = A(t)\cos(\omega t)$ and a DC component $i_{dc}(t)$:

$$i_{gs}(t) = A(t)\cos(\omega t) + i_{dc}(t) \tag{10.1}$$

$$i_{gs}(t) = \frac{P\sqrt{2}}{U\sqrt{3}} \left\{ \left[ \left( \frac{1}{X_d''} - \frac{1}{X_d'} \right) \exp\left( \frac{-t}{T_d''} \right) + \left( \frac{1}{X_d'} - \frac{1}{X_d} \right) \exp\left( \frac{-t}{T_d'} \right) + \frac{1}{X_d} \right] \right.$$

$$\left. \times \cos(\omega t) - \frac{1}{X_d''} \exp\left( \frac{-t}{T_a} \right) \right\} \tag{10.2}$$

with $P$ and $U$ being the rated generator active power and voltage respectively. Equation 10.2 describes the transient behaviour of generators of the network model that is shown in Figure 10.2. In the lower part, the unique characteristic of this type of fault is highlighted: "missing" current zeros or delayed current zeros. The first current zero can be delayed for several cycles. This implies that during this period, the generator circuit-breaker is not able to interrupt the fault current.

The circuit-breaker and the fault arcs reduce the circuit time constant because the effective arc resistances add to the circuit resistance [8]. The delayed current zero may, therefore, be advanced with respect to the situation without arc(s). In Figure 10.3 it is shown how the SF$_6$ circuit-breaker arc voltage effectively reduces the DC time constant of

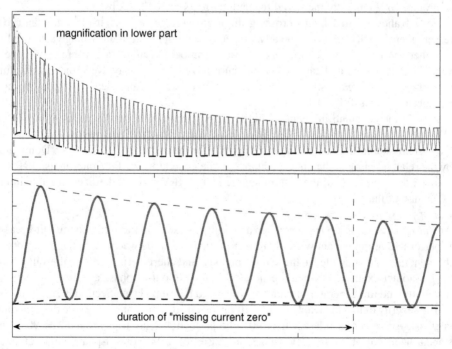

**Figure 10.2** Single-phase generator-source fault current with magnification of the period of "missing zeros".

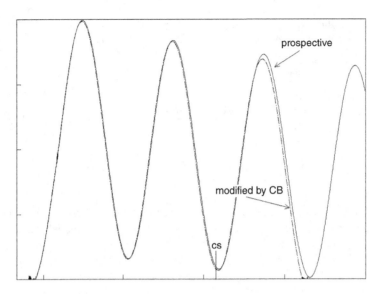

**Figure 10.3**  Test result of a prospective generator-source fault current (circuit-breaker remains closed) and current influenced by arc voltage ("modified by CB"); cs- contact separation. Time: 15 ms/div, current: 65 kA/div.

the generator-source fault current and shortens the arcing time by one major asymmetrical current loop [9]. The arc voltage of the fault arc also contributes to this reduction. Values of around 10 V cm$^{-1}$ are measured for fault currents up to 70 kA [8].

For advancing current zero crossing, high arc voltage (as for air-blast arcs and to a lesser extent for SF$_6$ arcs) can be advantageous, although a high arc voltage also implies a higher thermal stress for the interruption chamber. Vacuum circuit-breakers have a very low arc voltage and cannot advance current zero in this way. Nevertheless, vacuum generator circuit-breakers with significant generator-source fault clearing capability have been tested successfully [10].

A DC time constant, inherent to this situation circuit, is standardized as $T_a = 133$ ms [1]. The value of the generator reactance limits the short-circuit current to a value below the system-source-fault case. The same applies for the associated TRV stress. Dominated by the relatively large inherent capacitance of the generator, the TRV rate-of-rise (RRRV) is lower than the value of the system-source fault RRRV and standardized at a maximum of about half that value.

(d) *Out-of-phase fault*

An *out-of phase condition* occurs when a making operation of the generator circuit-breaker is performed at the instant when there is no synchronism between the voltage phasors of the generator at one side of the circuit-breaker and the external grid at the other side [11]. Another example is when a generator is running out-of-phase as a result of system instability and the generator circuit-breaker has to be tripped (see Section 3.7.2).

The severity of this interruption depends on the out-of-phase angle $\delta$. Since the generator is at risk for values of $\delta$ larger than 90°, the protection relay trips around $\delta = 90°$. The standardized out-of-phase TRV values are based on a 90° out-of-phase angle at rated voltage. For smaller generator units, a large out-of-phase angle can, however, occur [12].

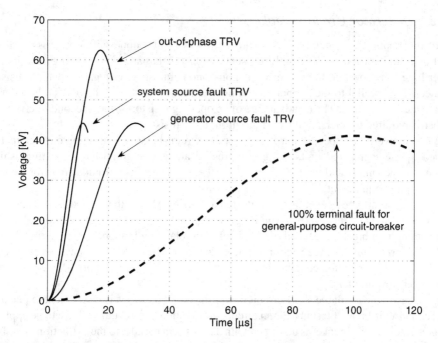

**Figure 10.4**   Standardized TRV shapes of various generator faults compared with TRV of a 100 % fault for a 24 kV general purpose circuit-breaker.

In the case of an out-of-phase angle of $\delta = 90°$, the current is about 50% of the fault current supplied by the system. On the voltage side, the generator circuit-breaker experiences a TRV with a RRRV roughly of the same order as in the system-source fault, but with a peak value nearly two times higher.

The out-of-phase current specified in the ANSI/IEEE standard [1] is half the system-source-fault current.

In Figure 10.4, TRV wave shapes based on the ANSI/IEEE standard [1] are shown for various interruption duties of a generator circuit-breaker, rated voltage 24 kV, connected to a generator having a rated power between 200 and 400 MVA. For comparison, a standardized TRV wave shape for the 100% short-circuit current interruption duty for a general-purpose 24 kV circuit-breaker, in accordance with IEC 62271-100 [13], is added. It is apparent that, apart from current stresses, the TRV requirements for generator circuit-breakers are far more severe than the TRV requirements for general-purpose circuit-breakers.

For lower generator-power ratings, even below 100 MVA, the RRRV of the system-source-fault TRV is still 2.6 times higher than the T10 duty standardized for general-purpose circuit-breakers [14].

In general, the generator neutral is unearthed. Therefore, the first-pole-to-clear faces a power-frequency recovery voltage of 1.5 times the phase-to-earth voltage of the system, see Section 3.3.2.1.

Since the generator circuit-breaker is located directly at the output of the power plant, the reliability of generator circuit-breakers must be extremely high [15].

## 10.1.2   Generator Circuit-Breakers

Current interruption in generator circuit-breakers was at first done with compressed air in extinction chambers assisted by a high-pressure air blast. Air as arc-extinguishing medium has a rather long time constant to recover to its non-conducting state. Often, an opening resistor is connected in parallel to the extinction chamber to reduce drastically the RRRV in order to facilitate the interruption. The disadvantage of an opening resistor is that a second interruption chamber is required to interrupt the current through the resistor.

The first delivery of a specific-purpose generator circuit-breaker, consisting of three metal-enclosed, phase-segregated units using compressed air as operating and arc-extinguishing medium, was accomplished in 1970. Since then, there has been a continuous development of this power-plant equipment.

The new generation of generator circuit-breakers using $SF_6$ as the arc extinction medium came on the market in the 1980s. Exploiting the thermal energy of the arc combined with a puffer action, high breaking capacity is realized with relatively low operating energy, the so-called *self-blast technology,* see Section 7.3.4. Reduction of the RRRV by capacitors connected in parallel to the interruption chamber (*surge capacitor*, see Figure 10.5 part 10) has a positive influence on current interruption.

In most cases, the available short-circuit power of test laboratories is insufficient to perform a generator circuit-breaker test in a direct circuit, and synthetic test methods have to be applied. If the generator circuit-breaker is equipped with a resistor in parallel to the extinction chamber, the energy in the HV capacitor-part of the synthetic circuit is usually insufficient to produce the TRV wave shape that a direct circuit would yield. Two-part test methods have to be used, that is, separate testing of the interruption behaviour during the thermal and dielectric period of the post-arc recovery [16–20].

In addition to the obvious requirement that a generator circuit-breaker must be able to carry the full load current of the generator and ensure a sufficient insulation level at all times, it must also be capable of performing the following functions:

- synchronize the generator with the main system;
- disconnect the generator from the main system;
- make, carry, and interrupt load currents up to the full load current of the generator;

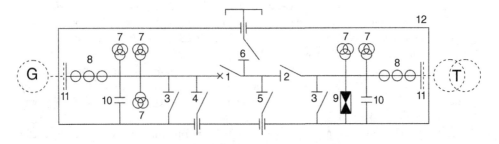

**Figure 10.5**   Generator circuit-breaker system [21]. 1 – Circuit-breaker; 2 – disconnector; 3 – earthing switches; 4 – starting switch; 5 – starting switch (back-to-back); 6 – short-circuiting switch / braking switch; 7 – voltage transformers; 8 – current transformers; 9 – surge arrester; 10 – surge capacitors; 11 – terminals; 12 – enclosure.

- make, carry, and interrupt the system-source short-circuit currents and generator-source short-circuit currents;
- make, carry, and interrupt currents under out-of-phase conditions.

All the associated items of switchgear can also be integrated into the generator-circuit-breaker enclosure as an option to their separate installation. Such items comprise a series disconnector, earthing switches, a short-circuiting switch, current transformers, voltage transformers, protective capacitors, and surge arresters.

Depending on the type of power plant, additional items, such as starting switches in gas-turbine and hydro power plants and braking switches in hydro power plants, can also be mounted in the generator circuit-breaker enclosure (see Figure 10.5).

In addition to the very large $SF_6$ generator circuit-breakers for the protection of (large) power-plant generators, generator vacuum circuit-breakers [22] are on the market for fault currents of 72 kA and higher [10].

## 10.2  Delayed Current Zero in Transmission Systems

Delayed current zeros occur for asymmetrical currents where the relative values of the AC and DC components are such that the current does not cross the zero line.

For transmission circuit-breakers close to generating stations, delayed current zeros require certain generator operating modes, underexcited, overexcited, full load or no-load, and fault type, usually three-phase simultaneous or non-simultaneous [23]. A study of 550 kV circuit-breakers at the Itaipu substation in Brazil assumed a theoretically highest DC component resulting from a two-phase line-to-line fault at zero line-to-line voltage that develops into a three-phase fault at zero voltage to earth on the third phase [24]. Subsequent short-circuit testing demonstrated the capability of the circuit-breaker to force current zero crossings due to the high arc voltage of the blown arc.

In Figure 10.6 an example from a test shows how the arc voltage of a $SF_6$ circuit-breaker advances current zero significantly.

A further study on the series compensated 735 kV system also showed that, for certain circuit-breaker locations and sequence of tripping, delayed current zeros could occur [25]. Later testing again demonstrated the positive influence of high arc voltages.

All in all, given the multiple contingencies required for delayed zero crossings, such actual events are conceiveable but not probable for transmission circuit-breakers.

## 10.3  Disconnector Switching

### 10.3.1  Introduction

*Disconnectors*, or *disconnecting switches* are switching devices that commonly fulfil the following functions:

- Isolation. Isolating system components from energized sections is a common no-load switching operation. Isolation is necessary for safe maintenance, repair, and replacement of power-system components. Only after isolation and earthing can personnel approach and touch the component. A visual break between live and workable parts is required in many countries.

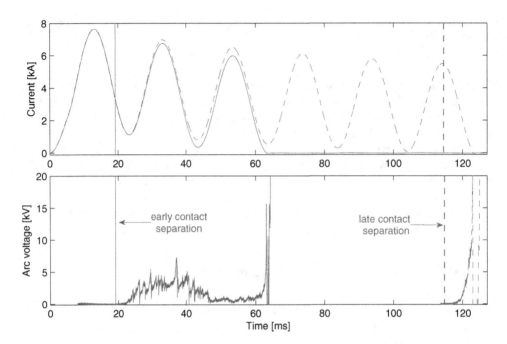

**Figure 10.6** Interruption of a 245 kV $SF_6$ circuit-breaker in a delayed current-zero test-situation with early contact separation (drawn traces) and late contact separation (dashed traces). Upper traces: current; lower traces: voltage across the breaker.

To reduce the probability of a dielectric breakdown to its absolute minimum, a large contact distance is necessary.

- Commutation. A typical topology of substations is the double busbar arrangement. In the case when the normal load current has to be diverted (or commutated) from one busbar to another, disconnectors normally transfer the current to the parallel busbar and the current will continue to flow uninterrupted. Because of the presence of the parallel busbar, the voltage across the disconnector remains low and current transfer is rather straightforward.

Disconnectors must ensure a firm connection even at short-circuit currents (see Section 2.3.1). Disconnectors operate in open air, as in outdoor substations, or in $SF_6$, as in gas-insulated substations (GIS).

## 10.3.2   No-Load-Current Switching

### 10.3.2.1   Introduction

Disconnectors are not designed for interruption of current since they are, in practice, used to isolate sections or components in the power systems. The intrinsic capacitance, however, of an unloaded section of a gas- or air-insulated substation causes a capacitive current to be switched. These currents may reach from a few tens of milliamperes in unloaded GIS sections, to a few hundred milliamperes in long busbars, circuit-breaker grading capacitors, or other energized parts such as short lengths of cables. Even such small currents can cause substantial switching overvoltage transients, in particular during GIS-disconnector operation at higher system voltages.

No-load-current switching by disconnectors is characterized by arcing between the slowly moving disconnector contacts. The movement can last from seconds in gas-insulated stations to some ten seconds in air-insulated stations. Unlike arcing at load currents or even fault conditions, the disconnector arc is a rapid succession of breakdown, arcing, interruption, restrike, and so forth. The parameters of this repetitive process, such as the breakdown voltage, duration of arcing, and frequency of occurrence of arcing, depend largely on the interruption medium and on the circuit topology in the vicinity of the disconnector.

### 10.3.2.2 Disconnector Switching in GIS

In GIS, the current to be interrupted is small, of the order of milliamperes, since only short sections of GIS busbar are involved. In practice, the no-load current in GIS does not exceed 0.5 A [26]. The small current to be switched causes a discharge between the arcing contacts to last only few microseconds, but this short time is sufficient to equalize the voltage at the load side with the source side. This process of the charge transfer generates a *very fast transient overvoltage* (VFTO). The amplitude of the VFTO is the difference between the instantaneous load- and source-side voltage, multiplied by an overshoot or peak factor.

Due to the relatively slow operating speed, a large number of pre-strikes and restrikes between the disconnector contacts occur at each closing and opening operation of the disconnector.

The whole switching process lasts at least 100 ms. At disconnector opening, the amplitude of the VFTOs increases, at closing a wave train of decreasing transients arises.

The impact of VFTOs on equipment can be divided into internal and external VFTOs [27, 28]:

- Internal VFTOs: Travelling waves that are initiated between the inner conductor and the enclosure, putting high stresses on the internal insulation.
- External VFTOs: At discontinuities of the GIS enclosure, such as windows and bushings, the electromagnetic wave is partially transmitted to the outside resulting in:
  - *Transient electromagnetic fields* (TEMF) causing stresses and *electromagnetic interference* (EMI) in connected primary equipment like transformers and instrument transformers.
  - Travelling waves on overhead lines causing stresses in the connected equipment (transformers, instrument transformers).
  - *Transient enclosure voltages* (TEV) causing stresses and electromagnetic interference in secondary equipment. According to IEC 62271-1 [29], switching of a small portion of (unloaded) GIS is the most severe situation with regard to the generation of electromagnetic disturbances in the secondary system [30, 31].

A typical sequence of events from a test of a 145 kV three-phase enclosed GIS disconnector can be seen in Figure 10.7 for the energization process, and in Figure 10.8 for the de-energization

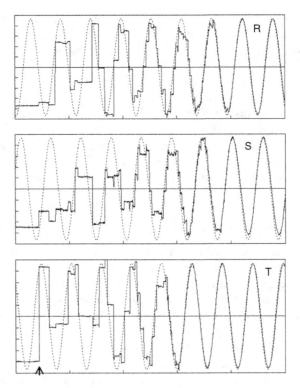

**Figure 10.7**    Three-phase GIS-disconnector closing operation. Drawn traces: load-side voltages; dotted traces: source-side voltages. Time: 15 ms/div; voltages: 25 kV/div.

process [32]. Every vertical voltage jump is a breakdown of the switching gap, immediately followed by arc extinguishing because of the very small current. During the de-energization, the interval between the breakdown events becomes gradually longer because of the increasing contact distance; the opposite is true for the pre-strikes during energization. These events trigger the VFTOs inside the GIS enclosure.

After de-energization of the unloaded GIS section, a certain charge can be trapped at the load-side capacitance, causing a maximum voltage value of $\hat{U}_t = -1$ p.u., if no capacitive coupling between the phases is taken into account. This value corresponds to the peak value of the source phase-to-earth rated voltage. A subsequent closing operation has a high probability to pre-strike at the opposite peak of the source voltage $\hat{U}_s \approx 1$ p.u., so that the load side will oscillate with a 2 p.u. excursion multiplied by an amplitude factor $A_f$ which has a standardized value of $A_f \geq 1.4$. This yields a peak value of the phase-to-earth voltage $\hat{U}_{p0,m}$:

$$\hat{U}_{p0,m} = |\hat{U}_s + (A_f - 1)(\hat{U}_s - \hat{U}_t)| = 1.8 \text{ p.u.} \tag{10.3}$$

whereas the phase-to-phase voltage $\hat{U}_{pp,m}$ is

$$\hat{U}_{pp,m} = |1 + \hat{U}_s + (A_f - 1)(\hat{U}_s - \hat{U}_t)| = 2.8 \text{ p.u.} \tag{10.4}$$

occurring when the trapped charge of one or both of the other phases has the value of 1 p.u.

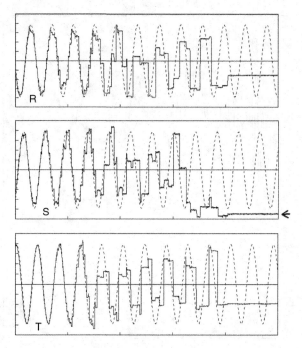

**Figure 10.8**   Three-phase GIS-disconnector opening operation. Drawn traces: load-side voltages; dotted traces: source-side voltages. Time: 15 ms/div; voltages: 25 kV/div.

The most severe VFTO is triggered when the pre-strike occurs between a GIS section, still charged after a previous disconnecting operation, and the source side at the maximum but opposite peak of the power-frequency phase-to-earth voltage. This situation is indicated in Figure 10.7 by an arrow. At first sight, the maximum jump in voltage in such a situation can be 2 p.u. However, in three-phase enclosed disconnectors, a voltage higher than 1 p.u. can remain trapped at the disconnected GIS section (as can be seen in Figure 10.8) due to phase-to-phase capacitive coupling. This voltage jump starts to travel as an electromagnetic wave in the GIS enclosure.

For certification purposes, IEC 62271-102, Annex F [26] prescribes the test duty no. 1 (*buscharging switching* duty) with the AC phase-to-earth voltage of 1.1 p.u. at the source side and the load side pre-charged with negative DC voltage of −1.1 p.u. prior to closing of the disconnector. The factor 1.1 is a safety-margin factor. In testing, the transient voltage of the conductor to enclosure, that is, the *transient voltage to earth* (TVE), shall be measured with a capacitive sensor located inside the GIS within 1 m of the contacts of the disconnector. Measurement of VFTO inside GIS is a major challenge because of the high voltages and the high frequencies involved [33–35].

A typical voltage waveform can be seen in Figure 10.9, showing in detail the VFTO initiated by the first voltage jump highlighted in Figure 10.7 by the arrow. The megahertz oscillations originate from the reflections of the wave, travelling in the GIS section, the lower frequencies originate from the external circuit.

VFTOs lead to a transient rise in potential of the discharge plasma with respect to neighbouring conductors. Due to the compact design of GIS, it is, therefore, essential that the

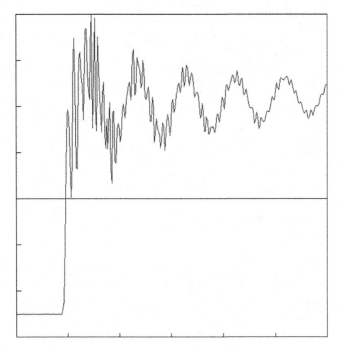

**Figure 10.9**  Voltage pattern of VFTO during closing operation of a GIS disconnector [32]. Time: 1 µs/div, voltage 50 kV/div.

disconnector discharge remains confined between the disconnector contacts and does not develop leaders branching off towards the enclosure or phase conductors, which would otherwise initiate an internal-arc fault. Tests, as specified in IEC 62271-102, Annex F, have to verify the absence of internal-arc faults due to disconnector action in GIS.

VFTOs in GIS are of concern at the highest rated voltages, for which the ratio of the lightning-impulse withstand voltage to the system voltage is relatively low [36]. This relationship is expressed in Figure 10.10, showing calculated VFTO magnitudes in three EHV and UHV projects in the world [37], compared with standardized GIS withstand voltages [38]. The expected VFTO level in GIS-disconnector switching can exceed the rated lightning-impulse withstand level at the highest rated voltage. Also, disconnector-switching overvoltages in GIS substations are higher than in *mixed-technology substations* (MTS), having hybrid SF$_6$ and air insulation. VFTO can become the limiting dielectric parameter which defines the dimensions at rated voltage levels above 800 kV.

Increasing the contact speed of disconnectors in GIS helps to reduce the magnitude of the VFTOs. In addition, the magnitude of VFTOs can be mitigated through the application of damping resistors [39] or magnetic rings installed coaxially on the conductors of each phase [40]. Application of resonators that dissipate the VFTO energy by allowing sparking in a safe region inside the GIS is considered [41].

A good overview on the backgrounds and experiences with GIS disconnector switching can be found in [27, 42, 43].

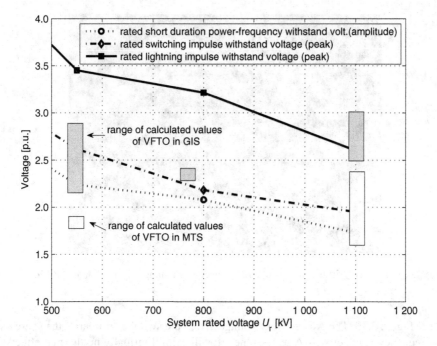

**Figure 10.10** Calculated magnitude of VFTO compared with various withstand voltage levels for GIS and mixed-technology substation designs [36].

### 10.3.2.3 Disconnector Switching in Open Air

Disconnector switching of a small current in open air results in a free burning arc in air between two slowly separating contacts situated at the tip of the *blades*. Since no active arc extinction is present, arcs can become long in length and in duration. Generally, both arc length and duration increase with the system voltage and arc current. Extended arcs, moving away from the shortest distance between the arc roots at the contacts, can get too close to energized parts of the system. In Figure 10.11, a long disconnector arc is shown just before its final extinction. This arc lasted 2 s with a current of 2.1 A at a driving voltage of 173 kV.

HV air-break disconnectors are commonly used by utilities to interrupt transformer magnetizing, capacitive, and bus transfer currents. The bus transfer currents are more often referred to as *loop currents* (see Section 10.3.3). Practices in North America have been based on the *arc-reach* approach based on tests in the 1940s that established empirical relationships between the current, the open-circuit voltage across the disconnector after current interruption, and the ultimate reach of the arc [44]. For most utilities worldwide, practice has tended to be based on trial and error. Recent research has shown that the associated free-burning arc behaviour is quite different than previously thought for each type of current and that reconsideration of past practices is warranted including a recommendation against the use of the arc-reach approach [45].

#### 10.3.2.3.1 Transformer Magnetizing-Current Switching
For today's low loss power transformers, the HV-side magnetizing current is typically 1 A or less. The transformer HV-side TRV is a critically damped oscillation of a few hundreds of

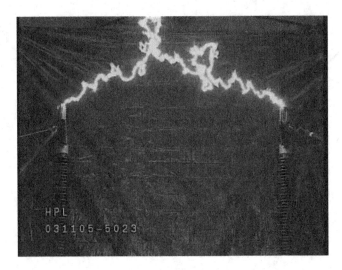

**Figure 10.11**   Erratic motion of a disconnector switching arc of 2.1 A at 173 kV.

hertz, see Figure 3.10. The AC recovery voltage across the switch is primarily the source-side power-frequency voltage. At 1 A or less, the arc will exhibit virtually no thermal effects and current interruption is in fact a dielectric event, which means that current interruption will occur when the increasing contact gap is sufficiently large to withstand the recovery voltage. The arc reach will be insignificant.

Generally, utilities recognize that there may be possible adverse effects on the transformer-insulation structure due to re-strike and pre-strike events during disconnector operation. Restrikes and pre-strikes may initiate inrush currents. The inrush current is usually limited because restrike or pre-strike occur close to the voltage peak, whereas the maximum inrush current arises at voltage zero (see Section 11.4.5.4). In addition, the short arc duration prevents increase of the arc reach. The effect of the inrush current prolongs the duration of the arc by the inrush-current duration.

Restrike can be prevented by using high-velocity devices, such as a spring-loaded whip, or *whip-type device* [46], which is first restrained as the disconnector main blades open arc-less and then released at a certain point of the blade travel to achieve a fast and restrike-free current interruption.

### 10.3.2.3.2  *Capacitive-Current Switching*

User requirements for capacitive current switching using air-insulated disconnectors frequently exceed the value of 0.5 A as stated for GIS disconnectors [47]. The relevant capacitive current ranges are from less than 1 A for bus section and instrument-transformer switching in substations up to about 10 A for short cable and transmission-line switching (see Section 4.2).

Interrupting of current by disconnectors is in fact a repetitive break–make event with numerous restrikes. This is demonstrated in Figure 10.12 showing current through and voltage across a disconnector arc. As can be seen in Figure 10.13, at each arc current zero (once per half-cycle) the arc current is interrupted and restrikes occur near the following peak of the

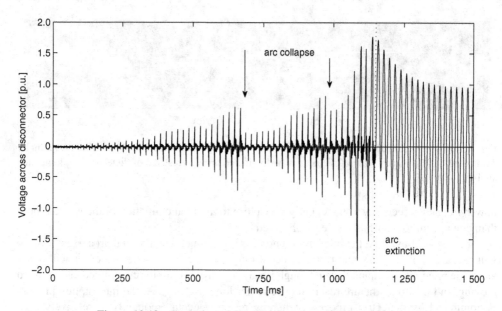

**Figure 10.12**   Evolution of the voltage across a disconnector arc.

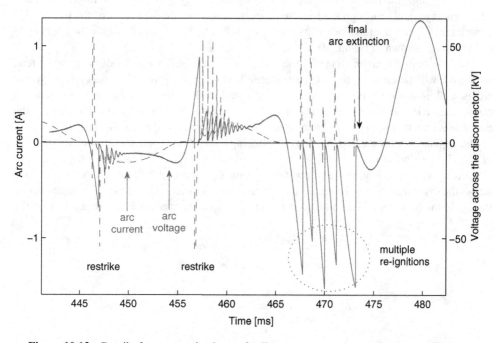

**Figure 10.13**   Detail of current and voltage of a disconnector arc close to final arc extinction.

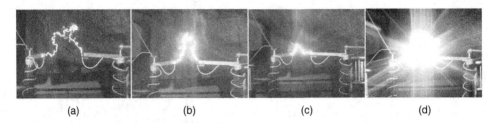

(a)                       (b)                       (c)                       (d)

**Figure 10.14** Disconnector arc: (a) steady state of thermally dominated arc; (b) restrike of thermally dominated arc; (c) steady state of dielectrically dominated arc; (d) restrike of dielectrically dominated arc [48].

power-frequency recovery voltage. Only very close to the final extinction of the arc can more than one re-ignition per half-cycle be observed.

For currents above 1 A, the interruption process is dominated by thermal processes: restrike voltages are relatively low and restrikes occur only once per half-cycle. Restrikes will occur along the previous arc channel even though it may be long, circuitous, convoluted, and upward moving. In this mode, the arc reach may be rather long. Below 1 A, the interruption process is dominated by dielectric processes whereby restrikes occur frequently at relatively high voltages. The arc is much more confined between the contacts.

Figure 10.14 shows the two modes and the visible differences in the restrike intensity.

The arc is sustained in the same arc channel by the energy input from the load current and by the energy supply through the restrike process, depending on the values of the source-side capacitance $C_S$ and the load-side capacitance $C_L$ [48,49]. The basic circuit for the analysis of this process is shown in Figure 10.15.

When $C_S$ is much larger than $C_L$, the equalization voltage to which both capacitors tend at restrike will be close to the source voltage. This situation gives low energy input to the arc as well as low overvoltages and short arc duration. This mode is the very fast transient case common in GIS disconnector operation. However, when $C_S$ is much smaller than $C_L$, the opposite is true, that is, high energy input to the arc, high overvoltages, and longer arc duration.

The explanation for the dependence of switching overvoltages and arc duration on the ratio of $C_S/C_L$ is as follows:

Current interruption leaves a trapped charge on $C_L$, and one half-cycle later the voltage across the disconnector will be 2 p.u. in the worst case. An eventual restrike takes place in two stages: in the first stage, the voltages at $C_S$ and $C_L$ equalize at a value $U_{eq}$ dependent on the

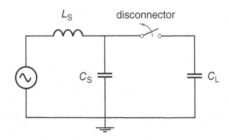

**Figure 10.15** Basic circuit for capacitive-current disconnector switching.

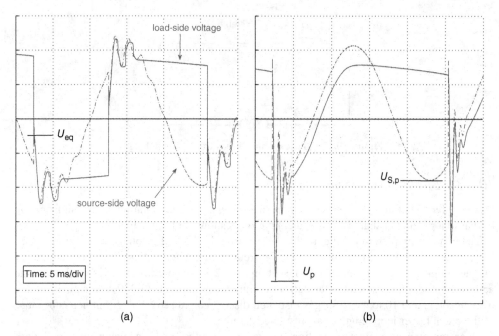

(a)                                                              (b)

**Figure 10.16** Capacitive-current interruption, detail of load- and source-side voltage equalization at restrike: (a) $C_S/C_L = 2.5$; (b) $C_S/C_L = 0.04$.

$C_S/C_L$ ratio; in the second stage, the voltage of $C_S$ and $C_L$ recovers to the source-side voltage by means of a main-circuit oscillation with an overvoltage peak $U_p$. Figure 10.16 shows actual oscillograms of the interruption of 2 A using a 300 kV centre-break disconnector. In Figure 10.16a, $C_S/C_L = 2.5$; it can be seen that the equalization voltage $U_{eq}$ is close to the source-side peak voltage $U_{S,p}$ and the subsequent recovery overvoltage $U_p$ is moderate.

In Figure 10.16b, $C_S/C_L = 0.04$ and now the equalization voltage is close to the load-side voltage and the subsequent recovery overvoltage $U_p$ is high. The high restrike voltage also implies a high restrike current, which injects energy into the arc, leading to longer arc duration than in the case of a high value of $C_S/C_L$.

A further phenomenon that contributes to sustaining the arc and prolonging the arcing time is the so-called *partial arc collapse*. Particularly for long arcs, two points on the arc come together, thereby short-circuiting a part of the arc between them, which then collapses, similarly to what happens to a lesser amount in circuit-breaker arcs (see Figure 3.21 in Section 3.6.1.6). The collapse shortens the arc and increases its sustainability since the power input needed to sustain the shorter arc is lower than a longer arc needs.

Whip-type devices are also widely used to interrupt capacitive currents. However, these devices are primarily dielectric-interrupting devices and the thermal effect of the arc will set a heretofore unrecognized limit on their use. In the past, there were no standards for testing of the devices and users mostly had to rely on catalogue information on the device capability. Such information, however, may not be based on realistic test circuits. The IEC Technical Report 62271-305 [47] describes how air-break disconnectors, both without and with auxiliary interrupting devices, should be tested for capacitive-current interrupting capability.

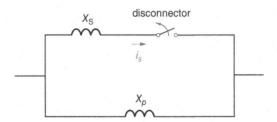

**Figure 10.17**   Basic circuit for bus-transfer switching.

## 10.3.3   Bus-Transfer Switching

Bus-transfer switching, also known as *loop switching*, is the commutation of current from one busbar to a parallel busbar. The equivalent circuit is shown in Figure 10.17.

The conducting path in parallel to the device strongly limits the voltage across the switching disconnector, allowing a much higher current to be switched than in the capacitive- or magnetizing-current case discussed in Section 10.3.2.3. Standardized values of the voltage across the open disconnector after current commutation are 100 to 330 V for air insulated disconnectors and 10 to 40 V for gas insulated disconnectors [26]. The difference is because of the much larger loop length in air-insulated stations.

Once initiated at contact separation, the arc propagates freely, achieving a long length through blade movement, thermal effects and the Lorentz force. As the arc elongates, its voltage increases, forcing the current into the parallel path, and this continues until the arc is extinguished. Extinction occurs when the arc voltage equals the initial current in the disconnector $i_S$ multiplied by the total impedance of the loop:

$$u_a = i_S(X_S + X_p) \tag{10.5}$$

The mechanism, however, is driven by the power being delivered to the arc: the arc will be sustained if the rate of change of the power supply is positive; if it becomes zero, the arc will stall and probably become unstable. If it becomes negative, the arc will collapse and extinguish. Arc duration tends to be a matter of seconds but the arc reach is limited by repeated partial arc collapses. Such collapses revive the arc, reducing its voltage and some current transfers back from the parallel path to the disconnector path.

This is illustrated by an actual laboratory measurement in Figure 10.18. The arc collapse clearly causes some of the current to shift back to the disconnector and the arc power drops because the arc voltage decrease is greater than the increase in the current. Ultimately, the rate of change of the power input becomes irreversibly negative and the final transfer of current is achieved.

There is an inverse relationship between the permissible current and the loop impedance. For transmission loops, currents up to 100 A and loop impedances of tens of ohms appear to be the industry practice. Bus transfer switching between busbars in substations is a special case in which the currents can be as high as 1600 A (the maximum standardized value) with loop impedances of less than 1 Ω. For disconnectors rated at 1100 and 1200 kV, this capability is now 80% of the rated normal current [26]. Disconnectors used for this purpose are sometimes provided with auxiliary commutating contacts or other suitable means to achieve current interruption.

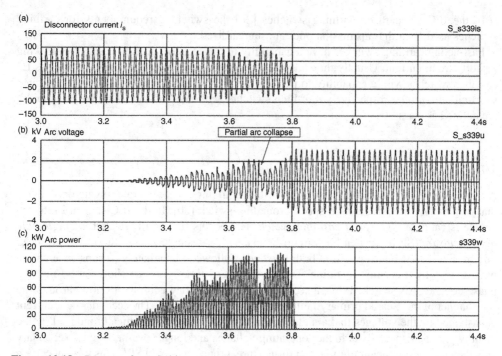

**Figure 10.18**  Bus-transfer switching process showing arc collapse. (a) Current in disconnector branch; (b) disconnector - arc voltage; (c) arc power.

## 10.4  Earthing

### 10.4.1  Earthing Switches

Earthing switches are mechanical switching devices for earthing parts of a circuit, capable of withstanding for a specified time currents under abnormal conditions such as those of short circuit, but they are not required to carry current under normal conditions of the circuit. An earthing switch may have short-circuit current making capacity.

In the case of multiple configurations of overhead transmission lines, such as tower configurations with more than one system being mounted on the same line tower, current may flow in de-energized and earthed lines as a result of capacitive and inductive coupling with adjacent energized lines. Earthing switches applied to earth these lines shall therefore be capable of performing the following switching operations:

- making and breaking of a capacitive current when the earth connection is open at one termination and earthing switching is performed at the other termination;
- making and breaking of an inductive current when the line is earthed at one termination and earthing switching is performed at the other termination;
- carrying continuously the capacitive and inductive currents.

The magnitude of the induced currents to be switched by the earthing switch depends on the capacitive and inductive coupling factors between the lines, and on the voltage, load and length of the parallel system.

In the IEC standard on earthing switches [26] the switching requirements for earthing switches used to earth transmission lines are standardized.

Fast-acting earthing switches have a capability to connect live parts to earth, thus making a short-circuit current. Usually they are spring-operated and defined as class E1 or class E2 earthing switches with a capability to perform two or five closing operations with the rated short-circuit current, respectively. Earthing switches without short-circuit making capability are defined as class E0.


## 10.4.2  High-Speed Earthing Switches

A special type of (fast-acting) earthing switch is a single-pole-operated switching device that originally was called a high-speed grounding switch [50]. In the IEC standard [26] the device is called a *high-speed earthing switch* (HSES). Its function is related to *single-pole auto-reclosure* (SPAR), when line circuit-breakers clear a single-phase fault on an overhead line by single-pole interruptions at both line ends, followed immediately (for instance within one second) by a closing operation. During the "dead time" (i.e. the short period that the phase conductor is switched off), a secondary arc current may continue to flow through the arc channel of the cleared primary fault current, see Section 6.2.4. The secondary arc current is sustained by the voltages that are electrostatically and electromagnetically induced by the healthy phases. Especially at higher operating voltages and longer overhead lines, secondary arcs may have a significant lifetime and form a threat for a successful SPAR, as it is not certain that the secondary arc current will be extinguished in due time by external means, such as elongation, wind cooling and so on. Unsuccessful SPAR will lead to three-phase interruptions at both line ends and forms a risk for load interruption and/or system instability.

Secondary arc currents up to a few tens of amperes tend to extinguish within some hundreds of milliseconds, but currents up to 100 A may last longer than one second.

A classical means to reduce the secondary arc current is the application of four-legged shunt reactors (i.e. with a neutral reactor) instead of three-limb shunt reactors. The neutral reactor is tuned to the zero sequence capacitance of the transposed overhead line circuit in such a way that the secondary current becomes minimal.

An alternative solution, especially for un-transposed lines, is to apply switched shunt reactors, where the shunt reactor in one of the healthy phases is switched off [51]. Four-legged shunt reactors and switched shunt reactors are perfect solutions for long single-circuit overhead lines, but rather expensive for short lines (without shunt reactors) and less effective for double-circuit overhead lines.

In such cases the high-speed earthing switch offers a very effective method to reduce the secondary arc current as it connects the faulty phase to earth to divert the secondary arc current from the primary fault location to the high-speed earthing switch. Usually a high-speed earthing switch is applied at both line ends, see Figure 10.19. In this figure, the upper circuit shows the undisturbed situation, or in the case of a double circuit, the undisturbed circuit. The lower circuit shows the case of a fault on the lower phase where, after the interruption of the fault current by the line breakers at either side of the fault location, the high-speed earthing switches in the lower phase close, bypass the secondary arc and extinguish it.

The order of operation is: (i) single-phase fault clearing by both line breakers followed by (ii) closing the corresponding pole of the high-speed earthing switch at both line ends within

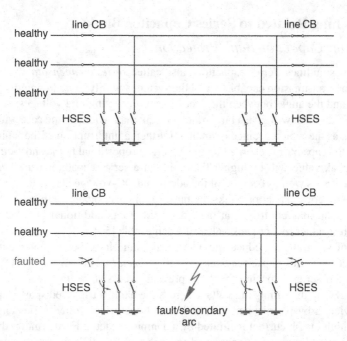

**Figure 10.19**  Principle of high-speed earthing switch (HSES) in single-phase auto-reclosure.

a few hundred milliseconds. About half a second later (iii) the high-speed earthing switches are opened and (iv) the line breakers close within the one second "dead time" of the SPAR operation.

High-speed earthing switches, as with circuit-breakers, are equipped with an arcing chamber and a fast operating mechanism. Electrically, the relevant duty is to interrupt the currents that are electrostatically and electromagnetically induced by the healthy phases and the phase currents of the parallel circuit. Most severe is the condition when a fault current is flowing though one of the previously healthy phases exactly at the moment of opening of the high-speed earthing switch. Such situations may occur during thunderstorms with lightning strokes hitting the line directly or in its vicinity. Thunderstorms with successive lightning strokes along the same path are quite normal and should be considered when specifying the duties for high-speed earthing switches.

Another speciality is the number of SPAR operations at the same time or, stated otherwise, the number of phases that may be switched off before system stability will be endangered. This is relevant with long-distance transmission, especially at 800 kV and above [52–55]. In case of three single-circuit overhead lines in parallel, one may consider to apply *three-phase auto-reclosure* (TPAR), so that secondary arcs are not an issue. However, in the case of double-circuit overhead lines SPAR will be applied, possibly with high-speed earthing switches, so that poles can be operated independently. In the latter case one gets six poles, two per phase, and multiple SPAR can be allowed at the same time, provided that at least two poles of two different phases remain closed to transport energy. Such a way of operating overhead lines is called *multi-phase auto-reclosure* (MPAR).

## 10.5   Switching Related to Series Capacitor Banks

### 10.5.1   Series Capacitor-Bank Protection

In long transmission lines, series capacitors, also called *series compensators*, are installed to increase the line transmission capability and the system stability by reducing the voltage drop along the line and the angle between the sending- and receiving-end voltages. A fault behind the series capacitor bank will give a larger fault current than would be the case without a series capacitor bank, as the total reactance is smaller. Further, at interruption of the fault current – at current zero – the capacitor is charged up to a voltage proportional to the short-circuit current. So, the TRV peak value will be higher than without a series capacitor bank because of the higher fault current as well as because of the additional DC-voltage across the series capacitor bank. Moreover, series capacitor banks, as far as applied in other overhead lines connected to the same busbar, may contribute at the busbar side to an additional DC voltage. Without countermeasures, rather higher peak values are achievable [56].

A number of solutions to reduce the TRV peak value have been implemented, most of them to protect the series capacitor bank as well. A *metal oxide varistor* (MOV) is applied in parallel to the series capacitor bank to prevent excessive voltage across the bank. In addition, a spark-gap that limits the voltage across the bank, a triggered spark gap [57], a fast protection device, a by-pass switch [58] and/or a thyristor-controlled series capacitor bank are used. The high inrush current is limited by a damping reactor. Furthermore, the TRV peak value can be limited by arresters connected to earth and by MOV across the circuit-breaker (see Section 11.3.3.2). The fast protection device is triggered either by the line protection that gives a tripping command to the line circuit-breaker or by the protection of the series bank. The series bank is thus by-passed at fault clearing and no longer influences the TRV peak value.

Examples of different configurations of series compensation banks are given in IEC 60143-1 [59].

Figure 10.20 shows a fully protected series compensation bank with a MOV, current limiting reactor, a spark gap, and a by-pass switch (see Section 10.5.2).

Arrangements without a MOV or spark gap are also possible. In some cases, though rarely found in EHV systems, only the by-pass switch has to protect the capacitor bank.

Extreme stresses arise in the case of a series capacitor-bank discharge current superimposed upon the power-frequency fault current. Arrester pressure relief tests have been reported at the

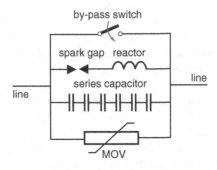

**Figure 10.20**   Basic diagram of series capacitor-bank protection circuitry.

**Figure 10.21**  By-pass switch in a series capacitor bank.

245 kV level with 31.5 kA, 60 Hz fault current, superimposed by the peak discharge current of 447 kA at 3 kHz from a series compensation bank [60].

## 10.5.2   By-Pass Switch

To protect the capacitor bank in cases of line faults, but also to adjust the level of compensation, a fast switching device, the *by-pass switch*, is needed (see Figure 10.21). The closed by-pass switch ensures a metallic short-circuit of the capacitor bank, having the MOV, the spark-gap or the faster protection devices in parallel, if any. Requirements for by-pass switches are given in IEC 62271-109 [61].

In order to reduce the effective insertion time of these protective devices, a short making time of the by-pass switch is required. Therefore, by-pass switches are normally equipped with operating mechanisms giving faster closing- than opening operations. The by-pass switch is also used to re-insert the capacitor into the transmission line after clearing the fault or for adapting the level of compensation.

### 10.5.2.1   Dielectric Stresses

The design of a by-pass switch is normally characterized by unequal air clearances to earth and across the by-pass units, because of the different insulation levels for these components (by-pass unit and pole column). The insulation level to earth depends on the rated voltage of the transmission line, which is typically above 300 kV.

The insulation level of the by-pass units has to match that of the voltage rating of the capacitor bank and the other protective devices. The rated voltage of the by-pass switch, derived from the insulation level of the capacitor bank single-phase value to a three-phase value and rounded to the next rated voltage according to IEC 62271-1 [29], is typically below 300 kV. So normally, a single-unit or smaller double-interrupter design for higher stresses is used.

### 10.5.2.2 By-Pass Current Making

By-pass switches have to switch on ('make') extremely high inrush currents associated with the series capacitor-bank discharge. Although the discharge current is limited by a series current-limiting reactor, peak values of inrush current above 100 kA are possible.

Moreover, the combination of the inrush-current frequency with the high peak value causes a considerable arcing stress during the *pre-arcing time* (arcing between closing contacts), since the inrush peak current is reached already during the pre-arcing time.

This very severe pre-arcing stress is far beyond the stress that general-purpose circuit-breakers face. The most severe making operation for general-purpose circuit-breakers is the energization of shunt capacitor banks in a back-to-back configuration (see Section 4.2.6), that is, when one capacitor bank is energized while other nearby parallel bank(s) are in service.

An appropriate design of the contact system and mechanism should prevent unacceptable contact welding and wear. The contact system must cope with the presence of forces originating from high magnetic fields, pressure rise from arcing, and frictional forces of (eroded) contacts [62]. Depending on the presence of a fault on the transmission line, the switching conditions while by-passing the capacitor bank differ in terms of the actual making current and the voltage applied to the by-pass unit. During closing at normal service condition, only the capacitor-bank-discharge current stresses the by-pass switch. The parameters of this current depend on the design of the by-pass platform equipment (capacitance and damping device). Standardized values are a peak value up to 100 kA and frequencies up to 1 kHz.

In the case of a fault on the transmission line, these currents are superimposed on the fault current from the line. Note that normally the by-pass switch will close before the line circuit-breaker can clear the fault current.

### 10.5.2.3 Series-Bank Re-Insertion

A by-pass switch has to (re-)insert a series capacitor bank under full-load conditions by interrupting the by-pass circuit, so that the load current is transferred from the by-pass branch back into the series capacitor. This is basically a load switching operation with a specific recovery voltage (RV) that has a capacitive (1 − cos) nature due to the presence of the capacitor. It differs from the *capacitive-current switching tests* defined in IEC 62271-100, clause 6.111, as the frequency of the RV is determined by a combination of power-frequency voltage and transients having (higher) frequencies, which arise from the interaction of the line inductance with the series capacitance. This results in a time-to-peak of the RV that can be shorter than the usual value of half the power-frequency period of a standard capacitive RV. To cover most 50 and 60 Hz applications with one single test-duty, the RV is standardized to have a time-to-peak of 5.6 ms and to follow a (1 − cos) curve [61]. The risk of restrike must be kept to an absolute minimum, in order to avoid restrike currents that can theoretically be even larger than the by-pass making current.

### 10.5.2.4 Technology

Caused by the high current during the closing operation and its high frequency, the energy of the pre-arc during the making operation imposes an extreme stress on the contact material of the by-pass switch. This stress is increased by physical effects, for example, skin effect, leading to a high current density in the surface area of the contacts. To reduce the wear of the contacts, three different measures can be taken:

- reduction of the pre-arcing time;
- optimization of the current distribution;
- improvement of the wear resistance of the contacts.

The pre-arcing time can be shortened by reducing the dielectric field to achieve smaller gap distances at the moment of the first discharge through the gap and/or by increasing the closing speed of the by-pass switch. To implement this, an insulating-nozzle by-pass switch with a double-motion arcing contact system can be used. By an additional motion of the contact pin, the relative closing speed of the arcing contacts is distinctly increased, compared with the speed of the main contacts.

The first discharge occurs between the contact pin and the contact tulip. The incipient pre-arc burns until the closing of this contact gap. Due to the small dimensions of the arcing contacts, the current density in the contact surface area increases highly, making an immediate closing of the main contacts necessary.

## 10.6   Switching Leading to Ferroresonance

In a series connection of a capacitance, an inductance and a small resistance series resonance may occur with a frequency that is mainly determined by the capacitance and the inductance. With excitation frequencies of the order of magnitude of the resonance frequency, and especially at resonance frequency, large voltages will appear across the capacitance and the inductance. The voltage phasors are of opposite direction and cancel out. If the magnetic circuit of the inductance is of ferromagnetic material, the high voltage will lead to saturation with considerable magnetizing currents. A consequence of the saturation is the distortion of the power frequency voltage and current. Moreover the non-linear characteristic of the ferromagnetic circuit introduces a wide range of possible resonance frequencies. Resonance phenomena influenced by ferromagnetic material are called *ferroresonance*. Care should be taken to avoid ferroresonance, as large overvoltages and large overcurrents may lead to damage to equipment and disturbances of the power system. As the best known method to avoid ferroresonance conditions, detuning of the series capacitance-inductance circuit is considered as well as additional damping in the primary or secondary circuits.

The quintessence though, is to recognize conditions that may lead to ferroresonance, as it is a very complex phenomenon [63]. Due to the non-linear characteristics ferroresonance may appear in different steady states. It is strongly dependent on the system parameters and very sensitive to the initial conditions [64]. The initial condition is the reason why switching can trigger it, although ferroresonance may also be initiated by other changes in the system (fault initiation, voltage jump, harmonics, etc.).

Some examples of system conditions that may lead to ferroresonance are the following:

- Voltage transformers (VTs) that are connected to a power frequency voltage source (for instance a busbar) through the grading capacitors of an open circuit-breaker. Problems can be avoided by an open disconnector, by a large capacitance in parallel to the VT (for instance an overhead line) or by a damping resistance at the secondary side of the VT.

- Voltage transformers connected to a non-effectively earthed neutral system with a certain zero sequence capacitance. Mitigation can be achieved by enlarging the zero sequence capacitance (adding cables and/or capacitors).
- Stuck poles of a three-phase circuit-breaker when (de)energizing transformers or shunt reactors. At single-pole or two-pole energization the zero-sequence capacitance and/or inter-phase capacitance (depending on the neutral treatment of the transformer) plays an important role. This situation can be avoided by detecting pole discrepancy followed by automatic switching of the other poles or other circuit-breakers. The risk of ferroresonance is limited when the transformers are loaded.
  Similar problems occur with stuck poles of switches or blown fuses in one or two phases [65].
- Single pole auto-reclosure of lines that are equipped with shunt reactors or that are directly connected to transformers. The risk of ferroresonance is limited when the magnetizing characteristic of the shunt reactors is linear up to 1.3 p.u. or more and when the transformers are loaded.
- Single-phase faults in non-effectively earthed neutral systems where the neutral voltage is capacitively coupled with the secondary side and its voltage transformers. A sustained neutral voltage after fault clearing enlarges the ferroresonance risk. Damping devices in the secondary side neutral and/or at the voltage transformer secondary sides will limit the risk of sustained ferroresonance.

It should be recognized that overvoltages due to ferroresonance may also occur at the healthy phases [66].

It is beyond the scope of this book to elaborate ferroresonance in more depth and to explain the physics of countermeasures in both the primary and secondary plants and protection systems. Reference is made to the literature on ferroresonance [67–69].

## 10.7 Fault-Current Interruption Near Shunt Capacitor Banks

Though not belonging to the capacitive current interruption duty, the presence of capacitance, and especially capacitor banks, close to a fault location has an impact on fault-current interruption that should not be neglected.

The general effect of capacitance on the TRV is a reduction of its rate-of-rise because the frequency of TRV is reduced in the simple approach in which the equivalent TRV frequency $f_{TRV}$ is:

$$f_{TRV} = \frac{1}{2\pi}\sqrt{\frac{1}{L_S C_p}} \qquad L_S = \frac{U_r}{\sqrt{3}\omega I_{sc}} \qquad (10.6)$$

This influence is sometimes used to reduce the RRRV, for example, if the RRRV in service exceeds the tested values of the RRRV, for example when current-limiting series reactors increase the RRRV to an unacceptable value, see Section 3.5.

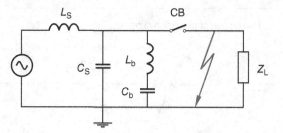

**Figure 10.22**  Basic circuit for TRV calculation in the case of a fault near a capacitor bank. $Z_L$ – load impedance; $L_b$ – inrush current limiting reactor; $C_b$ – capacitor bank $C_S \ll C_b$.

In the case of a short circuit close to a capacitor bank, the resulting TRV is strongly influenced by the capacitance ($C_b$) of the bank. The basic circuit is shown in Figure 10.22.

A comparison is shown in Figure 10.23, where a standardized TRV after fault-current interruption (in a 150 kV system with $C_S = 5$ nF of stray capacitance) is compared with the TRV after fault-current interruption near a capacitor bank with $C_b = 10$ μF. In this example, the time to reach the first TRV peak is increased from 20 to 820 μs. This is still a small value compared with the case of capacitive current interruption, where the recovery-voltage peak is reached after 10 ms at 50 Hz. Due to the mitigating effect of the capacitor bank, the arcing time can be small with a considerable risk of re-ignition due to the small contact gap while

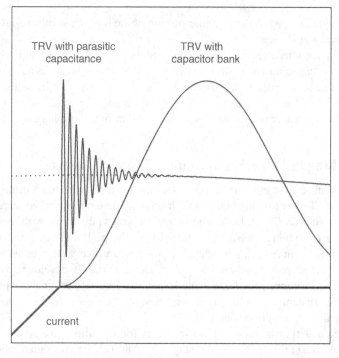

**Figure 10.23**  TRV with parasitic capacitance and a capacitor bank.

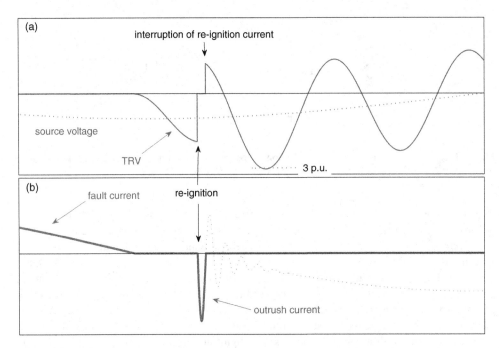

**Figure 10.24** Re-ignition after fault current interruption (a) and outrush current (b).

TRV develops. Studies have shown that interruption of the high-frequency re-ignition current can lead to severe overvoltages [70].

The electrical phenomena are demonstrated in Figure 10.24. As a result of re-ignition of the circuit-breaker, the capacitor bank contributes to the re-ignition current as an *outrush current*. This current can be interrupted by circuit-breakers able to interrupt high values of d$i$/d$t$, such as vacuum circuit-breakers and sometimes $SF_6$ circuit-breakers. Because of trapped charge in the capacitor bank, TRV can rise to very high values. Additional damping can reduce this [71].

## 10.8 Switching in Ultra-High-Voltage (UHV) Systems

The increasing demand for electric energy goes hand in hand with the worldwide need for sustainable power. This is true in countries with growing economies and an increasing use of electrical energy, such as China, India, and Brazil, as well as countries with more developed economies and a continuing transfer to electricity as the preferred energy source, such as Japan, Europe, and North America. Sustainable power sources as well as conventional power plants, including hydro-power and nuclear plants, are increasingly developed at locations far from load centers. Moreover, the large scale implementation of sustainable power plants at distributed and/or distant locations requires a strong interconnecting infrastructure for the exchange of the fluctuating power output.

Long-distance bulk transmission of electric energy forces utilities in a number of countries to implement electrical grids with a rated voltage above 800 kV for AC transmission and above 600 kV for DC transmission. According to IEC terminology, the rated voltage is the maximum continuous system voltage that is specified in discrete values. Therefore, systems with a rated voltage of 800 kV include systems designated as 735, 765, and 787 kV. These voltages belong to what is called the extra-high voltage (EHV) class. The ultra-high voltage (UHV) class is used for systems with a rated AC voltage above 800 kV, in practice 1100 and 1200 kV.

All these voltages are typical for long-distance bulk transmission of electric energy. The transmission corridors consist of two or three single-circuit overhead lines or double-circuit overhead lines. The transmission capacity is typically 6000 to 7000 MVA per circuit. The power-generation sites are 10 000 to 15 000 MW or even more. Three-phase transformer groups show for instance a capacity of 3000 to 4500 MVA.

### 10.8.1   Insulation Levels

For spatial, aesthetical, and economic reasons, the dimensions of the UHV infrastructure have to be kept as small as possible. Special measures have been taken to reduce as much as possible the transient switching surges and the effects of lightning strokes, while maintaining the highest reliability standards of the bulk-power transmission lines. By the application of modern surge arresters and very accurate calculation and simulation techniques, insulation coordination can be optimized to reduce insulation levels as well as margins.

The technical specification for systems of 800 kV level, which have been in use for over 45 years [72], is based on extrapolation from 420 and 550 kV levels. On the other hand, in the case of UHV systems, the new specifications are to a large extent based on up-to-date analysis techniques, taking advantage of possibilities to reduce transient and temporary overvoltages. This has resulted in recommended insulation levels for UHV equipment not far above those for 800 kV.

### 10.8.2   UHV System Characteristics Related to Switching

Because of the high voltage and the large transmission capacity, heavy multi-conductor bundles are applied for UHV, leading to less damping of travelling waves in comparison with lower voltages. Moreover, the radial network topology of the UHV systems leads to simple reflection patterns of the travelling waves and thus less damping due to refraction and distortion. Therefore, the amplitude factor $k_{af}$ (see Section 13.1.2), used to determine the TRV peak value $u_c$, will be relatively high. However, thehe first-pole-to-clear factor $k_{pp}$ (see Section 3.3.2.2), will be relatively low as a large part of the short-circuit current will be supplied through large power transformers with a low $X_0/X_1$ ratio. At the same time, it should be considered that in real service conditions, the effective *switching-impulse protection level* (SIPL, see Section 11.3.3.2) of the applied metal oxide surge arrester) will clip the peak value $u_c$ to values only marginally above SIPL.

The more compact designs of 800 kV and UHV lines lead to a lower inductance per km, a higher capacitance, and a lower surge impedance in comparison with line designs that are extrapolated from the lower rated voltages. At the higher frequencies, tens of kilohertz, applicable with travelling waves, surge impedances are somewhat lower than those at power frequency. Furthermore, the surge impedances of contracted bundle conductors are much higher than those of non-contracted or normal bundle conductors. For transient phenomena, however, bundle contraction can be neglected due to the stiffness of the UHV bundles and the short fault-clearing times. Duration of the bundle contraction from (90 to 124) ms have been measured in Japanese UHV studies with 41 to 53 kA of fault current [73].

From studies on the surge impedance of faulted overhead lines in Japan, Italy, and South Africa it appears that for the last pole-to-clear, the equivalent surge impedance of uncontracted lines is close to 300 Ω against up to 430 Ω in the contracted state [73].

IEC standards specify the following rated values of the equivalent surge impedance for the last-clearing pole for short-line faults:

- 450 Ω in case of rated voltages up to and including 800 kV; and
- 330 Ω for UHV systems.

The initial part of the TRV is strongly dependent on the equivalent surge impedance. The steepest TRVs come from the travelling waves excited by the clearing of short-line faults, where the rate-of-rise of TRV (RRRV) is determined by the line surge impedance and the short-circuit current (see Section 3.6.1). Due to capacitance between line and circuit-breaker, the steep TRV appears after a time delay (see Section 3.6.1.3), standardized to be represented in test circuits with an inherent value of 0.5 μs. With circuit-breakers applied in air-insulated substations, which are very large for UHV, travelling waves are generated within the substation busbars. This phenomenon is known as the initial transient recovery voltage (ITRV) (see Section 3.6.1.2). These travelling waves lead immediately to a steep rise in the TRV for the first 1.5 μs (after 1 to 2 μs). Negatively reflected travelling waves reduce the initially steep rise of the TRV. Typical values of the busbar surge impedance are 300 Ω (UHV) and 325 Ω (800 kV).

The surge impedance of the busbar in gas-insulated substations and hybrid-insulated substations is much lower, typically 90 Ω. Moreover, the lengths are much shorter and the capacitance of equipment, like bushings, is much higher. Therefore, the ITRV is not a relevant problem in such substations and can be neglected.

At rated voltages up to and including 800 kV, the circuit-breaker capability to cope with ITRV phenomena has to be verified by the short-line-fault test (with surge impedance 450 Ω and current 90% of the rated short-circuit breaking current) without a substantial time delay (delay time < 0.1 μs). Since at UHV levels, the surge impedance of an overhead line is close to that of a busbar, ITRV phenomena are only regarded as covered when the short-line-fault test has been performed with line surge impedance of 450 Ω. This should normally not be a problem for the two-chamber UHV circuit-breakers, each chamber being capable of clearing short-line faults at 550 kV.

A particularity for the large UHV substations – in AIS conductor lengths of several hundreds of meters – is the probability of a fault at some distance from the circuit-breaker. In that case, significant initial TRV will occur at both sides of the UHV circuit-breaker, leading to a doubling of the initial TRV across the circuit-breaker. Usually, such substations are designed with double-breaker or breaker-and-a-half schemes [74], two circuit-breakers in the substation have to clear the fault current simultaneously. Ultimately, one of them will be the last clearing one. By a short-line fault type test with 450 Ω and 90% of the rated short-circuit breaking current, the double-sided initial TRV, corresponding to a surge impedance of 2 × 300 Ω, is covered to up to 70% of the rated short-circuit current. Users of UHV AIS substations have to study carefully whether the double-sided ITRV has to be considered, as well as the surge impedance of the busbars and the maximum expected fault current level in the substation.

The heavy bundle conductors applied in UHV systems lead to DC time constants of the short-circuit current larger than the alternative value of 75 ms given in the IEC standard [13] for a rated voltage of 550 kV and above. For UHV systems, the DC time constant values between 100 and 150 ms have been reported [73]. The effect of the difference of a DC-time constant between 120 and 150 ms is negligible for the peak value of the fault current and for the major loop of the asymmetrical current prior to current zero. Therefore, IEC, for UHV systems, specifies the DC time constant as 120 ms.

A number of optimization techniques, as proposed for UHV, can be considered also for 800 kV – sometimes even for lower rated voltages. In particular, most characteristics and parameters that are not related to the length of busbars or bay connections show comparable values for UHV and 800 kV: surge impedances of lines and busbars in AIS, first-pole-to-clear and amplitude factors $k_{pp}$ and $k_{af}$, impact and optimization of surge arresters, long-line-fault conditions (see Section 3.6.2), DC components in short-circuit currents, capacitive-current-switching TRVs and inductive-load-current switching characteristics. Other parameters show similar tendencies but due to the differences in dimensions they differ essentially from UHV parameters. Examples are the time to ITRV peak $t_i$ or the surge capacitance of power transformers, which is an important parameter for transformer-limited faults.

On the other hand, since service experience with 800 kV systems is much greaterer than with UHV, valuable information from 800 kV has been used to adapt the standards for UHV. This has resulted in the inclusion of the UHV voltages 1100 and 1200 kV in the latest circuit-breaker standard [13].

## 10.9 High-Voltage AC Cable System Characteristics

### 10.9.1 Background

The application of HV cables in AC power grids is popular for extension of existing grids. People who live near overhead lines are reluctant to accept the construction of new HV overhead lines. The appearance of HV towers and lines in landscapes may have a large visual and ecological impact. Moreover, health concerns exist related to continuous exposure to low-frequency electromagnetic fields from HV overhead lines.

The highest electric-field strength occurs at the conductor surface of the line. The electric-field strength directly below 400 kV overhead lines is between 7 and 10 kV m$^{-1}$. At a distance of 25 m from the overhead line, the field strength can be expected to be less than 5 kV m$^{-1}$, which is an admissible level for human exposure. In cable systems, there exists only an electric field between the inner conductor and the screen but not outside the cable, since the screen conductor of the cable is usually connected to earth potential. In this way, the electric field is enclosed between the cable-core conductor and the screen. The magnetic-field strength around current-carrying cables is higher than with overhead lines, but decreases more rapidly with distance. There exist exposure limits for the magnetic-flux density outside the right-of-way of new overhead lines: 3 μT in Italy, 1 μT in Switzerland, and 0.4 μT in the Netherlands [75].

### 10.9.2 Current Situation

Several transmission utilities around the world are now studying the possibility of application of EHV AC cable lines in their power grids [76]. In Denmark an underground cabling is planned of the existing 132 and 150 kV grids by 2030. Cables can be applied to extend the grid or to replace existing overhead lines. Cables are already widely used for the lower (distribution) voltage levels. In 2005, a total length of almost 33 000 km of AC land cables was in service [77]. At voltage levels below 220 kV, more than 90% of the cables installed between 2001 and 2005 were of the XLPE insulated type. Above 220 kV, more than 40% are still of the *self-contained oil-filled* (SCOF) type.

Currently, network planners and system operators do not have much experience with the behaviour of power systems in which long EHV AC cables are integrated. Currently (2014), the longest EHV cable connection is in use in Japan; four circuits in parallel, each with a length of 40 km at 500 kV [78]. In Denmark, a 150 kV connection to an offshore wind farm has been installed and has a length of 100 km [77].

The application of cables in power grids and mixed line–cable–line sections will influence the behaviour of the power system as such in several aspects. The steady-state and transient response of cables for both small and large disturbances differ significantly from those of overhead lines. This has implications for the operation of power systems with cables in terms of grid stability and security of supply. Therefore, all steady-state and transient phenomena should be studied before the integration of cables is realized in practice.

From the installation point of view, HV cables longer than 1 to 2 km require cross-bonding schemes to reduce induced sheath currents. For shorter lengths, single-point bonding is generally applied. From the electrical point of view, overhead lines cannot be simply replaced by cables. The electrical behaviour of cables differs from that of overhead lines. An overhead line is a transmission line surrounded by air, has no other insulation than the air spacing, and has a dominant inductive behaviour. A cable consists of an inner and a sheath conductor with an insulating material with semiconductor layers in between and is, therefore, predominantly capacitive. HV-cable capacitance and inductance per unit length differ from the values of overhead lines. The series inductance of a cable is five times smaller and the capacitance to earth is 20 times larger than those of an overhead line. As a result, the surge impedance of cables is ten times smaller and the travelling wave velocity is about two times smaller. Thus, the cable surge-impedance value is between 30 and 70 $\Omega$. This means that the *surge impedance loading* (SIL) for a cable is several times larger as determined by the surge impedance and the voltage applied. Hence, a cable loaded below its SIL behaves like a capacitor to earth. A cable loaded above its SIL behaves as a series reactor. A large SIL will exceed the cable *ampacity* (i.e. normal current-carrying capacity), thus resulting in capacitive behaviour. When the cable load equals its SIL, there is not net reactive-power flow, resulting in a flat voltage profile along the cable.

A capacitance can be seen as a source of reactive power. Therefore, an energized cable injects reactive power into the grid. The magnitude of the capacitive current of a cable depends on the applied voltage and on the capacitance per unit length multiplied by the cable length. When the cable ampacity is consumed completely by the capacitive current, no active power can be transferred through the cable and the critical cable length is reached. Capacitive current is therefore the major limitation in the application of AC cables for long distances. The definition of the *maximum operable length at thermal limit* (MOLTL) is used [79–81] as a steady-state operational design criterion of cables. The large capacitive charging current also has consequences for the cable life time. Deterioration of cables caused by their intrinsic current is an important issue when using long EHV AC cables. HV equipment is dominantly inductive and, therefore, requires reactive power. Reactive power is indispensable for operating devices that make magnetic fields to perform their function, like transformers and motors. However, reactive power can only be transmitted along a short distance and needs to be injected and consumed locally. When long cables are applied in the system there can be an unbalance between the locations of produced and of required reactive power.

Reactive power surplus in any operating condition causes power-frequency voltage rise at the cable terminations and adjacent nodes in the grid. A step voltage change of 3% is usually allowed while connecting or disconnecting a cable [82]. The allowable voltage step change during switching of cables is prescribed in the grid code of the grid operator. To keep these stationary overvoltages below an acceptable level, compensation of reactive power is needed. Normally, cable connections longer than 30 km require compensation and this can be achieved by fixed or variable compensation using shunt reactors. Shunt reactors are usually installed at both cable ends or at the transformer tertiary winding.

Switching shunt reactors may result in an oscillating interchange of reactive power between system inductances and cable capacitance. These oscillations, superimposed to the system's natural frequency, can also lead to temporary voltage rise in the grid. Switching actions in shunt-compensated mixed line–cable–line connections may also result in overvoltages. In publication [83], simulations of de-energization operations were performed on a Danish 400 kV cable operating system with a length of 90 km. An overvoltage of 132% was shown compared with the voltage before switching. This overvoltage was caused by resonances between the cable and the shunt reactor. Another Danish study for a planned 60 kV cable of 18.5 km has shown no remarkable overvoltages after switching off [84]. Such cases are also reported in [56].

Compensation causes line resonances in situations when there is capacitive coupling between a disconnected phase and the phases remaining in service. When there is no capacitive coupling present between single-core insulated cables, the risk of line resonances is greatly reduced and the compensation rate can be 100%. The degree of compensation has an influence on the cable charging current and on the voltage profile along the cable. A study for a 400 kV cable system has shown that two shunt reactors placed at both cable ends reduce the charging current at both ends to half of the largest value.

System studies for long cables recommend installing shunt reactors with a distance of 15 to 40 km between them [85]. The influence of shunt compensation on the voltage profile along a 400 kV cable connection was investigated in a feasibility study [86]. Power-flow calculations were performed under no-load conditions when one cable end was opened while the voltage at the other end was set to 415 kV. The lowest compensation rate considered was 93.3% and it turned out that the voltage along the cable remained below 420 kV. For the highest (over)compensation rate of 111.9%, it was shown that the voltage at all locations along the line was below 415 kV.

In another study, carried out by CIGRE Working Group B1.05 [87], a cable end termination was placed in the cable-overhead-line joint. This termination was a linear variation of surge impedance from 90 to 30 $\Omega$. It was shown that an incoming wave with 2 µs front rise time raised the cable voltage by 2%, and for a 1 µs front rise time, the increase was 1%. The velocity of voltage and current waves along a cable is about 50% of the velocity of overhead lines and the waves propagate between the core and sheath.

The behaviour of mixed line–cable–line systems under lightning events is of importance in power-system transient studies. Lightning strokes can cause very high overvoltages, which can affect equipment insulation, so overvoltage protection like surge arresters may be applied.

The results of a study on a mixed overhead–cable 380 kV system have shown that shielding failure does not represent a critical event [88]. Furthermore, it was shown that installation of surge arresters at both cable ends reduces the tower-foot voltage by 10 to 15%.

Another issue related to shunt-compensated cables is the missing-zero phenomenon that normally occurs in shunt-compensated cables during the energization process of shunt-compensated cables. The decay of the current DC component depends on the cable- and reactors-resistances, and is generally small. This results in the current missing zero for several cycles, meaning that the interruption of current will be delayed. Therefore, countermeasures may be taken to avoid the failure of circuit-breakers. In practice, some methods have been proposed to minimize missing-zero phenomena by using a pre-insertion resistor [56, 89, 90].

Cables and overhead lines can create parallel paths. The lower impedance of cables can lead to an inequality in power flow, which could even result in overloading the cable connections. This implies that, during steady-state operation, compensation is needed in two different ways: for reactive-power surplus and for the difference in impedance to control power flows. The higher capacitance of a cable also has consequences for the system resonance frequencies, which are considerably lower. Installed long HV cables compensated by shunt reactors form a parallel-resonance circuit consisting of the cable capacitance and the shunt-reactor inductance. Series-resonance circuits are formed by the cable capacitance and the transformer leakage inductance. Both types of resonances can lead to temporary overvoltages.

Apart from affecting the steady-state operation, there are also implications for transient situations when using cables. In general, transients cause slow-front, fast-front, and very-fast-front overvoltages. Slow-front overvoltages occur for instance in the case of cable (de)-energization, line switching and fault clearing. Fast-front surges caused by lightning-induced current injections in overhead lines can lead to high overvoltages. Moreover, a restrike after switching a shunt reactor may cause overvoltages with a risk of ferroresonance (see Section 10.6).

Surge impedances of cables, cable joints and overhead lines differ from each other, meaning that there are impedance mismatches at line–cable–line transition points. Reflections of travelling-voltage and current waves will occur at junctions and may result in overvoltages. The surge impedance of cables and overhead lines are frequency dependent, which means that reflection coefficients at junctions are also frequency dependent. At high frequencies, reflections can lead to doubling of the voltage at those junctions. This can also happen at a junction formed by a transformer and cable, resulting in overstress of transformer insulation, which leads to accelerated ageing of the insulation.

Simulation studies of transients in cables are of great importance when studying cables in power grids. Accurate simulations of transients require a detailed cable model. Transients in power systems can result in steep voltage wave fronts. These steep voltages contain high-frequency oscillations. For a cable model in transient studies, the frequency dependence of cable parameters must be taken into account to obtain accurate simulation results. The modelling of a cable and the calculation of parameters for transient simulations is a rather complicated and delicate task, especially regarding the frequency dependence of the different parameters and the influence of the earth return path. The parameters need to be found for the different cable layers, the surrounding conditions, and the skin- and proximity effects. The total series impedance and shunt admittance of a cable has to be determined as well. The admittance is formed by insulating and semiconducting layers, and the material properties of the admittances are modelled by complex permittivities.

## 10.10   Switching in DC Systems

### 10.10.1   Introduction

Interruption of DC current is basically impossible because of the absence of a current zero. Therefore, all interruption strategies in DC systems that have been developed in the past are based on creating a current zero, either by the switching device itself, or by an auxiliary circuit [91].

An additional challenge to DC current interruption is the absorption of the energy stored in the circuit inductance. In an AC circuit this energy is equal to zero at current zero because the main circuit current is zero. In the DC interruption, the current at the circuit-breaker's location can be (forced to) zero, but the main circuit current still has a significant value. Therefore, at current interruption in DC systems, there is still a considerable energy stored in the system. DC breakers need an integrated provision to absorb this energy.

DC circuit-breakers have to act faster than equivalent AC breakers because, in many cases, the rise time of the short-circuit current (e.g. the discharge current of HV DC capacitance) is in the range of few milliseconds.

Another difference with respect to AC circuit-breakers is the interaction with the network. In the case of AC breakers, it is the network that determines the (transient-) recovery voltage that stresses the breaker. In DC systems it is the interrupting device itself that will determine the recovery voltage, rather independent from the system in which it is embedded.

The preferred strategy of interruption of a DC circuit is different for low- and medium voltage systems and high-voltage systems.

### 10.10.2   Low- and Medium Voltage DC Interruption

Low-voltage DC systems (roughly below 1.2 kV) are mostly applied for public and for mine traction, with various types of drives and converter systems. In these systems the method of arc elongation can be used, as explained in Section 6.2.2. By increasing the arc voltage to a value exceeding the source voltage, the arc current is forced to zero and the current can be interrupted (see Figure 6.7). The energy absorption is done by the arc.

In medium voltage DC systems in the range 1.2 to 3 kV, the method of arc elongation can also be used, but technically complicated measures, as shown in Section 6.2.3, are necessary to create the high arc voltage.

As an alternative, semiconductor based circuit-breakers can be applied that are equipped with auxiliary protection devices.

Mechanical and solid-state breakers have their positive and negative features, as summarized in Table 10.1 [92].

A combination of semiconductor devices with a mechanical breaker is called a *hybrid switching device*. The positive features of each technology are kept but without the drawbacks.

In a hybrid switching device, two different mechanical switches are combined: a main switch and an isolating disconnector, see Figure 10.25. The main switch has a mechanical contact set to provide very low losses when the device is in the closed position whereas the isolation disconnector provides the dielectric withstand capability after current interruption.

A commutation circuit is connected in parallel to the main switch. When the main switch opens and an arc appears, the arc voltage forces the current into the commutation circuit.

**Table 10.1**   Comparison of mechanical and semiconductor breaker features

| Feature | Mechanical breaker | Semiconductor breaker |
|---|---|---|
| Switching mechanism | metallic contact and arc | PN-junction |
| Contact resistance | $\mu\Omega$ to m$\Omega$ | few m$\Omega$ |
| Power loss | very small | relatively high |
| Voltage drop at normal current | less than 10 mV | 1 to 2 V |
| Galvanic isolation | yes | no |
| Isolation capability | very high | limited (sensitive for overvoltages) |
| Overload capacity | very high | limited by $I^2t$ value |
| Delay/response time | few ms to 20 ms | few $\mu$s |
| Life expectancy | limited by contact erosion | theoretically unlimited |
| Contact reliability | high | very high |
| Frequent switching ability | high | very high |
| Surge capabilities | high | limited (device-dependent) |
| Overvoltage protection | not necessary | snubber circuit/varistor |
| Size and volume | compact and small | relatively big due to cooling being necessary |
| Maintenance | necessary | not necessary |
| Cost | relatively low | relatively high |

The semiconductor switch operates and interrupts the current. A snubber circuit serves as a suppressor of transients for the protection of the semiconductor switch and a voltage limiting element absorbs the energy. During normal operation, when the breaker is in its closed position, the commutation circuit forms a very high impedance and only becomes active during the process of current interruption.

In addition to arc elongation and hybrid switching, active commutation is applied for DC current interruption at medium voltage and only mechanical switching devices are used, in general vacuum circuit-breakers [92]. A separate auxiliary circuit, consisting of a making switch, a pre-charged capacitor and an inductor in series, injects an oscillating current, with an amplitude higher than the DC current, into the DC current arc. The sum of the injected current and the DC current has to reach zero in order to extinguish the arc. As with hybrid breakers, energy absorption and voltage limitation is necessary. This is accomplished with

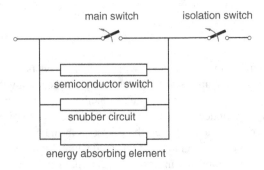

**Figure 10.25**   Schematic lay-out of a hybrid switching device.

parallel-connected snubbers and/or energy-absorbing elements. Very fast opening mechanical breakers must be applied in order to interrupt the fault current in the earliest possible stage of its development. This is challenging, since the short-circuit current rises in a very short time period [93, 94].

Topologies for full solid-state switching devices have also been proposed which make use of thyristors and/or *insulated-gate bipolar transistors* (IGBTs) [95].

## 10.10.3   High-Voltage DC Interruption

High-voltage DC systems are applied for transmission or used as back-to-back connection of AC systems. At present, the vast majority of HV DC connections have a point-to-point (single line) configuration. In the case of a fault on the DC line, the converter control is used to de-energize the faulted line, or an AC circuit-breaker at the AC side of the converter is operated. In the near future, however, multi-terminal, or meshed, HV DC systems are foreseen. In such grids, de-energization of a converter would also de-energize healthy DC connections. Therefore, also in HV DC grids, fault interrupting devices are essential, but are not available yet [96]. This is a major obstacle for the development of meshed DC grids.

Two technologies in HV DC transmission are used; they differ in the conversion technology [97]:

- *Current source conversion* (CSC) is a type of *line commutation conversion* (LCC). In this technology, thyristors [98] are used as switching elements in the converters and because thyristors can only be switched off when the conducted current passes zero, it requires the presence of line voltage for commutation. Large DC reactors realize a stiff current source. Considerable reactive power has to be provided at the terminals and large filters at the AC side of the converter have to attenuate the effects of low-order harmonics. In the case of a fault, the rate-of-rise of the DC current is limited by the large series reactor.

  The CSC technology is mature and has seen a continuous development since the 1950s. It has been employed up to ±800 kV for the transmission of power exceeding 7000 MW and is under development for ±1100 kV in China, with power capability over 10 000 MW.

- *Voltage source conversion* (VSC) is a more recent technology for lower voltage and power levels (typically up to ±320 kV and 800 MW). It is based on IGBTs as the semiconductor switching element [99] in multi-level *pulse-width modulation* (PWM) switching mode combined with large parallel DC capacitors to maintain a stiff voltage source. The key advantages of VSC with respect to CSC conversion are the possibility to control active and reactive power separately and to provide a stable voltage support. Compared with the CSC technology, the losses are higher, but because of the high switching frequency of the IGBTs the generation of harmonics is less and there is no need for large filters. This reduces overvoltages in the case of a fault. With a DC fault, the fault current is limited by the AC system and the rate-of-rise of the fault current is high compared with CSC-based systems. The main contribution to the DC fault current comes from the discharge of the DC capacitor and the cable capacitance [100]. In the case of a fault, there is no control of the converter as in the CSC technology and a breaker at the AC side must operate. As such, a VSC system is more vulnerable to a DC fault than a CSC system.

  Nevertheless, for meshed DC grid development, VSC is a very promising technology.

Current interruption in HV DC systems was studied intensively in the past [101], but no commercial product or even a unanimously accepted concept has yet emerged [96].

The principal candidate-technologies for HV DC current interruption are:

- Active commutation. This concept is similar to that applied in some medium voltage DC applications. An oscillating current from a pre-charged capacitor bank in a series $LC$ circuit is injected into an arcing gap to create a current zero and enable interruption. From that instant, the circuit current first commutates in the $LC$ branch, which causes the voltage to rise with a rate-of-rise $du/dt = I/C$ till a voltage level is reached that leads to commutation in a parallel-connected voltage-limiting element (usually a surge arrester) that is able to absorb the energy in the circuit [102]. Unlike for medium voltage, vacuum circuit-breakers cannot easily be applied in high-voltage, and keeping a large capacitor(-bank) charged continuously is a costly affair.
- Passive commutation. With this concept also, a series $LC$ circuit is connected in parallel to a conventional circuit-breaker. When arcing starts in the mechanical circuit-breaker, an oscillation is created in the circuit formed by the breaker and the $LC$ circuit.

  Arcs in gases have a negative current–voltage characteristic, see Section 3.6.1.6 Figure 3.23, resulting in a negative dynamic arc resistance $du/di$. As a result, any instability in the arc, like partial arc voltage collapse as shown in Figure 3.22, excites an oscillation with negative damping and increasing amplitude. This follows directly from the dynamic arc stability theory [103] that is also applied successfully to current chopping [23] and DC switching in airblast- [104] and $SF_6$ circuit-breakers [105]. After a certain time the increasing oscillation creates a current zero. As with active commutation, current commutates first into the $LC$ branch and soon thereafter into a parallel surge arrester.

  This interruption principle is applied in the HV DC system in various switching devices for load- and load-transfer-current switching but also for fault current interruption, for example in *metallic-return transfer breakers* (MRTBs). These devices have to eliminate neutral-to-earth faults by shunting a faulted section of the system and then interrupting the fault current [106].

  The principle of self-excited oscillation with increasing amplitude is demonstrated in Figure 10.26 in a 25 kV circuit, interrupting a 16.7 Hz current of 1 kA in an oscillatory manner far before its natural zero crossing [107]. The circuit-breaker used was a standard 245 kV $SF_6$ AC breaker.
- Hybrid technology. This technology, as explained in the text along with Figure 10.25, has been upscaled from medium- to high-voltage for a HV DC circuit-breaker that can interrupt 9 kA in a 320 kV system that consists of four modules in series, each having 20 IGBTs rated at 4.5 kV each [108]. In this particular design, the commutation from the main branch into the parallel branch is assisted by a semiconductor load break switch in series with the fast operating mechanical main switch.

## 10.11  Distributed Generation and Switching Transients

### 10.11.1  General Considerations

Distribution networks are currently changing from passive grids to active grids. In passive grids, the power flows from transformers down to the load, whereas the active grids are

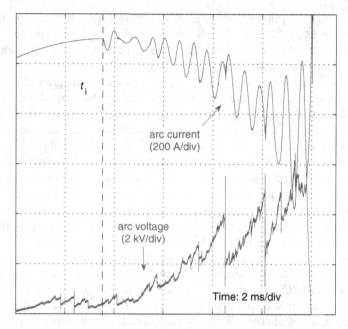

**Figure 10.26**   Arc current and voltage during current interruption by arc oscillations with increasing amplitude.

characterized by distributed generation and possible reverse power flow [109]. The design and operation of such distribution networks become more complicated and tend to be comparable to the transmission grids. The consequences of these developments are quite clear for the secondary equipment: more metering, more signalling, more remote control, fast and advanced protection, re-closure and automatic system restoration functions, extended telecommunication and dynamic loading. Apart from that, there are special requirements in active grids, such as a fault-ride-through capability of equipment, synchro-check and synchronization functions, system separation and islanding possibilities, voltage and frequency control, spinning reserve and load rejection.

Distribution circuit-breakers have to cope with these conditions and specifications of the circuit-breakers have to be assessed with respect to the prospective short-circuit current (AC and DC component, asymmetrical current peak, etc.) and the load current. Nowadays, just like in transmission networks, the operators of distribution networks have to cope with incomplete information about the development and operation of power plants. The lack of information and the uncertainty on information collected makes it more difficult to predict future load and fault current levels. Therefore, larger margins have to be applied in the specification of HV equipment. Flow patterns may change quite rapidly and protection systems have to cope with varying system conditions.

Care should also be taken for deviating operating voltages, harmonics, and non-standard power frequencies, especially under islanding conditions. If filter-bank switching is required, the main problem is the possibility of a higher peak value of the TRV than specified for capacitor-bank switching. Circuit-breakers have to be specified and tested for such special

TRV-waveforms, that is, shunt capacitor-bank switching with a higher capacitive voltage factor (see Section 13.2). Of course, in that case, the circuit-breakers are required to have a very low probability of restrike and extended mechanical endurance. This applies also to filter-bank switching in transmission networks [110].

Also, higher operating voltages and temporary overvoltages, as are probable in wind-power collection grids, pose more severe stresses to the switchgear involved. Appropriate specification of the rated voltage and/or additional testing for fault-current and load- current switching are strongly recommended. In addition, wind mills or wind mill farms may be equipped with capacitor banks to compensate for the reactive-power demand of the induction generators and power electronic convertors. Switching of the capacitors in combination with higher operating voltages also deserves special attention. The temporary and transient overvoltages are also harmful to other electrical equipment [111–115].

A fault current in a distribution grid depends strongly on the HV/MV transformer characteristics and the way the transformer is operated (e.g. transformers in parallel, tap position, tap control, load current, power factor, etc.). The length and type of cable(s) between the transformer and fault location considerably reduce the fault-current amplitude and asymmetrical peak, since both reactance and resistance per kilometre of cables is substantial compared with the reactance of a transformer and its $X/R$ ratio. On the other hand, distributed generators, by their individual size, show a large reactance compared with cables so that, in general, the impedance of the cables can be neglected when considering the fault-current contribution of such a generator.

Until the fault is cleared, the rotor will accelerate and, after short-circuit current interruption, the stator voltage will resume. With an induction machine (an induction generator), the resumption requires a large amount of reactive power that prevents the voltage reaching its normal value within a short time. Normally, before resumption, the stator current has almost vanished and the transient currents at resumption are comparable to fault currents.

Windmills are equipped with constant-speed generators or variable-speed drives.

The constant-speed generator units consist of a conventional synchronous or induction generator and a step-up transformer.

Variable-speed generators are available with a limited range of speed variation, so-called double-fed induction generators, or with a wide range of speed variation, for example synchronous generators with a converter. The transient and dynamic behaviour of variable-speed windmills is completely different from that of conventional generators. The fault-current contribution is usually reduced to values of a few p.u. peak value [116, 117] and the current after fault clearing is comparable with the rated current.

## 10.11.2 Out-of-Phase Conditions

System overloading and system stability issues will occur more frequently as systems run up to their limits. This applies to transmission systems where the number of corridors is scarce but may also be encountered in sub-transmission and distribution systems. As long as distributed generators are switched off in the case of faults and disturbances, the probability of out-of-phase conditions will be limited. However, since fault-ride-through requirements are imposed on smaller power plants to prevent problems with system stability, out-of-phase conditions will occur more frequently.

Synchronous generators powered by low inertia gas-turbines will accelerate rather fast under fault conditions and need to be tripped. Because of the fast acceleration, large out-of-phase angles can be present. The out-of-phase current is determined by almost double the system voltage divided by the transient reactance of the power unit plus the much smaller reactance of the system. The out-of-phase current may become larger than the generator's contribution to a busbar fault, which is approximately the system voltage divided by the subtransient reactance of the power unit. The out-of-phase current may endanger the equipment of the power plant, which has been specified according to its prospective short-circuit current only. The circuit-breaker, however, has to cope with the larger fault currents from the system and must therefore be capable to handle the dynamic forces of an out-of-phase current [113].

More severe for the circuit-breaker is the TRV-stress, as out-of-phase angles above approximately 100° will give higher peak values than those specified in the standards (for the out-of-phase test duty: 3.13 p.u. according to IEC 62271-100 [13] and 3.18 p.u. according to IEEE C37.013 [1]). The RRRV is also higher than specified in the standards for the out-of-phase test duty, but can be regarded as covered by T30 or T30S (Annex M of IEC 62271-100 [13]) for transformer-limited faults (see Section 3.4) [12].

In distribution systems, not much experience is yet available with out-of-phase conditions since distributed generators normally will be switched off at a serious disturbance of the network. Some examples of problems with conventional generators under out-of-phase conditions are given in [12]. Other power conversion technologies, like windmills with power electronic converters or PV-systems, behave rather stably, also under transient conditions. They adapt and limit their output very rapidly in order not to overstress the power electronic converters. Experience still has to be gained also with larger networks supplied by renewable energy.

## 10.12 Switching with Non-Mechanical Devices

### 10.12.1 Fault-Current Limitation

Increasing fault levels are a major concern for utilities. Quite apart from substation fault levels relative to the circuit-breaker short-circuit current ratings, increasing fault levels impact on all series-connected equipment, that is, buswork, disconnectors and CTs (see Section 2.3.1). Simply replacing the circuit-breakers with higher rated units does not solve the issue and an overall concept of fault current limitation is required.

Fault current limiting measures can be either passive or active [118, 119]. The common passive method as a standard practice to limit fault levels on distribution feeders as well as to limit fault levels at transmission voltages is the insertion of series reactors [120, 121]. The introduction of series reactors has consequences for the existing circuit-breakers, as discussed in Section 3.5.

Active fault current limiters are generally based on innovative concepts, including superconduction or solid-state solutions [118], but actual applications have been limited to R&D and demonstration projects because of the complexity of the devices and the high costs as compared with the use of series reactors [119].

### 10.12.2 Fuses

The fuse is the oldest and simplest of all protective devices. The idea is to insert a fusible element as the weakest link in a network and sacrifice it in the case of too large current. The early pioneers, such as Michael Faraday, had observed that a wire could be fused by an electric current and so the fuse was born. Primitive fuses consisted of an open wire between two terminals (Thomas Alva Edison for example used a lead wire) but soon improvement was sought and the wire was replaced by a strip made of different materials, usually of a lower melting point than copper, such as zinc. However, as the available power increased, the behaviour of the operating fuse became increasingly violent until attempts were made to screen the fuse element by a tube and fuse designers concentrated on how to limit the emission of flame from the ends of the tube. A useful degree of breaking capacity was achieved, but the introduction of the filled cartridge fuse marked the greatest innovation.

A fuse is a weak link in a circuit and as such has one important advantage over circuit-breakers. Because the element in the fuse has a much smaller cross-section than the cable it protects, the fuse element will melt before the cable. The larger the current, the quicker the fuse element melts. The fuse interrupts a very large current in a much shorter time than a circuit-breaker. So short in fact that the current will be cut off before it reaches its peak value, which in a 50 Hz system implies operation in less than 5 ms, hence serious overheating and electromechanical forces in the system are avoided. This current limiting action is an important characteristic that has application in many industrial low-voltage installations.

The single-shot feature of a fuse requires that a blown fuse has to be replaced before service can be restored. This means a delay, the need to have a spare fuse and qualified maintenance personnel who must go and replace the fuse in the field. This is a serious disadvantage of a fuse compared with a circuit-breaker.

Fuses for high-voltage applications require a high breaking capacity. The fuse body, the cartridge, is made of tough material, usually ceramic, and contains, apart from the fuse element, a filler, such as powdered quartz. The purpose of the quartz filler is to condense as quickly as possible the metal vapour that is produced when a large overcurrent blows the fuse element. The filler prevents a dangerous pressure rise in the hermetically sealed enclosure. The filler should be neither too fine nor too coarse: an intermediate grain size provides the optimum cooling.

The main classification of fuses is into *current-limiting* and *non-current-limiting* types [122–124]. *Current-limiting* fuses describe a class of fuses defined by the behaviour that occurs when the current is so high that the fuse element melts before the peak of the fault current. Upon melting, this type of fuse introduces a high arc voltage in the circuit, counteracting the source voltage so rapidly and effectively that the current stops rising and instead is forced quickly to zero before a natural current zero would occur.

The element is heated so rapidly that there is no time for the loss to the surroundings; there is a uniform temperature along the element and all parts reach their melting temperature simultaneously. The element thus becomes a liquid cylinder that becomes unstable and breaks up into a series of droplets. The fuse element has been replaced by a line of globules. The current that can pass through is limited, because the surrounding filler has a nearly infinite resistance. In consequence the voltage drop at the beginning of arcing becomes above the source voltage and causes rapid suppression of the current. The number of globules per centimetre is about 10 to 12, and as the voltage drop in a short arc is about 20 V, the voltage drop across the element is approximately 200 to 250 V cm$^{-1}$. The maximum voltage rise across a fuse depends on the length and design of the fuse element.

The principle of current limitation in fuses is the same as described in Section 6.2.2 and schematically explained in Figure 6.5. The difference with the mechanical device described there is that, in the case of a fuse, the arc is initiated very quickly by the melting of the fuse (instead of automatic contact separation) and the increase is arc voltage is by adding many arcs in series (instead of arc elongation and segmentation). In high-voltage fuses the conductors have a very large number of notches, each of which will turn into an arc. The addition of all the series arc voltages will reach close enough to the system voltage to limit the short-circuit current.

The fuse limits the current in magnitude as well as in duration hence the name: current limiting. Because of the rapid decrease in the fault current through the circuit's inductance, the current-limiting fuse introduces an overvoltage, called the *fuse switching voltage*, into the system during the current-limiting action.

In the case of very large current, as described above, there is little doubt that fuses melt. At low(er) overcurrent, there exists a critical range of current in which the fuse will not interrupt or the interruption will last (too) long. Depending on the extent of this range, three types of current limiting fuses are defined:

- *Back-up fuses:* for this type, the minimum breaking current is highest, 2.5 to 3 times the normal current. The minimum breaking current is the lowest current that fuses are able to interrupt. This type is most commonly used because of its fast reaction.

  Back-up fuses are commonly used in *switch-fuse combinations* [125]. Herein, the fuse extends the breaking capability of the load-break switch that is activated by a striker pin released upon fuse melting. Alternatively, at lower overcurrent, the switch can interrupt before thermal overload of the fuse. Switch-fuse combinations act considerably faster than circuit-breakers and are very often applied for MV transformer protection.
- *General purpose fuses:* this type is designed to interrupt all current values, starting from the current that causes the fuse to melt after one hour upwards, at 1.5 times the normal current, to the specified short-circuit breaking current.
- *Full-range fuses:* this type can interrupt all current from the smallest current that causes the fuse elements to melt up to the short-circuit breaking current.

*Non-current-limiting* fuses or *expulsion* fuses melt under the same circumstances but add only a low arc voltage into the circuit, so that the current continues to about the same peak as would occur if the fuse had not melted. An expulsion action (that is where gas is generated by the arc and expelled along with ionized material) produces a physical gap such that, at natural current zero, the arc does not re-ignite and the current is interrupted. The expulsion fuse limits the duration of the fault current, but not its magnitude.

## 10.12.3   $I_S$ Limiters

A disadvantage of fuses is heat generation by conduction of the normal current. Since the fuse wire has to melt at a current not too far beyond the rated current, the fuse conductor does not allow large load currents. In situations with a large load current, passing of current through a fuse should be avoided.

In the $I_s$ limiter, a current limiting fuse is placed in parallel with a solid conductor (carrying currents up to 5000 A), thus avoiding thin fuse elements in the circuit. When a short-circuit current is detected by its steep increase ($di/dt$) an explosive charge is detonated that ruptures the main conductor and the arc voltage will commutate the fault current into the parallel fuse that will in turn interrupt the current in less than 1 ms. Fault current interruption up to 210 kA has been demonstrated in 12 kV systems [126].

The $I_s$ limiters are applied when the available short-circuit current exceeds that of individual components or network sections. Examples of applications are:

- Parallel service of transformers. the secondary busbars are connected through an $I_s$ limiter. This device separates the two transformers and the contribution to the fault current at the secondary side is limited to one transformer only.
- In parallel with a current limiting reactor. In the undisturbed situation, the normal current is carried by the $I_s$ limiter and losses across the reactor are avoided. In the case of a fault, the $I_s$ limiter blows and the fault current is limited by the reactor.
- In the connection from the network to local generation centres.

# References

[1] ANSI/IEEE C37.013-1997 (1997) Standard for AC-High Voltage Generator Circuit-Breakers Rated on a Symmetrical Current Basis, IEEE New York; IEC 62271-37-013 (2014) High-voltage switchgear and controlgear. Alternating-current generator circuit-breakers. (In 1997 a revision was published including generators rated 10-100 MVA as publication IEEE Std C37.013a-2007. In 2014, an updated revision will be issued as a dual logo IEC-IEEE publication IEC 62271-37-013.)

[2] Dufournet, D. and Willieme, J.M. (2002) Generator Circuit-Breakers: SF6 Breaking Chamber – Interruption of Current with non-zero passage – Influence of Cable Connection on TRV of System Fed Faults. CIGRE Conference, Paper 13–101.

[3] Canay, M. and Werren, L. (1969) Interrupting sudden asymmetric short-circuit currents without zero transition. *Brown Boveri Rev.*, **56**(10), 484–493.

[4] Owen, R.E. and Lewis, W.A. (1971) Asymmetric characteristics of progressive short circuit on large synchronous generators. *IEEE Trans. Power Ap. Syst.*, **90**, 587–596.

[5] Canay, M. and Klein, H. (1974) Asymmetric short-circuit currents from generators and the effect of the breaking arc. *Brown Boveri Rev.*, **61**(5), 199–206,.

[6] Lim, L.S. and Smith, I.R. (1977) Turbogenerator short circuits with delayed current zeros. *Proc. IEE*, **124**(12), 1163–1169.

[7] Boldea, I. (2005) *The Electric Generators Handbook*, CRC Press, ISBN 7949314810808.

[8] Kohyama, H., Kamei, K., Yoshida, D. *et al.* (2013) Study of Interrupting Duties of Delayed Zero Crossing Current in Generator Main Circuit. CIGRE A3/B2 Colloqiuum, Paper 421, Auckland.

[9] Smeets, R.P.P. and van der Linden, W.A. (1998) The testing of generator circuit-breakers. *IEEE Trans. Power Deliver.*, **13**(4), 1188–1193.

[10] Smeets, R.P.P., te Paske, L.H., Kuivenhoven, S. *et al.* (2009) The Testing of Vacuum Generator Circuit-Breakers. CIRED Conference, Paper No. 393, Prague.

[11] Task Force 13.00.2 of CIGRE SC 13 (1989) Generator circuit-breaker. transient recovery voltages under load current and out-of-phase switching conditions. *Electra* **126**, pp. 43–50.

[12] Jansen, A.L.J., Smeets, R.P.P., van der Linden, W.A. and van Riet, M.J.M. (2002) Distributed Generation in Relation to Phase Opposition and Short-Circuits. 10th Int. Symp. on Short-Circuit Currents in Pow. Syst., Lodz, Oct. 28–29, 2002.

[13] IEC 62271-100 (2012) High-voltage switchgear and controlgear – Part 100: Alternating-current circuit-breakers, Ed. 2.1.

[14] Dufournet, D. and Montillet, G. (2002) Transient recovery voltages requirements for system source fault interrupting by small generator circuit-breakers. *IEEE Trans. Power Deliver.*, **17**(2), 472–478.

[15] Palazzo, M., Braun, D., Cavaliere, G. *et al.* (2012) Reliability Analysis of Generator Circuit-Breakers. CIGRE Conference, Paper A3-206.

[16] van der Sluis, L. and van der Linden, W.A. (1985) Short circuit testing methods for generator circuit-breakers with a parallel resistor. *IEEE Trans. Power Ap. Syst.*, **PAS-104**(10), 2713–2720.

[17] Fröhlich, K.J. (1985) Synthetic testing of circuit-breakers equipped with a low ohmic parallel resistor (with special respect to generator circuit-breakers). *IEEE Trans. Power Ap. Syst.*, **PAS-104**(2), 2283–2288.

[18] Thuries, E., Van Doan, P., Dayet, J. and Joyeux-Bouillon, B. (1986) Synthetic testing method for generator circuit-breakers. *IEEE Trans. Power Deliver.*, **1**(1), 179–184.

[19] Braun, A., Huber, H. and Suiter, H. (1976) Determination of the Transient Recovery Voltages across Generator Circuit-Breakers in Large Power Stations. CIGRE Conference, paper 13-03.

[20] Task Force 13.00.2 of CIGRE SC 13 (Switching Equipment), (1987) Generator circuit-breaker. Transient recovery voltages in most severe short-circuit conditions. *Electra* **113**, 43–50.

[21] Smeets, R.P.P., Barts, H.D. and Zehnder, L. (2006) Extreme Stresses of Generator Circuit-Breakers. CIGRE Conference, Paper A3-306.

[22] Smith, R.K., Long, R.W. and Burmingham, D.L. (2003) Vacuum Interrupters for Generator Circuit-Breakers they're not just for Distribution Breakers Anymore. 17th CIRED Conference, Session 1, Paper 5.

[23] IEC 62271-306 (2012) High-voltage switchgear and controlgear – Part 306: Guide to IEC 62271-100, IEC 62271-1, and IEC standards related to alternating current circuit-breakers.

[24] Kulicke, B. and Schramm, H-H. (1980) Clearance of short-circuits with delayed current zeros in the Itaipu 550 kV substation. *IEEE Trans. Power Ap. Syst.*, **PAS-99**, 1406–1414.

[25] Bui-Van, Q., Khodabakhchian, B., Landry, M. *et al.* (1997) Performance of series-compensated line circuit breakers under delayed current-zero conditions. *IEEE Trans. Power Deliver.*, **12**, 227–233.

[26] IEC 62271-102, High-voltage switchgear and controlgear – Part 102: Alternating current disconnectors and earthing switches, Ed. 1.0 (2001); amendment 1 (2011), amendment 2 (2013).

[27] CIGRE Working Group 33/13.09 (2005) Monograph on GIS Very Fast Transients. CIGRE Technical Brochure 35.

[28] CIGRE Working Group A3.22 (2011) Background of Technical Specifications for Substation Equipment Exceeding 800 kV AC. CIGRE Technical Brochure 456.

[29] IEC 62271-1 (2007) High-voltage switchgear and controlgear – Part 1: Common specifications.

[30] CIGRE Working Group C4.208 (2013) EMC within Power Plants and Substations. CIGRE Technical Brochure 535.

[31] Lee, C.H., Hsu, S.C., Hsi, P.H. and Chen, S.L. (2011) Transferring of VFTO from EHV to MV system as observed in Taiwan's No. 3 nuclear power plant. *IEEE Trans. Power Deliver.*, **26**(2), 1008–1016.

[32] Smeets, R.P.P., van der Linden, W.A., Achterkamp, M. *et al.* (2000) Disconnector switching in GIS: three-phase testing and – phenomena. *IEEE Trans. Power Deliver.*, **15**(1), 122–127.

[33] Damstra, G.C., Nolson, T. and Matyáš, Z. (1996) Test Circuit for GIS Disconnector Fast Transient Measurements. 9th Int. Symp. on High-Voltage Eng., Graz, Paper 6794, pp. 245–248.

[34] Damstra, G.C., Eenink, A.H., Smallegang, C. and Smeenk, W. (1996) A New 50 MHz Multi-channel Digitizing System. 9th Int. Symp. on High-Voltage Eng., Graz, Paper 4528, pp. 206–209.

[35] Damstra, G.C. and Matyáš, Z. (1998) Improvement of Dividers for Fast and Very Fast HV Transients, Measurement and Calibration in High-Voltage Testing. ERA Conference, Report 98-1098, pp. 2.4.1–2.4.10.

[36] Riechert, U., Krüsi, U. and Sologuren-Sanchez, D. (2010) Very Fast Transient Overvoltages during Switching of Bus-Charging Currents by 1 100 kV Disconnector. CIGRE Conference, Paper A3-107.

[37] CIGRE Working Group D1.03 (2012) Very Fast Transient Overvoltages (VFTO) in Gas-Insulated UHV Substations. CIGRE Technical Brochure 519.

[38] IEC 62271-203 (2011) High-voltage switchgear and controlgear – Part 203: Gas-insulated metal-enclosed switchgear for rated voltages above 52 kV, Ed. 2.0.

[39] CIGRE Working Group A3.22 (2008) Background of Technical Specifications for Substation Equipment Exceeding 800 kV AC. CIGRE Technical Brochure 362.

[40] Liu, W. D., Jin, L. J. and Qian, J. L. (2001) Simulation Test of Suppressing VFT in GIS by Ferrite Rings. Proc. of Int. Symp. on Electr. Insul. Materials.

[41] Smajic, J., Holaus, W., Troeger, A. *et al.* (2011) HF Resonators for Damping of VFTs in GIS. Proc. Int. Conf. on Pow. Syst. Transients, Delft, paper No. 185.

[42] Riechert, U., Bösch, J., Smajic, J. *et al.* (2012) Mitigation of Very Fast Transient Overvoltages in Gas Insulated UHV Substations. CIGRE Conference, Paper A3-110.

[43] Liu, H., Sun Y.-Q., Lie, J.-P. *et al.* (2012) Testing Technology on Very Fast Transient Overvoltage in 500 kV HGIS Intelligent Substation. 2nd Int. Conf. on Instrumentation, Measurement, Computer, Communication and Control (IMCCC), Harbin, China.

[44] Andrews, F.E., Janes, L.R. and Anderson, M.A. (1950) Interrupting ability of horn-gap switches. *AIEE Trans.*, **69**, 1016–1027.

[45] Peelo, D.F. (2004) Current interruption using high voltage air-break disconnectors. Ph.D. Thesis, Eindhoven University, ISBN 90-386-1533-7, available through http://alexandria.tue.nl/extra2/200410772.pdf.

[46] Chai, Y., Wouters, P.A.A.F. and Smeets, R.P.P. (2011) Capacitive current interruption by HV air-break disconnectors with high-velocity opening auxiliary contacts. *IEEE Trans. Power Deliver.*, **26**(4), 2668–2675.

[47] IEC/TR 62271-305 (2009) High-voltage switchgear and controlgear – Part 305: Capacitive current switching capability of air-insulated disconnectors for rated voltages above 52 kV, Ed. 1.0.

[48] Chai, Y. (2012) Capacitive Current Interruption with High Voltage Air-Break Disconnectors. Ph.D. Thesis, Eindhoven University, ISBN 978-90-386-3097-7 (available through http://alexandria.tue.nl/extra2/728755.pdf).

[49] Peelo, D.F., Smeets, R.P.P., Kuivenhoven, S. and Krone, J.G. (2005) Capacitive Current Interruption in Atmospheric Air. CIGRE SC A3&B3 Joint Colloquium, Tokyo, Paper No. 106.

[50] Hasibar, R.M., Legate, A.C., Brunke, J. and Peterson, W.G. (1981) The application of high-speed grounding switches for single-pole reclosing on 500 kV power systems. *IEEE Trans. Power Ap. Syst.*, **PAS-100**(4), 1512–1515.

[51] Shperling, B.R. and Fakheri, A.J. (1979) Single phase switching parameters for untransposed EHV transmission lines. *IEEE Trans. Power Ap. Syst.*, **PAS-98**(2), 643–654.

[52] Kobayashi, A., Yamagata, Y., Yoshizumi, T. and Tsubaki, T. (1996) Development of 1,100 kV GIS – Gas Circuit Breakers, Disconnectors and High-Speed Grounding Switches. CIGRE Conference, paper 13-304.

[53] Agafonov, G.E., Babkin, I.V., Berlin, B.E. *et al.* (2004) High Speed Grounding Switch for Extra-High Voltage Lines. CIGRE Conference, paper A3-308.

[54] Toyoda, M., Yamagata, Y., Jaenicke, L.-R. *et al.* (2011) Considerations for the Standardization of high-speed earthing switches for secondary arc extinction on transmission lines (part 2). CIGRE SC A3 Colloquium, Vienna, paper A3-103.

[55] Mizoguchi, H., Hioki, I., Yokota, T. *et al.* (1998) Development of an interrupting chamber for 1000 kV high-speed grounding switch. *IEEE Trans. Power Deliver.*, **13**(2), 495–502.

[56] CIGRE Working Group A3.13 (2007) Changing Network Conditions and System Requirements, Part II: The impact of long distance transmission on HV equipment. CIGRE Technical Brochure 336.

[57] Ebbers, L., Hänninen, T., Pöyhönen, J. and Riffon, P. (2012) New Type Test Requirements for Forced Triggered Spark Gap in Series Capacitor Bank Applications. IET AC DC Conference, London.

[58] Smeets, R.P.P., Hofstee, A.B., Jänicke, L.-R and Punger, M. (2006) High-current Switching Protective Equipment in Capacitor Banks for Series Compensation of Very Long Overhead Lines. CIGRE Conference, paper A3-304.

[59] IEC 60143 (2010) Series capacitors for power systems – Part 4: Thyristor controlled series capacitors.

[60] Dubé, J.-F., Goehler, R., Hanninen, T. *et al.* (2012) New Achievements in Pressure-Relief Tests for Polymeric-Housed Varistors used on Series Compensated Capacitor Banks. IEEE PES General Meeting, pp. 1–8.

[61] IEC 62271-109 (2008) High-voltage switchgear and controlgear – Part 109: Alternating-current series capacitor by-pass switches.

[62] Smeets, R.P.P. and van der Linden, W.A. (2001) Verification of the short-circuit current making capability of high-voltage switching devices. *IEEE Trans. Power Deliver.*, **16**(4), 611–618.

[63] Iravani, M.R., Chaudhary, A.K.S., Giesbrecht, W.J. *et al.* (2000) Modeling and analysis guidelines for slow transients – part III: The study of ferroresonance. *IEEE Trans. on Pow. Del.*, **15**(1), 255–265.

[64] Ferracci, P. (1998) Ferroresonance, Cahier technique n° 190, Groupe Schneider.

[65] Greenwood, A. (1991) *Electrical Transients in Power Systems*, John Wiley and Sons Ltd, ISBN 0-471-62058-0.

[66] Rüdenberg, R. (1974) *Elektrische Schaltvorgänge*, 5th edn, Springer Verlag, ISBN 978-3642503344.

[67] CIGRE Working Group 33.10 (2000) Temporary overvoltages test case results, *Electra*, **188**, 71–87.

[68] CIGRE Working Group C4.307 (2014) Resonance and Ferroresonance in Power Networks, CIGRE Technical Brochure 569.

[69] Emin, Z., Martinez Duro, M. and Val Escudero, M. (2013) An overview of Resonance and Ferroresonance in Power Systems. CIGRE SC A2 & C4 Symposium, Zürich, Report 1-1.

[70] Janssen, A.L.J. and van der Sluis, L. (1990) Clearing faults near shunt capacitor banks. *IEEE Trans. on Pow. Del.*, **5**(3), 1346–1354.

[71] Martin, F. and Joncquel, E. (2006) Circuit-Breaker Tripping near Capacitor Bank. CIGRE Conference, paper A3-203.

[72] Ito, H., Janssen, A.L.J., Merwe, van der, C. *et al.* (2009) Comparison of UHV and 800 kV Specifications for Substation Equipment. CIGRE 6th Southern Africa Regional Conference, paper P401.

[73] CIGRE Working Group A3.22 (2008) Background of Technical Specifications for Substation Equipment Exceeding 800 kV AC. CIGRE Brochure No. 362.

[74] McDonald, J. (2007) *Electric Power Substations Engineering*, 2nd edn, CRC Press, ISBN 0-8493-7383-2.

[75] Noack, F. (2008) Comparison between overhead lines (OHL) and underground cables (UGC) as 400 kV transmission lines for the Woodland-Kingscourt-Turleenan Project, ASKON Consultinggroup, Ilmenau University of Technology, Germany.

[76] CIGRE Working Group C4.502 (2013) Power System Technical Performance Issues Related to the Application of Long HVAC Cables. CIGRE Technical Brochure 556.

[77] CIGRE Working Group B1.10 (2009) *Update of Service Experience of HV Underground and Submarine Cable Systems*, CIGRE, ISBN 978-2-85873-066-71.

[78] Ohki, Y. and Yasufuku, S. (2002) The worlds first long-distance 500 kV-XLPE cable line, part 2: joints and after-installation test. *IEEE Electr. Insul. Mag.*, **18**(3), 57–58.

[79] Colla, L., Gatta, F.M., Geri, A. *et al.* (2009) Steady-state Operation of Very Long EHV AC Cable Lines. Proc. of IEEE Pow. Tech Conf., Bucharest, June–July 2009.

[80] Gatta, F.M. and Lauria, S. (2005) Very Long EHV Cables and Mixed Overhead-Cable Lines. Steady-State Operation. Proc. of IEEE Pow. Tech Conf., St. Petersburg, June 2005.

[81] Colla, L., Gatta, F.M., Iliceto, F. and Lauria, S. (2005) Design and operation of EHV transmission lines including long insulated cable and overhead Sections. Proc. of IEEE Pow. Eng. Conf., Nov.–Dec. 2005.

[82] Lauria, S., Gatta, F.M. and Colla, L. (2007) Shunt compensation of EHV Cables and Mixed Overhead-Cable Lines. Proc. of IEEE Lausanne Pow. Tech Conf., July 2007.

[83] Bak, C.L., Wiechowski, W., Sogaard, K. and Mikkelsen, S.D. (2007) Analysis and simulation of switching surge generation when disconnecting a combined 400 kV cable/overhead line with shunt reactor. Proc. of IPST Conf., Lyon, France, June 2007.

[84] Bak, C.L., Baldursson, H. and Oumarou, A.M. Switching Overvoltages in 60 kV Reactor Compensated Cable Grid due to Resonance after Disconnection. Inst. of Energy Technology, Aalborg Univ., Denmark.

[85] Burges, K., Bömer, J., Nabe, C. and Papaefthymiou, G. (2008) Study on the comparative merits of overhead electricity transmission lines versus underground cables, Ecofys Germany GmbH.

[86] Tokyo Electric Power Company (2008) Joint Feasibility Study on the 400 kV Cable Line Endrup-Idomlund, Final Report, April 2008.

[87] CIGRE Working Group B1.05 (2005) Transient voltages affecting long cables. CIGRE Technical Brochure 268.

[88] Massaro, F., Morana, G. and Musca, R. (2009) Transient behavior of a mixed overhead-cable EHV line under lightning events. *Proceedings of the 44th International Universities Power Engineering Conference (UPEC)*, IEEE.

[89] Faria de Silva, F., Bak, C.L., Gudmundsdottir, U.S. *et al.* (2010) Methods to Minimize Zero-Missing Phenomenon. *IEEE Trans. Power Deliver.*, **25**(4), 2923–2930.

[90] Faria de Silva, F., Bak, C.L., Gudmundsdottir, U.S. *et al.* (2009) Use of a Pre-Insertion Resistor to Minimize Zero-Missing Phenomenon and Switching Overvoltages. IEEE Power Eng. Soc, General Meeting, Calgary, AB, Canada.

[91] Pugliese, H. and von Kannewurff, M. (2013) Discovering DC: a primer on DC circuit breakers, their advantages, and design. *IEEE Ind. Appl. Mag.*, **19**(5), 22–28.

[92] Atmadji, A. (2000) Direct Current Hybrid Breakers. A Design and Its Realization. Ph.D. thesis, Eindhoven Univ. of Techn., ISBN 90-386-1740-2 (available through http://alexandria.tue.nl/extra2/200001242.pdf).

[93] Niwa, Y., Yokokura, K. and Matsuzaki, J. (2010) Fundamental Investigation and Application of High-speed VCB for DC Power System of Railway. XXIVth Int. Symp. on Disch. and Elec. Insul. in Vac., Braunschweig.

[94] Meyer, J.-M. and Rufer, A. (2006) A DC Hybrid Circuit Breaker With Ultra-Fast Contact Opening and Integrated Gate-Commutated Thyristors (IGCTs). *IEEE Trans. Power Deliver.*, **21**(2), 646–651.

[95] Meyer, C. and de Doncker, R.W. (2006) Solid-state circuit breaker based on active thyristor topologies. *IEEE Trans. Power Electr.*, **21**(2), 450–458.

[96] Franck, C.M. (2011) HVDC circuit breakers: a review identifying future research needs. *IEEE Trans. Power Deliver.*, **26**(2), 998–1007.

[97] Kim, C.-K., Sood, V.K., Jang, G.-S. *et al.* (2009) *HV DC Transmission. Power Conversion Applications in Power Systems*, IEEE Press, and John Wiley & Sons (Asia) Pte Ltd, ISBN 978-0-470-82295.

[98] Huang, H. and Uder, M. (2008) Application of high Power Thyristors in HVDC and FACTS Systems. Int. Conf. on Elec. Pow. Suppy Ind., (CEPSI) paper 262.

[99] Perret, R. (2009) *Power Electronics Semiconductor Devices*, ISTE Ltd and John Wiley & Sons Inc, ISBN 978-1-84821-064-6.

[100] Bucher, M.K. and Franck, C.M. (2013) Contribution of fault current sources in multiterminal HVDC cable networks. *IEEE Trans. Power Deliver.*, **28**(3), 1796–1803.

[101] Pucher, W. (1968) Fundamentals of HVDC interruption, *Electra*, **5**, 24–39.

[102] Tokuyama, S., Arimatsu, K., Yoshioka, Y. *et al.* (1985) Development and interrupting tests on 250 kV 8 kA HVDC circuit breaker. *IEEE Trans. Power App Syst.*, **PAS-104**(9), 2453–2458.

[103] Rizk, F. (1963) Interruption of Small Inductive Currents with Air-Blast Circuit-Breakers; Time-Constant, Instability and Dielectric Strength. Diss. Chalmer's Inst. of Techn., No. 37, Gøteborg.

[104] Bachmann, B., Mauthe, G., Ruoss, E. and Lips, H.E. (1985) Development of a 500 kV airblast HVDC circuit-breaker. *IEEE Trans. Power Ap. Syst.*, **PAS-104**(9), 2460–2466.

[105] Ito, H., Hamano, S., Ibuki, K. *et al.* (1997) Instability of DC arc in SF6 circuit breaker. *IEEE Trans. Power Deliver.*, **12**(4), 1508–1513.

[106] Nakao, H., Nakagoshi, Y., Hatano, M. *et al.* (2001) DC Current interruption in HVDC SF6 Gas MRTB by means of self-excited oscillation superimposition. *IEEE Trans. Power Deliver.*, **16**(4), 687–693.

[107] Smeets, R.P.P., Kertész, V. and Yanushkevich, A. (2014) Modelling and Experimental Verification of DC Current Interruption Phenomena and Associated Test-Circuits. CIGRE Conference, paper A3-114.

[108] Callavik, M., Blomberg, A., Häffner, J. and Jacobson, B. (2013) Breakthrough! ABB's hybrid HVDC breaker, an innovation enabling reliable HVDC grids, ABB Review, No. 2, pp. 7–13.

[109] Badrzadeh, B., Høgdahl, M.and Isabegovic, E. (2011) Transients in wind powerplants – Part I: Modeling methodology and validation. *Proceedings of IAS Annual Meeting*, IEEE.

[110] CIGRE Working Group A3.12 (2007) Changing Network Conditions and System Requirements, Part II: The impact of long distance transmission on HV equipment. CIGRE Technical Brochure 336.

[111] Reza, M., Breder, H., Liljestrand, L. *et al.* (2009) An Experimental Study of Switching Transients in a Wind-Collection Grid Scale Model in a Cable System Laboratory. 20th Int. Conf. on Electr. Distr. (CIRED), Paper 0364, Prague.

[112] Chennamadhavuni, A., Munji, K.K. and Bhimasingu, R. (2012) Investigation of transient and temporary overvoltages in a wind farm. 2012 IEEE Int. Conf. on Pow. Syst. Techn. (POWERCON 2012), Auckland, New Zealand.

[113] Sallam, A. and Malik, O.P. (2010) *Electric Distribution Systems*, John Wiley & Sons, Inc., ISBN 978-0-470-27682-2.

[114] Badrzadeh, B., Høgdahl, M., Singh, N. *et al.* (2011) Transients in wind power plants – Part II: Case studies. *Proceedings of IAS Annual Meeting*, IEEE.

[115] CIGRE Working Group A3.12 (2007) Changing Network Conditions and System Requirements, Part I: The impact of distributed generation on equipment rated above 1 kV. CIGRE Technical Brochure 335.

[116] Janssen, A.L.J., van Riet, M., Bozelie, J. and Au-yeung, J. (2011) Fault current contribution from state of the art DG's and its limitation. Int. Conf. on Pow. Syst. Transients, IPST 2011, Delft, the Netherlands, Report 113.

[117] Janssen, A.L.J., van Riet, M., Smeets, R.P.P. *et al.* (2012) Prospective Single and Multi-Phase Short-Circuit Current Levels in the Dutch Transmission, Sub-Transmission and Distribution Grids. CIGRE Conference, Paper A3-103.

[118] CIGRE Working Group A3.10 (2003) Fault Current Limiters in Electrical Medium and High Voltage Systems. Technical Brochure 239.

[119] CIGRE Working Group A3.23 (2012) Application and Feasibility of Fault Current Limiters in Power Systems. Technical Brochure 497.

[120] Peelo, D.F., Polovick, G.S., Sawada, J.H. *et al.* (1996) Mitigation of Circuit-Breaker Transient Recovery Voltages Associated with Current Limiting Reactors. *IEEE Trans. Power Deliver.*, **11**(2), 865–871.

[121] Amon Filho, J., Fernandez, P.C., Rose, E.H. *et al.* (2009) Brazilian Successful Experience in the Usage of Current Limiting Reactors for Short-Circuit Limitation. XIth Symposium of Specialists in Electric Operational and Expansion Planning, Belem, Brazil.

[122] IEC/TR 62655 (2013) Tutorial and application guide for high-voltage fuses, ed. 1.0.

[123] Wright, A. and Newbery, P.G. (2004) *Electric Fuses*, 3rd edn, IEEE, IEEE Power & Energy Series **49**, ISBN 0 86341 339 4.

[124] IEEE Std C37.48.1-2012 (2012) IEEE Guide for Operation, Classification, Application, and Coordination of Current-Limiting Fuses with Rated Voltages 1 - 38 kV.

[125] IEC 62271-105 (2012) High-voltage and controlgear – Part 105: Alternating current switch-fuse combinations for rated voltages above 1 kV up to 52 kV, ed. 2.

[126] Hartung, K-.H. and Schmidt, V. (2009) Limitation of Short-Circuit Current by an Is-limiter. 10th Int. Conf. on Electr. Pow. Qual. and Util., EPQU.

# 11

# Switching Overvoltages and Their Mitigation

## 11.1 Overvoltages

Overvoltages, stressing a power system, can generally be classified into two categories regarding their origin:

- External overvoltages, generated by lightning strokes, which are the most common and severe atmospheric disturbances.
- Internal overvoltages, generated by changes in the operating conditions of the network, like switching.

Voltages and overvoltages are classified in IEC 60071-1 [1] according to their shape and duration as follows:

*Low-frequency voltages and overvoltages* (of constant r.m.s. value):

- *Continuous voltages* having a power frequency of 50 or 60 Hz and duration of at least 1 h.
- *Temporary overvoltages* (TOV) are power-frequency and harmonic overvoltages of relatively long duration, from 20 ms to 1 h. They may be undamped or weakly damped. Frequencies in practice are in the range from 10 to 500 Hz. The standardized voltage shape for testing is 48 to 52 Hz with duration of 60 s.

  The most common temporary overvoltages occur on the healthy phases of a system during phase-to-earth faults. Some other well-known events leading to the generation of temporary overvoltages are load rejection, Ferranti rise, resonance and ferroresonance phenomena (see Section 10.6). The temporary overvoltages may lead to overstressing of surge arresters and magnetic saturation of transformers and shunt reactors. Temporary overvoltages are considerably lower than transient overvoltages.

*Switching in Electrical Transmission and Distribution Systems*, First Edition.
René Smeets, Lou van der Sluis, Mirsad Kapetanović, David F. Peelo, and Anton Janssen.
© 2015 John Wiley & Sons, Ltd. Published 2015 by John Wiley & Sons, Ltd.

The following means can be used to limit temporary overvoltages:

- adequate earthing and neutral treatment;
- compensation by shunt capacitor banks and/or shunt reactors;
- switching off cables and overhead lines to reduce reactive power generation;
- adapting or blocking the voltage regulation of generators and transformers;
- application of switchable, potentially sacrificial surge arresters [2];
- series compensation with series capacitor banks in very long AC lines.

*Transient overvoltages* (of a few milliseconds or less, usually highly damped):

- *Switching overvoltages* or *slow-front overvoltages* (SFO) can be caused by switching operations as well as by faults occurring on the system. Circuit-breakers do not generate SFOs directly; they just initiate them by changing the topology of a circuit. If present, re-ignition, restrike, inrush current, chopping, multiple re-ignitions, and NSDD (non-sustained disruptive discharge) phenomena can also be responsible for generation of SFOs. In practice, the SFOs have time-to-peak between 20 μs and 5 ms and tail duration less than 20 ms. Testing is performed with standardized *switching impulses* (SI) having *time to peak* $T_p = 250$ μs and *decay time to half-value* $T_2 = 2.5$ ms [3].
  Switching overvoltages are treated in detail in Section 11.2.
- *Lightning overvoltages* or *fast-front overvoltages* (FFO) can appear in a substation either due to a lightning stroke directly to the substation or a strike to the transmission line feeding the substation [4]. Alternatively, flash-overs, re-ignitions, and restrikes at a short distance may cause the FFOs. Typical FFOs have time to peak between 100 ns and 20 μs and tail duration of 300 μs. Testing is performed with standard *lightning impulses* (LI) having a *front time* $T_1 = 1.2$ μs and *decay time to half-value* $T_2 = 50$ μs, referred to as 1.2/50 impulse [3].
- *Very-fast transient overvoltages* (VFTO) (see Section 10.3.2.2) belong to the highest frequency range, 30 kHz up to 100 MHz, with time to peak between 3 and 100 ns and duration shorter than 3 ms. They are mainly produced by *gas-insulated switchgear*, especially by switching with disconnectors. They may have a direct impact on primary equipment, transformers, or on secondary control systems due to their strong electromagnetic interference potential.

Figure 11.1 shows a comparison of the relevant phase-to-earth insulation levels[1] for various rated voltages $U_r$ (r.m.s. values). The insulation levels refer to:

- amplitude of the short-duration withstand voltage $U_{d,p}$ [p.u.];
- peak of switching-impulse voltages $U_s$ [p.u.]; and
- peak of lightning-impulse voltages $U_p$ [p.u.].

In each case, 1 p.u. $= U_r \sqrt{2} / \sqrt{3}$.

---

[1] These rated insulation levels are specified by IEC 62271-1 (2007), Table 1a and 2a, and in Amendment 1 (2011), Table 2a. The value of the short-duration withstand voltages $U_{d,p}$ have been calculated from the specified r.m.s. rated insulation levels $U_d$, that is, $U_{d,p} = U_d \sqrt{2}$.

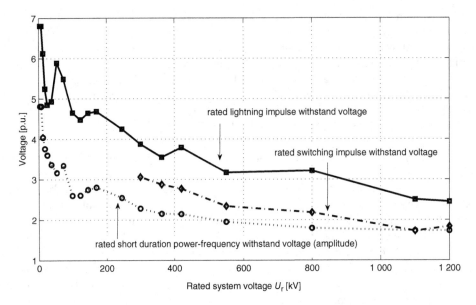

**Figure 11.1** Rated short-duration power-frequency-, switching impulse- and lightning impulse with-stand voltages versus rated system voltage [5].

One of the oldest means of protection against overvoltages that is still used widely is the spark gap. By allowing a gap to flash over, the propagation of undesired voltages in the network can be prevented. However, the spark gap is not a very efficient overvoltage protection device. The main drawbacks are the variation in the breakdown voltage with the polarity and environmental conditions, and the fact that once an arc is initiated it continues even after the overvoltage has disappeared, initiating a phase-to-earth fault. Moreover, the flashover itself generates switching overvoltages with a very steep front.

## 11.2  Switching Overvoltages

Apart from being caused by dielectric faults or flashover, switching overvoltages appear in the power systems due to switching of load and/or fault currents, and they cannot be avoided. In addition, the network is also stressed by lightning overvoltages. Although components in the system are selected with regard to expected internal and external overvoltages, it is impossible to exclude stresses exceeding the insulation level of the equipment. Therefore, substations have to be protected from dangerous overvoltages by preventing their penetration into parts of the substation where they can cause severe damage. Design criteria are usually based on overvoltages with levels that cover 98% of the expected overvoltages in service [6].

Individual components in substations have different insulation levels achieved by proper insulation coordination. In order to protect power transformers, being the most valuable and difficult to replace component in a substation, insulation coordination is directed towards creating several lines of defence by which other components may fail instead of the power transformers. For GIS similar considerations are applicable. The intention is to direct any

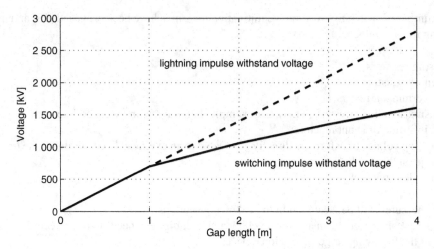

**Figure 11.2** Lightning- and switching-impulse withstand voltage versus gap distance for a rod–rod gap [7].

disruptive discharge caused by voltage surges to components of the substation which are less valuable or easily replaceable.

Switching overvoltages are the primary dimensioning parameter for air-clearances in EHV and UHV systems. The reason is that, while clearances for lightning impulse withstand voltages increase linearly with gap distance, those for switching impulse withstand level tend to saturate with increasing gap distances (see Figure 11.2). Clearly, there is great merit in mitigating switching overvoltage levels.

The historical context of this subject is interesting. In Europe, the first EHV system was the 400 kV system developed in Sweden in the mid-1950s. No account was taken of switching overvoltages because their influence remained to be identified and, obviously, no mitigation was considered. The development of the 400 kV system in Italy in the 1960s was the first real recognition of switching overvoltages. A very large number of switching impulse tests were performed and resulted in the now well-known $U$-curves and the notion of gap factors [8, 9].

The explanation for the $U$-curve was provided by research work at Electricité de France [10]. Switching-impulse breakdown involves both streamer and leader development with the latter being the main driver, whereas lightning-impulse breakdown involves streamers only. The minimum value of the $U$-curves thus represents optimal leader development and occurs at front times in the range of 100 to 400 μs, leading to a selection of a front time of 250 μs for standardization purposes. Detailed descriptions of electric discharges and breakdown in air can be found in [11, 12].

## 11.3   Switching-Voltage Mitigation

### 11.3.1   Principles of Mitigation

By limiting the power-frequency temporary overvoltages, the switching overvoltages are usually also reduced as, with limited temporary overvoltages, the jump from a steady state to another steady state will be smaller.

A number of switching-overvoltage mitigation measures have been proposed [13] and used, starting in the 1960s:

- closing resistors;
- opening resistors;
- line-terminal arresters;
- transmission line arresters (TLA), for example, halfway on an overhead line;
- fast insertion of shunt reactors;
- installing voltage transformers directly connected to an overhead line to drain off the trapped charge on the line before reclosure;
- controlled switching.

These measures will be discussed in the following sections.

If surges can exceed the insulation level, then vital equipment needs to be protected by using some of the following surge-protection devices:

- spark gaps (see an example in Section 10.5.1);
- surge arresters (see Section 11.3.3);
- capacitance–resistance-type (*C–R*) surge suppressors (see the example in Section 4.3.3.4).

## 11.3.2   Mitigation by Closing Resistors

A closing resistor is a resistor (*R*) that is inserted temporarily by an auxiliary switch in parallel to the interrupter and in series with a transmission line of surge impedance *Z* at a predetermined time during the closing process. After a short time (*pre-insertion time*), the main contacts of the circuit-breaker close, bypassing the resistor. This principle is outlined in Figure 11.3.

Once the significance of switching overvoltages was recognized, closing resistors were applied on EHV airblast circuit-breakers, starting in the 1960s [14–16].

The principle behind the use of closing resistors is as follows. Assume a voltage $U_D$ across the open circuit-breaker at the instant of closing. Closing of the circuit-breaker produces a travelling wave of magnitude $U_D$ which divides between the resistor *R* and the line surge impedance *Z*. The amplitude of the voltage wave ($U_{oL}$) incident wave at the open end of the line is thus:

$$U_{oL} = \frac{U_D Z}{R + Z}$$

(11.1)

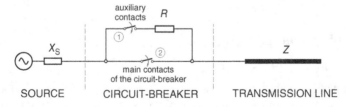

**Figure 11.3**   Diagram of circuit-breaker equipped with single-step closing resistor.

and after voltage doubling the line end voltage ($U_{eL}$) becomes:

$$U_{eL} = \frac{2U_D Z}{R + Z} - U_{TC} \tag{11.2}$$

where $U_{TC}$ is the voltage due to the initially trapped charge on the line.

Taking, for example, $U_D = 2$ p.u. (i.e. 1 p.u. positive source voltage and 1 p.u. negative voltage due to trapped charge)

$$U_{eL} = \frac{4Z}{R + Z} - 1 \tag{11.3}$$

If $R = 0$, then $U_{eL} = 3$ p.u.

Clearly, the higher the resistance $R$, the lower the value of $U_{eL}$. However, the resistor is inserted only momentarily and the bypassing generates a travelling wave, the magnitude of which is given by:

$$U_R = \frac{U_{R,D} R}{R + Z} \tag{11.4}$$

where $U_{R,D}$ is the voltage across the resistor at the instant of bypassing. At the line end with voltage doubling, the voltage is:

$$U_{R,eL} = \frac{2U_{R,D} R}{R + Z} \pm U_{pf} \tag{11.5}$$

where $U_{pf}$ is the power-frequency voltage at the instant of arrival of the travelling wave. An upper limit of the value of $R$ is therefore required. The value of $R$ is usually optimized through computer studies and is typically in the range of 250 to 600 $\Omega$, depending on the application.

Generally, the aim of closing resistors is to limit the voltage at the receiving end of the line to 2 p.u. However, closing resistors do little for the voltage profile along the line, and higher overvoltages than 2 p.u. occur at points where incident and reflected waves superimpose.

Besides its resistance value, the insertion time is the second important parameter of a closing resistor. The resistor insertion time is determined by two contradictory requirements:

- the time should be sufficiently long to attenuate the transients;
- the time should be short enough not to exceed the thermal capability of the resistor.

If the insertion time is shorter than the circuit-breaker pole scatter plus twice the transmission-line travelling-wave transit time, an increase in the overvoltage results. The insertion time should be longer than 8 ms for a closing resistor to be effective.

Since switching overvoltages during closing of long no-load transmission lines very much depend on the moment of closing, *controlled* or *synchronized*, or *point-on-wave closing* is increasingly used instead of closing resistors, see Section 11.4. Also, the mechanical complexity of the closing resistor, with its auxiliary switch and its inherent concerns of reliability and maintenance, leads to other solutions, such as controlled switching.

### 11.3.3   Mitigation by Surge Arresters

#### 11.3.3.1   Introduction

Surge arresters are used to limit overvoltages to a specified protection level which is, in principle, below the withstand voltage of the equipment.

The ideal surge arrester would be one that starts to conduct at a specified voltage level, at a certain margin above its rated voltage, holds that voltage level without variation for the duration of the overvoltage, and ceases to conduct as soon as the voltage across the surge arrester returns to a value below the specified voltage level. Therefore, such an arrester would only have to absorb the energy that is associated with the overvoltage.

The design and operation of surge arresters has radically changed over the last 30 years, from valve or spark-gap type silicon carbide (SiC) surge arresters to the gapless metal oxide (MO) or zinc oxide (ZnO) surge arresters. Major steps in the development of surge arresters have been made and modern surge arresters fulfil the present day requirements [17].

#### 11.3.3.2   Metal Oxide Surge Arresters

As an alternative to the use of closing resistors, *metal oxide surge arresters* were applied on line terminals starting in the late 1980s [18–21]. Ever since, this has become common practice at (extra-)high voltage applications, mainly for lightning protection.

Modern (gapless) surge arresters are based on MO resistors, which have an extremely non-linear voltage–current characteristic and a high energy-absorption capability. They are known as metal–oxide surge arresters. The metal oxide material is a ceramic made by mixing ZnO with small amounts of additives, such as $Bi_2O_3$, $CoO$, $Cr_2O_3$, $MnO$, and $Sb_2O_3$, granulating the mixture, then drying it, pressing it into discs, and finally sintering it. The ZnO grains of roughly 10 μm diameter have a low resistivity of $10^{-2}$ Ω m, but they are surrounded by a granular oxide layer 0.1 μm thick, the resistivity of which changes non-linearly from $10^8$ Ω m for low electric-field stresses to just below $10^{-2}$ Ω m for high stresses.

The MO disc can be represented by the equivalent circuit shown in Figure 11.4 [22]. $R_{ZnO}$ is the resistance of the ZnO grains, $L$ stands for the inductance of the MO disc and is determined by the geometry of the current path. $R$ is the non-linear resistance of the granular layers. The granular layer has a constant relative dielectric permittivity between 500 and 1200, depending on the manufacturing process. Its capacitance is indicated by $C$ in the circuit.

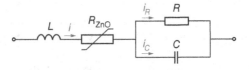

**Figure 11.4**   Equivalent circuit of a MO surge arrester disc.

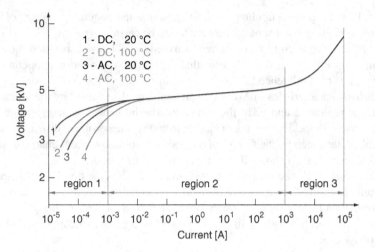

**Figure 11.5**  Voltage–current characteristic of a MO disc (diameter 80 mm, height 20 mm).

The current-voltage characteristic of the MO disc is shown in Figure 11.5.

Based on the conduction mechanism of the disc-material microstructure, the characteristic is divided into three regions:

- Low electric-field region (region 1)
  The conduction mechanism in this region is effective through the existence of the potential barrier of the granular layer. The barrier prevents electrons from moving from one grain to another. The presence of an electric field has the effect of lowering this barrier, known as the *Schottky effect*, enabling a certain number of electrons to pass through the barrier thermally. The resulting current is rather small, in the milliampere range. A higher temperature of the disc increases the energy of the electrons and they can pass through the potential barrier more easily.
- Medium electric-field region (region 2)
  When the electric field in the granular layer reaches a certain value, approximately $100$ kV mm$^{-1}$, electrons move through the potential barriers by the quantum mechanical *tunnel effect*. In this region, the voltage changes very slowly over a wide range of currents.
- High electric-field region (region 3)
  In this region, the voltage drop at the potential barrier due to the tunnel effect is small, and the voltage drop across the resistance of the ZnO grains dominates. The current then gradually approaches the linear relation with the voltage.

A MO surge arrester is essentially a collection of billions of microscopic junctions of MO grains that turn on and off in microseconds to create a current path from the top terminal to the earth terminal of the arrester. It can be regarded as a very fast-acting electronic switch, which is open to operating voltages and closed to switching and lightning overvoltages.

An important parameter of an arrester is the *switching-impulse protection level* (SIPL) defined as *the maximum permissible peak voltage on the terminals of a surge arrester subjected to switching impulses under specific conditions.*

In order to keep the power supplied to a MO arrester at the system operating voltage small, the continuous operating voltage of the arrester has to be chosen in region 1. In this region, the peak value of the resistive-current component is usually well below 1 mA and the capacitive-current component is dominant. This means that the voltage distribution at operating voltage is capacitive and is thus influenced by stray capacitance.

The protective characteristics of MO surge arresters are determined by the voltage–current characteristic in regions 2 and 3. In these regions, the influences of temperature and stray capacitances have disappeared and the deviation from the linear voltage distribution along the arrester is only determined by the scatter of the resistive voltage–current characteristic, which is small and, therefore, the voltage distribution is practically linear.

The voltage–current characteristic of the MO material offers the non-linearity necessary to fulfil the mutually contradicting requirements of an adequate protection level at overvoltages and low current, that is, low energy dissipation at the system operating voltage.

Metal oxide surge arresters are available for protection against switching overvoltages at all operating voltages.

Traditionally, porcelain-housed MO surge arresters (Figure 11.6) are used.

For satisfactory performance, it is important that the units are hermetically sealed for the lifetime of the arrester discs. The sealing arrangement at each end of the arrester consists of a stainless-steel plate with a rubber gasket. This plate exerts a continuous pressure on the gasket, against the surface of the insulator. It also serves to fix the column of the MO discs in place by springs. The sealing plate is designed to act as an overpressure relief system. Should the arrester be stressed in excess of its design capability, an internal arc is established. The ionized gases cause a rapid increase in the internal pressure, which in turn causes the sealing plate to open and the gases to flow out through the venting ducts. Since the ducts are directed towards each other, this results in an external arc, thus relieving the internal pressure and preventing a violent shattering of the insulator [23] (see Section 2.3.1 Figure 2.10).

However, porcelain-housed distribution arresters have tended to fail due to problems with sealing. The benefits of a leak tight design, using polymers, have been generally accepted, leading to the almost wholesale changeover from porcelain to polymers.

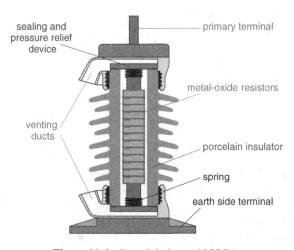

**Figure 11.6** Porcelain-housed MOSA.

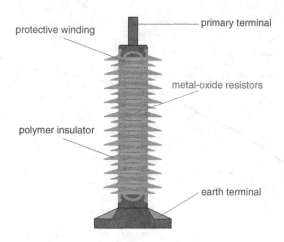

protective winding

primary terminal

metal-oxide resistors

polymer insulator

earth terminal

**Figure 11.7**   Polymer-housed MOSA.

Polymer-housed arresters (Figure 11.7) have a very reliable bond of the silicone rubber with the active parts. Hence, gaskets or sealing rings are not required. Should the arrester be electrically stressed in excess of its design capability, an internal arc is established, leading to rupture of the enclosure, instead of explosion. The arc will easily burn through the soft silicone material, permitting the resultant gases to escape quickly and directly. Hence, special pressure relief vents, with the aim of avoiding explosion of porcelain housing, are not required for this design.

Moreover, polymer-housed distribution arresters are cheaper than porcelain-enclosed ones.

For transmission-system voltages, the porcelain housing is also being replaced increasingly by polymer insulation.

### 11.3.3.3   Effect of Transmission Line Arresters (TLA)

The effect of line surge arresters is shown in Figure 11.8 for a typical EHV application, such as described in [21]. With arresters at the line ends only, the overvoltages at those points will be limited to the protective level of the arresters. However, the voltage profile along the line shows much higher overvoltage levels and suggests a need to flatten the profile. This is achieved by adding arresters at intermediate points along the line, as also shown in Figure 11.8.

The surge arresters applied for this purpose are special in the sense that their characteristics are derived from simulation system studies. A balance has to be achieved between the protective level and the energy absorbed, which are contradictory requirements, that is, decreasing the protective level increases the energy absorbed and vice versa.

Ultimately, further reduction in the overvoltage level requires more proactive mitigation measures.

### 11.3.4   Fast Insertion of Shunt Reactors

The primary purpose of the application of shunt reactors is compensation of the transmission-line capacitance. In addition, as a secondary feature, they provide some reduction of

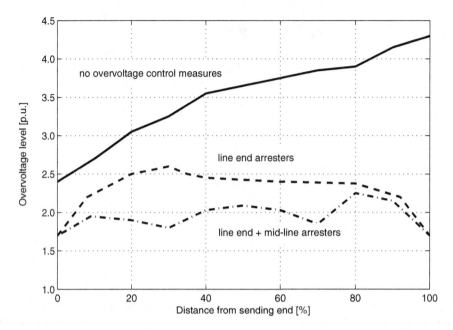

**Figure 11.8**   Effect of line surge arresters on switching-overvoltage levels along the line length.

switching-overvoltage levels. Shunt reactors, however, are never installed by utility companies specifically for this secondary effect, although shunt reactors contribute to a reduction of temporary overvoltages and discharge the line before reclosure when connected directly to the overhead line.

The shunt reactors, if present, are switched on preferably prior to energizing or re-energizing the line [24]. In comparison with other switching overvoltage mitigation means, shunt reactors are less significant.

## 11.4   Mitigation by Controlled Switching

### 11.4.1   Principles of Controlled Switching

All network switching operations result in a certain level of transient overvoltages that propagate through the system. The severity of the transients caused by operation of circuit-breakers strongly depends on the instant of switching.

Normally, this instant is random and there are several important circuit-breaker applications where random closing or opening may lead to severe voltage and current transients. This particularly applies to energization and de-energization of shunt-reactor banks, power transformers, shunt-capacitor banks and overhead lines. Very steep voltage surges or very high inrush currents can occur, causing high dielectric and mechanical stresses on circuit-breakers and other components. The stresses may cause immediate or gradual damage to equipment.

Thus, for example, the energization of no-load overhead transmission lines, and especially their re-closing, may result in very high overvoltages due to the presence of a trapped charge and travelling voltage waves on the isolated section of the overhead line. These switching surges have to be taken into account in the design of the line insulation. However, if the switching consists of a reclosing operation, the surge may lead to a re-initiation of the fault and unsuccessful reclosure.

Switching transients occur in the primary HV circuits, but they may induce transients in control and auxiliary circuits. Induced transients may lead to a variety of disturbances in control, protection and telecommunication systems and processors.

Historically, the protection methods presented in Sections 11.2 and 11.3.1 have been implemented to mitigate the transients. These methods – though generally accepted in practice – do not tackle the root of the problem.

The transients generated by switching at random instants can be eliminated by switching operations synchronized with regard to the instantaneous voltage and current waveforms – the so-called *controlled switching* [25, 26]. This method is also known as *point-on-wave*, *synchronized* or *intelligent switching*. Such intelligently controlled switching has the potential of avoiding virtually all transients instead of only reducing their effects, achievable by protection methods described in Sections 11.3.2 and 11.3.3.

Controlled switching is not a recent concept. Controlled closing of circuit-breakers to limit switching overvoltages was suggested as early as 1966 [27] but was viewed as less than practical until 1969 [15]. It was proposed as part of a combined closing resistor and controlled closing solution in 1970 and the method was actually tested on a power system in 1976 [28]. Only by the 1990s controlled switching had become a practically accepted reality [29, 30, 33–35]. The reason for this delay was the lack of availability of reliable control devices.

It was the intensive development of electronics that brought controlled switching into standard practice. Due to the high efficiency, reliability and low investment costs, there is currently a rapidly increasing demand from utility companies for controlled switching.

The term controlled switching applies to both *controlled opening* and *controlled closing*. Test requirements are formulated in IEC Technical Report 62271-302 [31]. The reference electrical signal can be either current (for opening operations) or voltage (for closing operations), as illustrated in Figure 11.9.

The term controlled opening refers to the technique of controlled contact separation of each pole of a switching device with respect to the phase angle of the current. In this way, the arcing times can be controlled in order to minimize stresses on the switching device and other components of the power system. More precisely, by avoiding small contact separation at the appearance of TRV, the probability of re-ignition and restrike can be reduced greatly.

Similarly, stresses can be minimized by the use of intelligent closing to control the instant of making, or current initiation, with respect to the phase angle of the voltage across the switching device.

## 11.4.2 Controlled Opening

To achieve controlled opening, the current through the circuit-breaker is monitored. The arcing time for each pole is controlled by setting the instant of contact separation with respect to the current waveform.

The schematic timing sequence for controlled opening is given in Figure 11.10.

For explanation of the terminology, all instants are identified by lowercase letter $t$ whereas intervals are designated by uppercase letter $T$. The *initial opening command* is issued randomly

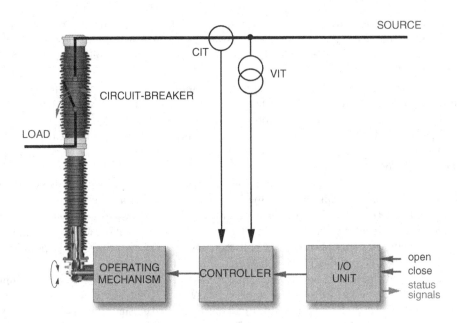

**Figure 11.9**   Diagram illustrating the principle of circuit-breaker control for controlled switching.

at some instant $t_{\text{command}}$, regardless of the phase angle of the reference signal. This command is delayed by the switching controller. The *total delay* $T_{\text{total}}$ is the sum of a *waiting time* $T_{\text{waiting}}$ used for calculations by the controller and an intentional *synchronizing delay* $T_{\text{controller}}$ introduced with respect to the relevant current zero crossing:

$$T_{\text{total}} = T_{\text{waiting}} + T_{\text{controller}} \qquad (11.6)$$

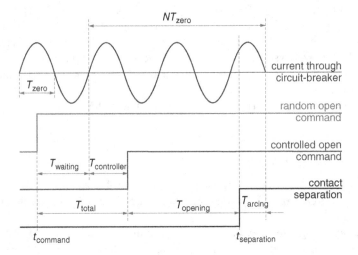

**Figure 11.10**   Schematic timing sequence for controlled opening.

$T_{\text{controller}}$ is determined by the *mechanical opening time* $T_{\text{opening}}$ and the targeted *arcing time* $T_{\text{arcing}}$.

$$T_{\text{controller}} = NT_{\text{zero}} - T_{\text{arcing}} - T_{\text{opening}} \tag{11.7}$$

$T_{\text{zero}}$ is the duration of a power-frequency half-cycle. The *opening time* $T_{\text{opening}}$ is the interval of time between the instant of energizing the opening release and the instant when the arcing contacts separate. $NT_{\text{zero}}$ is the time interval, made up of an integer number $N$ of half-cycles, which is required to achieve a positive value of $T_{\text{controller}}$.

Hence, accurate control of the *time instant of contact separation* $t_{\text{separation}}$ effectively defines the arcing time.

## 11.4.3   Controlled Closing

For controlled closing, the source voltage is monitored by the switching controller. Again, the *initial closing command* is issued randomly at some instant $t_{\text{command}}$ regardless of the phase angle of the reference signal. This command is delayed accordingly by the switching controller. Here, an intentional *synchronizing delay* $T_{\text{controller}}$ is determined by the *mechanical closing time* $T_{\text{closing}}$, the *pre-arcing time* $T_{\text{prearcing}}$, and the actual phase angle of the targeted instant for making. The schematic timing sequence for controlled closing is given in Figure 11.11. The example relates to a purely inductive load, where the optimum instant for making is at the voltage peak.

The controller calculates $T_{\text{controller}}$ by assuming a closing time and a pre-arcing time from:

$$T_{\text{total}} = T_{\text{waiting}} + T_{\text{controller}} \tag{11.8}$$

$$T_{\text{controller}} = NT_{\text{zero}} - \frac{T_{\text{zero}}}{2} - (T_{\text{closing}} - T_{\text{prearcing}}) = NT_{\text{zero}} - \frac{T_{\text{zero}}}{2} - T_{\text{making}} \tag{11.9}$$

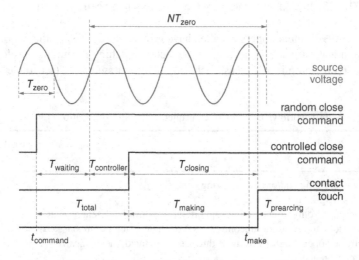

**Figure 11.11**   Schematic timing sequence for controlled closing.

The *making instant* $t_{make}$ is the instant where the current starts to flow. The *mechanical closing time* $T_{closing}$ is the time interval between energizing the closing release and the instant when the contacts touch. The *pre-arcing time* $T_{prearcing}$ is the interval of time between initiation of current flow (i.e. the instant of pre-strike) and the instant when the contacts touch. The *make time* $T_{making}$ is the interval of time between energizing the closing release and the instant when the current begins to flow.

## 11.4.4   Staggered Pole Closing

Staggered pole closing is a complementary switching overvoltage mitigation measure rather than a fundamental measure [32–35]. The principle is simply closing the individual poles one-half cycle apart in the expectation that transients in the closed phase will have greatly attenuated before the next poles close. The effect is to reduce the coupling of transients in any one phase from the other two phases. This measure is inexpensive, easily implemented and reliable.

In the case of closing into a short-circuit care must be taken that the pole delay does not lead to asymmetrical current peak values beyond the specified ones, see Section 2.2.3.

## 11.4.5   Applications of Controlled Switching

The potential benefits from controlled switching are dependent on the specific nature of the circuit to be switched. The benefits range from pure technical aspects to significant economic factors, including assessment of the costs of application and the avoided cost of maintenance.

A number of controlled load-switching applications have been developed and implemented so far. Controlled switching of shunt capacitor and reactor banks are the most common applications reported to date. Table 11.1 summarizes the percentage of conventional controlled-switching applications reported by CIGRE WG A3.07, based on a survey of installations from 1984 until 2001 [25, 36]. An estimated 2500 controllers had been installed around the world at that time.

Since the late 1990s, the numbers installed have grown rapidly. Today, the use of controlled switching in applications such as capacitor and reactor switching is routine; developments are focused on the more complicated areas, such as transformer switching (see Table 11.1).

**Table 11.1**   1984–2001 worldwide survey of controlled-switching applications [36].

| Application | Total applications reported [%] |
| --- | --- |
| Shunt-capacitor energizing and/or de-energizing | 64 |
| Shunt-reactor de-energizing and/or energizing | 17 |
| Transformer energizing only or energizing supported by controlled de-energizing | 17 |
| Line energizing and auto-reclosure (uncompensated or shunt-reactor compensated) | 2 |
| Combined controlled opening and closing of three-pole operated, mechanically staggered circuit-breakers | 7 |

There are potential technical benefits to be obtained in applying controlled switching beyond the widely accepted cases, such as capacitive and inductive switching. In practice, theoretically, almost all switching conditions can benefit from appropriate use of controlled switching.

### 11.4.5.1 Controlled Switching of Shunt Capacitor Banks

The application of controlled switching is most common in shunt capacitor bank switching. This is due to the well-defined transient behaviour of the capacitive load.

The energization of single capacitor banks causes an inrush current leading to a voltage dip at the shunt-capacitor-bank bus. This voltage dip is not a problem for the circuit-breaker but rather for the power quality, see Section 4.2.6. The optimum instant of making, that minimizes the inrush current, for a single-phase capacitor-bank current with no trapped charge is the zero crossing of the system voltage. Since earthed-neutral capacitor banks can be treated as three single-phase banks, the ideal energization instants are the zero crossings of the respective phase voltages. All three phases may be closed within 120 electric degrees, as shown in Figure 11.12a.

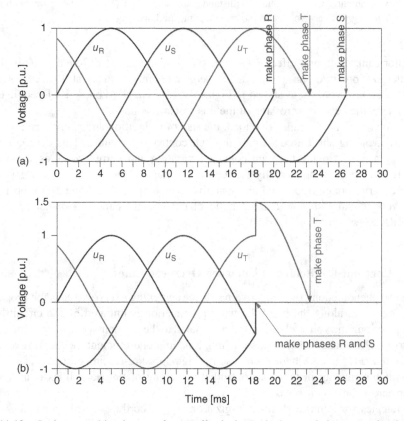

**Figure 11.12** Optimum making instants for (a) effectively-earthed-neutral shunt capacitor banks and (b) non-effectively-earthed shunt capacitor banks.

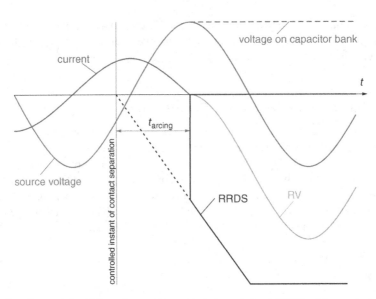

**Figure 11.13** Comparison of the voltage withstand characteristics (RRDS) of the contact gap with the capacitive-switching recovery voltage in the case of controlled opening.

Capacitor banks with non-effectively earthed neutrals require a different strategy for controlled energization. Two phases should be energized simultaneously at their corresponding phase-to-phase zero voltage, as shown in Figure 11.12b. The third phase is closed 90 electric degrees later at the next zero crossing of the line-to-earth voltage.

For the interruption of a capacitive load, the risk of re-ignition and restrike can be greatly reduced by avoiding short arcing times through controlled opening. This can be achieved by allowing a pre-set delay between contact separation and current zero, as visualized in Figure 11.13. It shows schematically the recovery voltage of the circuit-breaker gap and the system recovery voltage impressed at capacitive switching with sufficient margin between recovery voltage and withstand characteristic, characterized by the *rate-of-rise of dielectric strength* (RRDS, see Section 11.4.8.2).

### 11.4.5.2    Controlled Switching of Unloaded Overhead Lines

The physical phenomenon governing overhead-line switching overvoltages is the propagation of travelling waves along the line. The wave propagation is initiated by the circuit-breaker making operation. The voltage level is directly related to the momentary voltage at the moment of pre-strike. This makes controlled switching a natural and efficient means for overvoltage control for unloaded line switching and also for high-speed re-closing events.

The optimum making instant for controlled switching of unloaded lines is at a voltage minimum across the circuit-breaker pole.

The strategies for controlled line energization vary according to whether there is shunt compensation of the line or not and also whether there exists any trapped charge on the line, as in the case of re-closing.

Energization of unloaded lines without trapped charge is rather trivial. In that case, the line is treated in a manner similar to a shunt capacitor bank and the targeted instant of making is the zero crossing of the circuit-breaker's source-side voltage.

When an unloaded transmission line is switched off, the remaining voltage on the line will be a DC voltage, if the line is not compensated, or an oscillatory voltage, if the line is compensated, the frequency being dependent on the degree of compensation. The intent of controlled switching is then to close the circuit-breaker when the difference between the power-frequency voltage of the source and the line voltage is at a minimum. For the uncompensated line, the minimum will occur once every cycle when the polarities of the source and line voltages are the same as is shown in Figure 4.21. For the compensated line, the differential voltage minimums are frequency dependent.

Actual on-site measurements are shown in Figure 11.14. It shows that the line-side oscillation is more complicated with a higher degree of compensation. In reality, three oscillation modes are involved in the 40% compensation case and the minimums become even less distinct.

It is clear that this mitigation measure requires a dynamic controller that can analyse the differential voltage across the circuit-breaker, locate the minimums, predict the expected minimums, and close the circuit-breaker accordingly when the close signal is applied – all

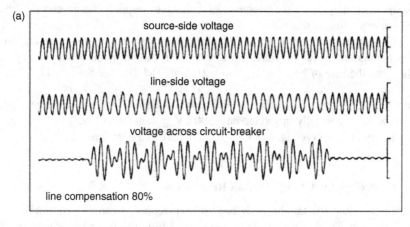

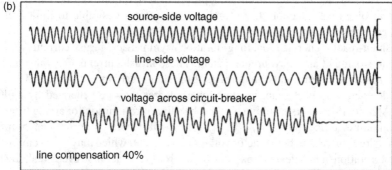

**Figure 11.14** Source-side, line-side and circuit-breaker voltage in case of de-energizing a 500 kV line compensated 80% (a) and 40% (b).

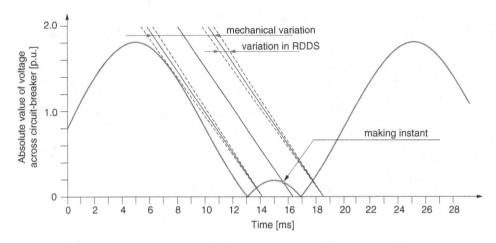

**Figure 11.15**   Controlled closing strategy under 0.8 p.u. trapped charge condition.

within 0.5 s or less. Figure 11.15 demonstrates the modification of the voltage across the circuit-breaker resulting from the presence of a 0.8 p.u. trapped charge and the optimum strategy for energization in that case.

The sophistication of the up-to-date control devices obviously goes far beyond the zero crossing detectors and sequence timers available when this measure was first proposed. In this application, the *rate-of-decay of dielectric strength* (RDDS, see Section 11.4.8.2) of the circuit-breaker is a factor.

Such a mitigation measure, combined with line-end and mid-line installed arresters rated at 372 kV, has been successfully in service on a 500 kV system since 1995 [30, 35]. The intent was to limit switching overvoltages to 1.7 p.u. anywhere along the line.

### 11.4.5.3   Controlled Switching of Shunt Reactors

The switching of shunt reactors has long been recognized as a source of current and voltage transients. Overvoltages due to current chopping and re-ignitions can be dangerous for all equipment involved (see Section 4.3) – and it is, therefore, desirable to limit them to the greatest degree possible.

During shunt-reactor energization, long-duration asymmetrical inrush currents may be generated due to closing at an unfavourable instant. These inrush currents may have undesirable effects on protection circuits and transformers if the shunt reactors are connected to buses.

During de-energization of shunt reactors, overvoltages may be generated by multiple re-ignitions. All circuit-breakers exhibit a high probability of re-ignition for arcing times shorter than a minimum (see Section 4.3.3.2). Re-ignitions will generate high-frequency transients, typically hundreds of kHz, in both reactor voltage and current, which may affect circuit-breaker elements: perforation of nozzles, arcs external to the arcing contacts and even breakdown across the external insulators of live tank circuit-breaker interrupters. The very steep voltage transients caused by re-ignitions will be unevenly distributed across the reactor winding, with the highest stress on the initial turns. There is at least a certain risk that the voltage stress will lead to

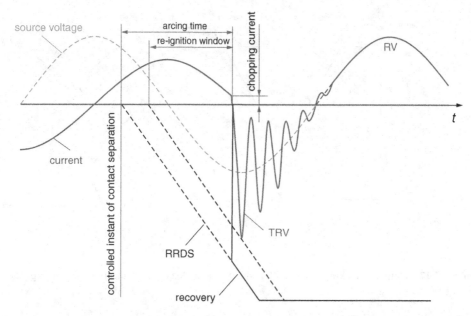

**Figure 11.16** Avoidance of re-ignition with contact separation to the left side of the re-ignition window at controlled switching off of an inductive load.

puncture of the winding insulation. Even surge arresters cannot protect the winding against these steep voltage stresses. Therefore, it is desirable to eliminate re-ignitions. Longer arcing times of course mean higher chopping overvoltages, but – since re-ignition overvoltages are normally more severe than chopping overvoltages – the use of controlled switching to increase arcing times to avoid re-ignition is highly advisable. Keeping the moment of contact separation out of the re-ignition window, see Figure 11.16, is the main guidance for avoidance of re-ignitions.

The targeted or controlled instant of making which minimizes reactor inrush currents is at the moment of power-frequency voltage maximum. For an ideal single-pole operated circuit-breaker the optimum making instants are as shown in Figure 11.17:

- making of poles R and S at the maximum voltage between the two phases;
- making of the pole T 90 electrical degrees later.

This sequence results in symmetrical energization currents in all three phases, so that the maximum inrush current is 1 p.u.

In this case, the associated switching overvoltage is generally low, but this strategy stresses the reactor insulation with a steep voltage wave front caused by the breakdown of the contact gap at the making instant. Thus, it is not possible to achieve reduction of both inrush current and transient stresses on the reactor simultaneously and, hence, a compromise solution must be reached. In practice, it is necessary for the user to decide which approach suits the requirements best.

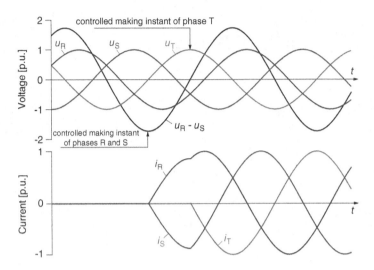

**Figure 11.17** Optimum making instants of a shunt reactor load for an ideal single-pole operated circuit-breaker.

#### 11.4.5.4 Controlled Switching of Transformers

Power transformers are vital components in electrical networks and are generally the most expensive equipment in substations. They are used in a variety of configurations and for various purposes and they are typically switched seldom, possibly once or twice per year, compared with shunt reactor or capacitor banks, which are usually switched daily. In addition, switching of unloaded transformers is a less severe duty than the switching of shunt reactors. The natural oscillations due to the transformer windings are much less pronounced and more strongly damped. The overvoltages generated during de-energization of unloaded transformers are low and normally have few consequences. Thus, there is no need for limitation of overvoltages by controlled opening.

The primary focus for controlled switching of power transformers is on energization control. The aim is to minimize current inrush transients in order to:

- reduce electromechanical stress on the windings;
- prevent temporary harmonic voltages, which may cause degradation of the power-supply quality; and
- avoid interference in secondary circuits from high zero-sequence current and avoid relay protection maloperation.

While energization of unloaded power transformers [37] is similar in many respects to the energization strategies for shunt reactors, there are also some differences. Namely, a *residual* or *remanent flux*[2] may remain in the transformer core after its de-energizing and it may influence the subsequent making conditions.

---

[2] Transformer de-energization results in a *remanent flux density* or *remanent magnetization* of the core due to *hysteresis* of the magnetic material. The value of the remanent flux in a transformer core is known as the "residual flux".

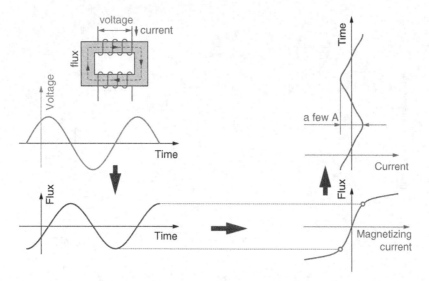

**Figure 11.18**   Relationship between applied voltage, flux and current in a power transformer in steady-state no-load condition.

The residual flux tends to be a more critical factor for controlled closing of unloaded transformers than for reactors, since reactors are either air-cored or with a gap in the core. Due to the air gap, the iron core cannot be saturated. For reasons of economy, transformers are designed with operational magnetic flux as close as possible to saturation; the core material is then optimally utilized.

The magnetic flux in the transformer core is proportional to the integral of the voltage across the winding. The resulting steady-state magnetizing current (i.e. transformer no-load current) is determined by the non-linear magnetizing characteristic. Therefore, even under steady-state conditions, the magnetizing current has a pronounced non-sinusoidal shape, as exemplified in Figure 11.18.

Due to saturation and *magnetic hysteresis*, a high flux asymmetry may be introduced if a transformer is energized at unfavourable instants, even in the case when the transformer has no residual flux. Without a residual flux, the most unfavourable instant is at the voltage zero. In this case, the flux is initially zero and the maximum developed flux will be twice the normal operational level. Since the operational level is already close to saturation, an increase in flux to double this value corresponds to extreme core saturation. Consequently, the inductance drops and the current rises rapidly to high values as in Figure 11.19a.

If a residual flux, resulting from the previous opening operation, is of the same polarity as the flux change, then the inrush current can reach even higher values, as this situation provokes still greater saturation of the magnetic circuit. Thus, the optimal making instants depend highly on the level and polarity of the residual flux. The optimum targeted instants for making are those where the prospective flux equals the residual flux. This is the way to create flux symmetry and an immediate steady-state condition, as illustrated in Figure 11.19b.

The residual flux resulting from the previous opening operation, prior to controlled closing, can be determined by measurement and integration of the voltage of each phase across the

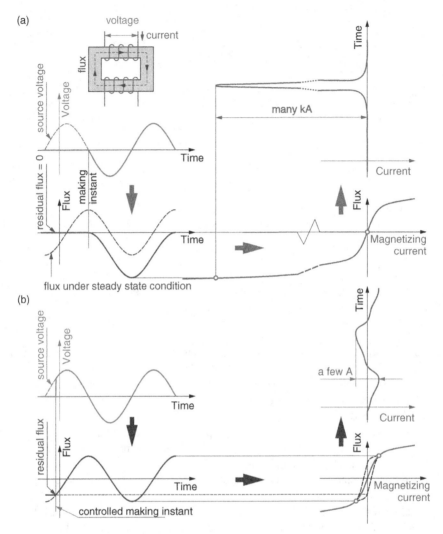

**Figure 11.19** Inrush current dependence on the moment of energization of a power transformer: (a) High inrush current resulting from the energization at voltage zero; (b) ideal making instant of energization taking into account the residual flux.

transformer terminals just before and during steady-state de-energization. This signal is more easily accessible than a direct measurement of the residual flux. Based on the calculated residual flux, the subsequent closing operation is then controlled in such a manner that the inrush current is minimized by optimizing the time instant of making in relation to the source voltage.

Residual flux in transformer cores is usually not taken into account for the choice of the controlled instant of closing, meaning that controlled closing is then not truly optimized. Normally residual flux prior to closing is taken for granted.

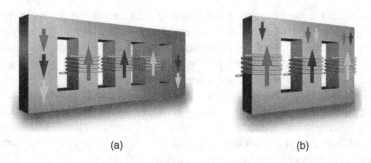

**Figure 11.20**  (a) Five-legged transformer core; (b) three-legged transformer core.

Since interruption at natural current zero leads to the lowest residual flux, controlled opening might be used in an attempt to control a level of residual flux, (i.e. its magnitude and polarity) for the next controlled closing operation. Thus, in situations where there may be residual flux, the inrush current can be limited even further by using a controlled opening technique as support for controlled closing.

The optimum targeted control of switching operations depends also on the core and winding arrangement. Depending on the arrangements, the individual phases may or may not influence each other during switching operations. In some cases, energization of the first phase results in a transient flux in the other core sections or core legs, which has to be considered when controlled switching is applied. This transient flux is termed the *dynamic core-flux*. The influence exists in the following cases of arrangements:

- a transformer with unearthed neutral on the switched side; and/or
- a three-legged transformer core; and/or
- a transformer with three separate cores with a delta-connected secondary or tertiary winding.

There is no mutual influence in the following cases:

- a transformer with a solidly earthed neutral; and/or
- a five-legged transformer core.

Five-legged cores in Figure 11.20 make the three phases magnetically independent, while three-legged cores, Figure 11.20b, enable magnetic coupling amongst the phases, as illustrated by the vertical lines, which indicate the magnetic flux of each phase. The type of core therefore determines the switching sequences for controlled switching.

For controlled switching of unloaded transformers, the circuit-breaker should be single-pole operated. Three-pole operation is unsuitable since relevant mechanical staggering is not available.

### 11.4.5.5  Controlled Fault Interruption

In recent years, investigations have been initiated to apply controlled switching to fault-current interruption. The main idea behind such a development is to limit the arcing time to the smallest possible value, thereby reducing the degradation of the interruption chamber and/or the contact system. At a random moment of contact separation, the arc duration is always longer than the minimum arcing time. If it were possible to ensure always a minimum (or slightly longer) arcing time by a controlled contact separation, the impact of the arc on the circuit-breaker chamber could be minimized. However, each type of fault (in fact each switching operation) needs, in principle, its own minimum arcing time.

Benefits of controlled fault interruption may extend beyond reduction of interrupter electrical wear; it can provide a basis for optimization of future circuit-breaker designs, including reduction in operating-energy requirements. Such potential benefits have been investigated in conjunction with high-power experiments conducted to measure the consistency and stability of minimum arcing times of HV SF$_6$ self-blast circuit-breakers [38].

Most challenging for such an application is the prediction of the moments of current zero, based on the fault-current wave shapes as measured by the protection system. Given the large variety of fault-current magnitudes and degree of asymmetry, very fast algorithms, operating in real time, have to predict current zero exactly by extrapolating the fault-current wave shape within a few milliseconds [39].

### 11.4.6  Comparison of Various Measures

Figure 11.21 shows the cumulative frequency distribution of the overvoltages at the open end of a transposed 330 km long 500 kV line following high-speed reclosure for the various

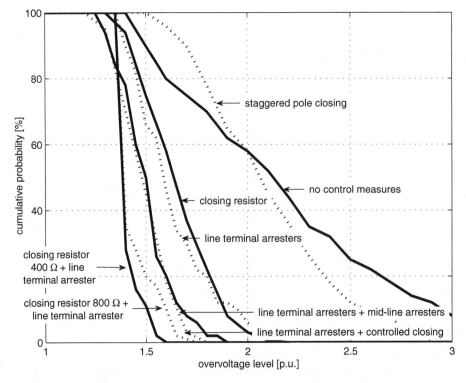

**Figure 11.21**  Cumulative distribution of frequency of occurrence of overvoltages for various switching overvoltage mitigation measures.

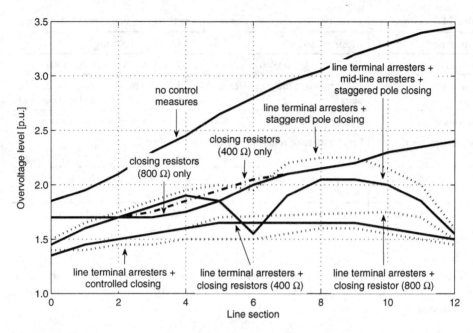

**Figure 11.22** Overvoltage profile along 330 km long, 500 kV transposed line for various overvoltage mitigation measures.

overvoltage limitation measures. The arresters considered were rated at 372 kV. Clearly, staggered pole closing has only incremental value and cannot be considered as a stand-alone option. Effective overvoltage mitigation measures are closing resistors, line-terminal arresters (meaning arresters at both line terminals), mid-line arresters, and controlled closing combined with line-terminal arresters.

The effect of the various options on the voltage profile along the line is of greater interest, and this is shown in Figure 11.22 for a 330 km long 500 kV line.

The two options that provide the most significant overvoltage limitation along the line profile are closing resistors or controlled closing, both combined with line-terminal arresters. At 800 kV and below, closing resistors are now rarely used; instead, either only the arresters option or the controlled closing option is adopted. At the UHV levels of 1000 kV and above, the situation is different, as discussed below.

Circuit-breakers applied at 1000 and 1200 kV will incorporate opening resistors in order to meet transient recovery voltage (TRV) requirements during fault clearing. The resistors, though probably rated resistance-wise for opening, can double as closing resistors. However, as shown in Figures 11.21 and 11.22, the lower the resistance value, the greater the overvoltage limitation. The choice between the closing resistor versus controlled closing is a decision for the user; a comparison is given in Table 11.2.

The above discussion and results of the study performed for a transposed 500 kV transmission line illustrate the combined contribution of the line-end arresters and closing resistors, or controlled closing, to switching overvoltage limitation. The results can certainly be viewed as indicative of performance but to determine absolute design parameters the user needs to conduct studies focused on the actual application details and considerations. Accurate

**Table 11.2**   Comparison of overvoltage limitation options: closing resistor and controlled closing.

| Attribute | Closing resistor | Controlled closing |
|---|---|---|
| Proven technology | Yes, in use since 1960 | Yes, in use since early 1990s |
| Complexity | High; multiple moving mechanical parts | Low; no moving parts, circuit boards and associated software only |
| Location | At line potential | At earth level in control room |
| Maintainability | Low; requires circuit-breaker outage | High; does not require circuit-breaker outage |
| Provision of spares | Yes; parts only rather than complete module | Yes; complete module or circuit boards |
| Future improvement potential | Limited if at all | Yes; component and software advances |

representation of arrester and controller characteristics, resistor insertion and line configuration is essential in order to achieve a valid design basis for lines meeting dependability requirements.

## 11.4.7   Influence of Metal-Oxide Surge Arresters on Circuit-Breaker TRVs

Standardized TRVs are inherent test-circuit values to be achieved without regard to either possible circuit-breaker and test-circuit interaction or network-component influence. One such component is the metal oxide surge arrester (MOSA). As AC system voltages have increased from HV through EHV 345 to 800 kV to UHV 1100 or 1200 kV, the ratio between insulation levels and system voltages has decreased, and, likewise, arrester protective levels have decreased with increased potential to limit TRV magnitudes.

The notion of TRV limitation by MOSA is not new. MOSAs are applied on EHV line terminals where the lines are series compensated with metal oxide varistors (MOV) protected series capacitor banks for the purpose of limiting the TRV seen by the line circuit-breaker [40]. It must be recognized, that application of MOSA at one side of the circuit-breaker does not automatically reduce TRV, since TRV is composed of transient contributions from the network at both sides of the device. In the Turkish 400 kV system, limitation was achieved by connecting the MOSA across the line circuit-breakers [41].

For general purpose circuit-breaker applications, the possible influence of MOSAs (and other network components) is treated as having added value rather than fundamental value. Figure 11.23 shows the IEC TRV envelopes (see Section 13.1.2) for circuit-breakers rated at 550 and 1100 kV in comparison with the typical switching-impulse protective levels (SIPL) of arresters applied.

Review of Figure 11.23 shows that the arresters will limit the peak TRV values for all the test duties, under the condition that the TRV is not composed of transient parts at either side of the circuit-breakers, as with transformer secondary fault, long line fault, out-of-phase fault, etc.

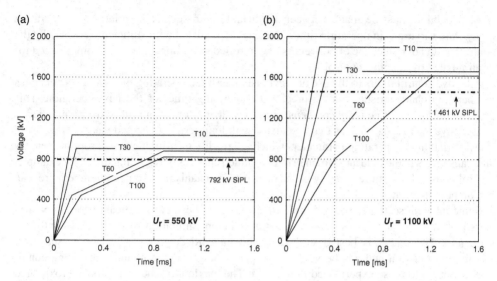

**Figure 11.23** Circuit-breaker TRVs: (a) circuit-breaker rated voltage 550 kV and surge arrester rated 396 kV SIPL; (b) circuit-breaker rated voltage 1100 kV and surge arrester rated 828 kV SIPL.

The importance of MOSAs in limiting TRVs, particularly for UHV systems, is evident and the following can be concluded:

- The influence of MOSAs on mitigating or limiting the TRV peak values of circuit-breakers has been presently viewed as a latent hidden value.
- Consideration of this influence has considerable merit, but generalization is difficult because MOSA ratings – and hence protective levels – are system-specific, being dependent on expected *temporary overvoltage* (TOV) magnitudes and durations.
- Users should recognize the technical advantage of the TRV peak limitation and the associated economic value that the approach offers by allowing the use of circuit-breakers with fewer interrupters in series.
- If the MOSA TRV limiting approach is pursued, then the user is advised to ensure that the actually applied TRVs are covered by available circuit-breaker-type test reports. The most recent amendment to the IEC circuit-breaker standard addresses the TRV requirements for 1100 and 1200 kV circuits and the calculations of the test TRVs as influenced by opening resistors [42]. The additional influence of MOSAs can be calculated similarly.

## 11.4.8   Functional Requirements for Circuit-Breakers

### 11.4.8.1   Mechanical Characteristics

With respect to controlled switching, an ideal circuit-breaker should show no variation of its operating times. Real circuit-breakers will exhibit some variation of operating, opening, and closing times. The absolute value of the variation of the closing time is usually larger than that of the opening time because opening times are often less than half the closing times. If these variations are relevant, then different approaches for corrections may be used.

A distinction must be made between predictable and stochastic variations in operating times. Any variations in operating times, that can be predicted with sufficient accuracy by the controller, do not reduce the effectiveness of controlled switching and can be compensated for or eliminated by adaptive control.

Adaptive control refers to the use of previously measured operating times to detect changes in operating characteristics and to predict the operating time for the next operation. This type of control can effectively compensate for any drift in operating times over a number of consecutive operations, such as that associated with long-term ageing and wear.

All variations in operating parameters which are readily measurable by appropriate sensors and transducers in the field and which result in defined changes of operating times, for example, control voltage level, stored energy of the operating mechanism, or ambient temperature, can be compensated for.

Some inherent scatter of the operating times will occur even at identical operating parameters and ambient conditions; this presents an inherent limitation to the use of controlled switching. The scatter is best described by the standard deviation $\sigma_{mech}$ of its statistical distribution function and can be assessed by performing operations under operating conditions identical to those experienced in the field. The maximum scatter may be approximated by $3\sigma_{mech}$.

In terms of controlled switching, there is a major difference between single-pole operated and three-pole operated circuit-breakers because the currents and the voltages in the three phases of the three-phase power systems are 120 electrical degrees apart. Controlled switching can be optimized only in the case of single-pole operated circuit-breakers, since each pole of the breaker must be controlled and operated independently. In the case of so-called "gang-operated" circuit-breakers with a single mechanism, a fixed mechanical stagger between poles is possible. However, for gang-operated circuit-breakers, which achieve a phase staggering by mechanical means, a variation in circuit-breaker characteristics cannot be compensated, and this type of controlled switching is rarely used.

### 11.4.8.2   Electrical Characteristics

An ideal circuit-breaker should have the following electrical characteristics with respect to controlled switching:

- During closing, the dielectric withstand characteristic of the contact gap is infinite as long as the contacts do not touch and, consequently, there is no pre-strike and no pre-arcing time. The making instant is always the same as the closing instant.
- During opening, the probability of re-ignition or restrike after initial interruption is zero. The current is interrupted at the first current zero.

For a given contact gap during circuit-breaker opening operation, there is a certain voltage level at which the gap will break down and current will flow. Without considering the effect of arcing, a first approximation is a linear increase in breakdown voltage with the gap spacing after the contact separation. Knowing the travel characteristics of the circuit-breaker contacts, this is related to time as a *rate-of-rise of dielectric strength* (RRDS), which is a very important electrical characteristic for controlled opening.

When interrupting capacitive currents, there is a certain probability of restrike for particular circuit-breakers, depending upon the contact gap at current zero and the RRDS. The dielectric stresses can be decreased if the recovery voltage is applied across a larger gap. Controlled opening facilitates current interruption at a relatively large contact gap by avoiding short arcing times. In this way, dielectric stresses and re-ignition probabilities can be markedly reduced. Controlled opening is, in this context, more common on filter banks rather than capacitor banks.

When interrupting inductive currents there is again a certain probability of re-ignition since the contact gap at current zero may not be sufficient to withstand the transient recovery voltages, which are determined by the multiple re-ignitions behaviour of the particular type of circuit-breakers and the load characteristics. The arcing time, and therefore the contact gap at current zero, should be such as to ensure interruption without re-ignitions and the application of controlled opening is an appropriate method of achieving this.

The key electrical characteristic for controlled closing is *rate-of-decay of dielectric strength* (RDDS). When the contacts are close to touching during the closing operation, the mean value of the decrease in the withstand voltage can be approximated by a linear function of time. The slope $S_0$ is proportional to the mean value of the closing velocity and the gas pressure in the case of gas circuit-breakers.

The making instant $t_{make}$ is when the voltage across the circuit-breaker exceeds the withstand voltage of the contact gap. The voltage across the circuit-breaker $u_{12}$ is given as an absolute value, as shown in Figure 11.24, since polarity effects may be neglected. Pre-strike will occur at time $t = t_{make}$, from the following equation:

$$|u_{12}(t)| = -S_0(t - t_{target}) \quad \text{for } t \leq t_{target} \tag{11.10}$$

where

$t_{target}$ is the targeted instant of the contact touch;

$|u_{12}(t)|$ is the absolute value of the voltage across the circuit-breaker;

$S_0$ is the RDDS.

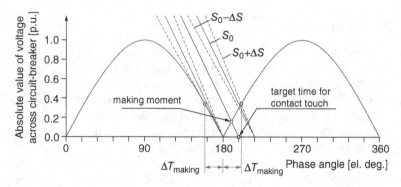

**Figure 11.24**   Switching strategy for closing targeted at voltage zero and the influence of mechanical and electrical variations.

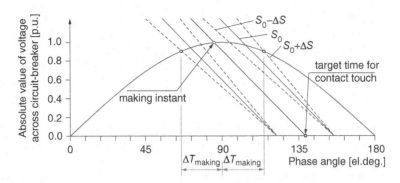

**Figure 11.25**  Switching strategy for closing targeted at voltage peak and the influence of mechanical and electrical variations.

The statistical nature of dielectric breakdown has a significant influence on the RDDS. Therefore, the actual RDDS exhibits some scatter. The maximum scatter may be approximated by $\Delta S = 3\sigma_{electrical}$.

Figure 11.24 shows targeting at voltage zero, for example, when energizing a capacitor bank. The making instants which result from a mechanical variation of the closing time and the pre-strike behaviour are indicated. The broken lines are due to variation of RDDS, the solid lines are due to the mechanical variations. Any value of the voltage that lies within the range bounded by the outer lines can occur as the actual making voltage. The corresponding making-time window is defined by $\pm\Delta T_{making}$ around voltage zero. The target point should be set such that symmetry of the making voltages on both falling and rising slope of the voltage waveform is achieved.

If the making target is at the voltage peak, for example, in order to obtain symmetrical current or inrush-current free transformer energization, the target point is optimized as in Figure 11.25. The corresponding making-time window is defined by $\pm\Delta T_{making}$ around the voltage peak.

It should be noted that a very small value of the RDDS ensures making near the voltage peak, no matter how large the mechanical scatter may be.

## 11.4.9   Reliability Aspects

Reliability of controlled-switching systems must be considered as a whole, that is, circuit-breaker, controller, sensors and auxiliaries. As with any electronic device, there is a risk of malfunction of the controller itself. The mean time between failures of electronic equipment is much shorter than that of $SF_6$ circuit-breakers [43]. It may be assumed that this also applies to the controllers. Consequently, the reliability of the combination of circuit-breaker and controller will be determined by the reliability of the controller and, consequently, it will be lower than the reliability of the circuit-breaker itself.

To some extent, this may be compensated by an improved reliability and lifetime of the equipment to be switched, that is, shunt capacitors, reactors, or transformers. It can be expected that the reduced stresses due to controlled switching will also reduce the failure rate of this equipment.

In addition, the ambient conditions under which controllers have to operate can hardly be controlled. Those components incorporated into circuit-breakers are exposed to critical environmental conditions due to local climate, mechanical vibrations, as well as electric and magnetic fields.

If errors are detected in the controller, switching operations can still be performed as uncontrolled. For this reason, the circuit-breaker must be designed to meet existing standards and has to be proven suitable for its basic duties. Moreover, conventional measures are required for protection of transformers, reactors and other equipment in the power system. Reliability of a controlled switching system is of paramount importance but, in addition, the consequences of failure are of particular importance. These consequences should be taken into consideration in the specification of the switching system. In general, each switching system requires a default mode for a possible failure of the controller to be able to still execute a required duty.

## 11.5   Practical Values of Switching Overvoltages

### 11.5.1   Overhead Lines

There is extensive literature and field experience on switching overvoltages of overhead lines and on most of the applicable means for mitigation [44]. Many utilities design overhead lines of rated voltages up to 420 kV with a 3 p.u. of the *switching-impulse withstand voltage* (SIWL). This implies a very high probability to withstand 3 p.u. because three times standard deviation gives 99.86% withstand probability in Gaussian statistics. With this approach, the only applied overvoltage mitigation measure up to 420 kV overhead lines is the specification of the pole discrepancy for closing of circuit-breakers to be less than 5 ms.

Overhead lines with a rated voltage of 550 kV and above are usually designed with lower per unit values of SIWLs and therefore they require means for limitation of the switching overvoltages. Note that the switching overvoltages appear at the line side of the circuit-breaker. The effects of various mitigation measures can be summarized as follows:

- Pre-insertion (closing) resistors reduce the switching overvoltages to less than 2.0 to 2.2 p.u. with one resistor per interruption chamber and to less than 1.5 to 1.6 p.u. with two resistor steps per interruption chamber.
- Controlled switching leads to closing at the busbar voltage zero and reduces the switching overvoltage to less than 2 p.u.
- Metal oxide surge arresters may reduce the switching overvoltage to values between 1.7 and 2.2 p.u., depending on their switching impulse protection level with the matching energy absorption capacity.
- Staggering the closure of circuit-breaker poles will reduce the transient overvoltages to less than 2.4 p.u.

Shunt reactors, when applied, are usually pre-connected to the line (preferably at the receiving end) before energizing; they provide limitation of the temporary overvoltages as well as the transient energizing and re-energizing overvoltages.

The overvoltages caused by the single-pole automatic reclosure (SPAR) are usually less than 2.4 p.u., and up to 2.7 p.u. or even 3.0 p.u. with very unusual line and grid configurations. In any case, they do not exceed the no-load energizing overvoltages as regards the above-mentioned limitation means. When no mitigation measures are taken, three-pole auto-reclosure (TPAR) may cause overvoltages as high as 3.8 to 4.0 p.u. due to the trapped charge on the healthy-phase conductors.

With TPAR the following limiting means may be applied:

- Shunt reactors connected to the overhead line will discharge the line in the dead time between fault clearing and reclosure. The TPAR overvoltage is practically reduced to the switching overvoltages at no-load line energizing.
- Inductive type voltage transformers directly connected to the line terminals discharge the line in the dead time. The TPAR overvoltage is practically reduced to the case of no-load energizing.
- MOSA connected at the line terminals will limit overvoltages to the SIPL.
- Controlled switching at circuit-breaker closure, preferably when the voltage across the open contacts of each pole is close to zero, may reduce the overvoltage to less than 2.0 p.u. When this solution is applied, measurement of the trapped charge DC voltage requires special apparatus, usually not available in substations. Practical solutions are the following:
  - If the line has shunt reactor(s), close each pole when the oscillatory discharge-voltage on the line side of the relevant phase is minimal.
  - If shunt reactors and/or inductive type potential transformers are not available, synchronize the pole reclosure of the healthy phases at the peak of the busbar voltage (or 50% of it) with the same polarity as the trapped charge. When this polarity is determined to be the same as the phase voltage at the instant of current interruption, the faulty phase(s) can be reclosed when the busbar voltage is zero. By assuming a variation of the circuit-breaker poles at closing of ±1.5 ms, the TPAR overvoltages are limited to less than 3 p.u.
- One step closing resistors reduce the switching overvoltages to less than 2.5 p.u.

## 11.5.2   Shunt Capacitor Banks and Shunt Reactors

Energizing shunt capacitor banks may result in a severe voltage dip, followed by a voltage overshoot at the natural frequency (typically 300 to 900 Hz) of the $LC$ circuit, determined by its short-circuit reactance and the bank's capacitance (see Section 4.2.6, Figure 4.23). Switching overvoltages are up to 1.6 p.u., but voltage magnification elsewhere may cause much higher overvoltages (for instance at remote capacitor banks, at open-ended lines, at radially-fed transformers, or due to capacitive coupling between transformer windings [45]). Under such circumstances, uncontrolled energization of capacitor banks presents an unacceptable risk of flashovers so that measures such as pre-insertion resistors, series reactors or controlled switching are necessary. The topic of overvoltage protection by means of surge arresters is complicated; it is considered in detail in [46].

Circuit-breakers for shunt capacitor banks are specially designed to interrupt capacitive currents with a minimum probability of restrikes. In the case of a restrike, the current starts to flow again, and the shunt capacitor bank is energized while it is still charged with voltage of opposite polarity. The fast-fronted overvoltage may show values twice as large as the slow-fronted overvoltage described in Section 4.2.2, and even more in the case of multiple restrikes. It is therefore strongly recommended to apply only circuit-breakers with a very low restrike probability, termed class C2 in IEC 62271-100 [42]. In any case, controlled switching is recommended.

The inrush current of transformers and shunt reactors is usually expressed in per unit value related to the peak value of the rated current. Energizing large power transformers leads to high inrush currents (up to typically 4 to 4.5 p.u. when energizing the outer winding). The inrush currents are rich in harmonic content and lead to temporary overvoltages rather than transient overvoltages, with peaks as high as 2.0 to 2.5 p.u. The interruption of transformer magnetizing current is normally not a problem, although high overvoltages have been reported under some exceptional conditions [47].

Shunt reactors are designed such that the inrush current does not exceed 2.8 to 3.0 p.u [48]. When de-energizing shunt reactors, multiple re-ignitions may occur, as described in Section 4.3.3.2. High, fast-fronted overvoltages may appear. Switching of shunt reactors may be quite frequent (daily) and therefore deserves due attention because re-ignitions can be dangerous for circuit-breakers and shunt reactors. Interruption chamber explosions, causing serious danger and damage to HV-equipment in the vicinity, have occasionally occurred with circuit-breakers not tested, not protected or not controlled for the specific duty.

IEC 62271-110 [49] defines tests and describes evaluation methods for circuit-breakers used for switching shunt reactors. However, it advises that laboratory tests using an actual shunt reactor will not necessarily be valid for other cases. On the other hand, it is not easy or practical to perform field tests for specific shunt-reactor applications. Thus, other countermeasures are advisable.

Limitation methods of the overvoltages to acceptable values for circuit-breakers used to switch shunt reactors are the following:

1. *Use of MOSA*
   If the current-chopping overvoltage causes the operation of MOSA, the phase-to-earth overvoltage is limited on the shunt-reactor side to the switching-impulse protection level of the MOSA; however, the risk of re-ignition, although somewhat reduced, is not eliminated. MOSAs are in any case applied to limit the switching and lightning overvoltages, which can stress the shunt reactors. The limitation is generally within a margin of not less than 25%.
2. *Use of opening resistors*
   Resistors with a resistance of the same order as the shunt-reactor surge impedance (for instance 4000 $\Omega$) provide limitation of the overvoltages caused by current chopping. They also reduce the probability and severity of re-ignitions. Opening resistors have been applied with air blast circuit-breaker technology, but they have been abandoned with single-pressure $SF_6$ technology because they bring about added complexity and increased risk of failure.
3. *Use of metal-oxide varistors connected across the circuit-breaker main contacts*
   These MOVs limit the TRVs across the circuit-breaker as well as the re-ignition overvoltages, thus reducing also the stresses on the shunt reactors.
4. *Use of synchronizing relays for controlled opening of circuit-breaker poles.*
   Controlled opening eliminates the risk of re-ignition because it permits contact separation in each pole at a point on the current wave that ensures a sufficiently long arcing time and thus an adequate contact gap for successful interruption at the first current zero. The sensor of the synchronizing controller is supplied by the secondary winding of the inductive or capacitive voltage transformers installed at the same busbar or line terminal to which the shunt reactor is connected.

In summary, the controlled pole opening of circuit-breakers used for switching shunt reactors is the most effective means for eliminating the risk of circuit-breaker damage. Combined with the MOSAs connected close to the shunt reactor terminals for protection against the switching and lightning overvoltages, it ensures trouble-free switching of shunt reactors. However, keeping in mind that the synchronizing device (a digital controller) might occasionally malfunction or be out of service, it is advisable to install circuit-breakers that have been type tested for switching-off shunt reactors in accordance with IEC 62271-110 [49].

# References

[1] IEC 60071-1 (2006) Insulation co-ordination – Part 1: Definitions, principles and rules, Ed. 8.0.
[2] CIGRE Working Group A3.12 (2007) Changing Network Conditions and System Requirements, Part II: The impact of long distance transmission on HV equipment. CIGRE Technical Brochure 336.
[3] IEC 60060-1 (2010) High-voltage test techniques – Part 1: General definitions and test requirements, Ed. 3.0.
[4] CIGRE Working Group C4.407 (2013) Lightning parameters for engineering applications. CIGRE Technical Brochure 549.
[5] IEC 62271-1 (2007) High-voltage switchgear and controlgear – Part 1: Common specifications, Ed. 1.0.
[6] IEC 62271-306 (2012) High-voltage switchgear and controlgear: Part 306 Guide to IEC 62271-100, IEC 62271-1 and other IEC standards related to alternating current circuit-breakers.
[7] CIGRE Working Group C4.306 (2013) Insulation coordination for UHV AC systems, CIGRE Technical Brochure 542.
[8] Carrara, G. (1964) Investigation on impulse sparkover characteristics of long rod/rod and rod/plane air gaps. CIGRE Report No. 328.
[9] Paris, L. and Cortina, R. (1968) Switching and lightning impulse characteristics of large air gaps and long insulator strings. *IEEE Trans. Power Ap. Syst.* (87), 947–957.
[10] Schneider, K.H. on behalf of CIGRE SC 33 (1977) Positive discharges in long air gap at Les Renardieres, *Electra*, **53**, 33–153.
[11] Allen, N.L. (2004) Mechanism of air breakdown, Chapter 1, in *Advances in High-Voltage Engineering* (eds M. Haddad and D. Warne), IEE Press, ISBN 0 85296 158 8.
[12] Cooray, V. (2003) Mechanism of electrical discharges, Chapter 3, in *The Lightning Flash*, IEE Press, ISBN 0 85296 780 2.
[13] IEEE Std. C57.142-2010 (2011) IEEE Guide to Describe the Occurrence and Mitigation of Switching Transients Induced by Transformers, Switching Device, and System Interaction.
[14] Wagner, C.L. and Bankoske, J.W. (1967) Evaluation of surge suppression resistors in high-voltage circuit-breakers. *Trans. Power Ap. Syst.*, **PAS-86**, 698–707.
[15] Baltensperger, P.A. and Djurdjevic, P. (1969) Damping of switching overvoltages in EHV networks – new economic aspects and solutions. *IEEE Trans. Power Ap. Syst.*, **PAS-88**, 1014–1022.
[16] Colclasser, R.G., Wagner, C.L. and Donohue, E.P. (1969) Multistep resistor control of switching surges. *IEEE Trans. Power Ap. Syst.*, **PAS-88**, 1022–1028.
[17] CIGRE Working Group A3.17 (2013) MO Surge Arresters. Stresses and Test Procedures. CIGRE Technical Brochure 544.
[18] Ribeiro, J.R. and McCallum, M.E. (1989) An application of metal oxide surge arresters in the elimination of need for closing resistors in EHV circuit-breakers. *IEEE Trans. Power Deliver.*, **4**, 282–291.
[19] Blakow, J.K. and Weaver, T.L. (1990) Switching Surge Control for the 500 kV California-Oregon Transmission Project. CIGRE Conference, paper 13-304.
[20] Eriksson, E., Grandl, J. and Knudsen, O. (1990) Optimized Line Switching Surge Control Using Circuit-Breakers Without Closing Resistors. CIGRE Conference, paper 13-305.
[21] Musa, Y.I., Keri, A.F.J., Halladay, J.A. *et al.* (2002) Application of 800 kV SF$_6$ dead tank circuit-breaker with transmission line surge arrester to control switching transient overvoltages. *IEEE Trans. Power Deliver.*, **17**, 957–962.

[22] CIGRE Working Group 33.06 (1990) Metal Oxide Surge Arresters in AC systems, Electra, No. 133.

[23] Smeets, R.P.P., Barts, H., Linden, van der, W., and Stenström, L. (2004) Modern ZnO Surge Arresters under Short-circuit Current Stresses: Test Experiences and Critical Review of the IEC Standard. CIGRE Conference, paper A3–105.

[24] Thoren, H.B. (1971) Reduction of switching overvoltages in EHV and UHV systems. *IEEE Trans. Power Ap. Syst.*, **PAS-90**, 1321–1326.

[25] CIGRE Working Group 13.07 (1999) Controlled switching of HVAC circuit-breakers – Guide for application lines, reactors, capacitors, transformers. 1st Part: *Electra* **183**, 43–73; 2nd Part: *Electra* **185**, 37–57.

[26] CIGRE Working Group A3.07 (2004) Controlled Switching of HV AC Circuit-Breakers – Part 1: Benefits and Economic Aspects, CIGRE Technical Brochure 262; Part 2: Guidance for further Applications including Unloaded Transformer Switching, Load and Fault Interruption and Circuit-Breaker Uprating, CIGRE Technical Brochure 263 (2004); Part 3: Planning, Specification and Testing of Controlled Switching Systems, CIGRE Technical Brochure 264 (2004).

[27] Maury, E. (1966) Synchronous Closing of 500 and 765 kV Circuit-Breakers: a Means of Reducing Switching Surges on Unloaded Lines. CIGRE Conference, paper 143.

[28] Konkel, H.E., Legate, A.C. and Ramberg, H.E. (1977) Limiting switching surge overvoltages with conventional power circuit-breakers. *IEEE Trans. Power Ap. Syst.*, **PAS-96**, 535–542.

[29] Khan, A.H., Johnson, D.S., Brunke, J.H. and Goldsworthy, D.L. (1996) Synchronous Closing Application in Utility Transmission Systems. CIGRE Conference, paper 13-306.

[30] Avent, B.L., Peelo, D.F. and Sawada, J. (2002) Application of 500 kV Circuit-Breakers on Transmission Line with MOV Protected Series Capacitor Bank. CIGRE Conference, paper 13-107.

[31] IEC Technical Brochure 62271-302 (2010) High-voltage switchgear and controlgear – Part 302: Alternating current circuit-breakers with intentionally non-simultaneous pole operation.

[32] Legate, A.C., Brunke, J.H., Ray, J.J. and Yasuda, E.J. (1988) Elimination of closing resistors on EHV circuit-breakers. *IEEE Trans. Power Deliver.*, **3**, 223–231.

[33] Avent, B. and Sawada, J. (1995) BC Hydro's Experience with Controlled Circuit-Breaker Closing on a 500 kV Line. Canadian Electrical Association, Engineering and Operating Division Meeting.

[34] Froehlich, K., Hoelzl, C., Carvalho, A.C. and Hofbauer, W. (1995) Transmission Line Controlled Switching. Canadian Electrical Association, Engineering and Operating Division Meeting.

[35] Froehlich, K., Hoelzl, C., Stanek, M. *et al.* (1997) Controlled closing on shunt reactor compensated transmission lines: Part 1 closing control device development and Part 2 application of closing control device for high-speed autoreclosing on BC hydro 500 kV transmission line. *IEEE Trans. Power Deliver.*, **12**, 734–746.

[36] CIGRE Working Group A3.07 (2004) Controlled switching: Non-conventional applications, *Electra*, **214**, 28–39.

[37] CIGRE Working Group C4.307 (2014) Transformer energization in power systems: a study guide, CIGRE Technical Brochure 568.

[38] Thomas, R. and Sölver, C.-E. (2007) Application of Controlled Switching for High Voltage Fault Current Interruption. CIGRE SC A3 Coll., Rio de Janeiro.

[39] Pöltl, A. and Fröhlich, K. (2003) A new algorithm enabling controlled short circuit interruption. *IEEE Trans. Power Deliver.*, **18**(3), 802–808.

[40] Avent, B.L., Peelo, D.F. and Sawada, J.H. (1995) Circuit-Breaker TRV Requirements for a Series Compensated 500 kV Line with MOV Protected Series Capacitors. Coll. of CIGRE SC 13, Florianopolos, Brazil.

[41] Gatta, F.M., Illiceto, F., Lauria, S. and Dilli, B. (2002) TRVs Across Circuit-Breakers of Series Compensated Lines. Analysis, Design and Operational Experience in the 420 kV Turkish Grid. CIGRE Conference, paper 13-109.

[42] IEC 62271-100 (2012) High-voltage switchgear and controlgear – Part 100: Alternating-current circuit-breakers, Ed. 2.1.

[43] Suiter, H., Degen, W. and Eggert, H. (1990) Consequences of Controlled Switching for System Operation and Circuit-Breaker Behaviour. CIGRE Conference, paper 13-202, Paris.

[44] CIGRE Working Group A3.13 (2007) Changing Network Conditions and System Requirements, Part II: The impact of long distance transmission on HV equipment, CIGRE Technical Brochure 336.

[45] CIGRE Working Group 13.04 (1999) Shunt capacitor bank switching – Stresses and Test Methods,1st Part, Electra No. 182, pp. 165–189.

[46] Stenström, L. and Mobedjina, M. (1995) Guidelines for the selection of surge arresters for shunt capacitor banks. *Electra*, **159**, 11–24.

[47] Shirato, T., Yokutsu, K., Yonezawa, H. *et al.* (2006) Severe Stresses on Switching Equipment of 500 kV Transmission System in Japan. CIGRE Conference, Paris, paper A3-303.

[48] CIGRE Working Group B5.37 (2013) Protection, Monitoring and Control of Shunt Reactors. CIGRE Technical Brochure 546.

[49] IEC 62271-110 (2012) High-voltage switchgear and controlgear – Part 110: Inductive load switching.

# 12

# Reliability Studies of Switchgear

## 12.1 CIGRE Studies on Reliability of Switchgear

### 12.1.1 Reliability

Reliability of equipment is usually approached by its opposite: the probability of the occurrence of failures. The failures and failure rate can be investigated with a focus on the consequences for the power system or with a focus on the performance of the equipment itself. Looking at the power system, the network topology and the service condition play a dominant role in assessing the consequences of malfunctioning equipment. However, looking at the switchgear itself, the system approach gives more or less underweighted information. Therefore, in this chapter, the focus will be on the switchgear, especially on the circuit-breaker.

Following the definitions in the IEC standards [1], failures and defects in a circuit-breaker will be distinguished into:

- *major failures* (MF) – failures that cause the cessation of one or more fundamental functions of a breaker;
- *minor failures (mf)* – other failures; and
- *defects* – imperfections or inherent weakness of the state of an item, which may result in one or more failures of the device itself or of another component under the specific service, environmental, or maintenance conditions in a stated period of time.

An MF will result in an immediate change of the power-system operating conditions, requiring the back-up protective equipment to remove the fault, or it will result in mandatory removal from service within 30 minutes for non-scheduled maintenance.

There are the following possible failures of a circuit-breaker to perform its fundamental functions (i.e. the case of an MF):

- does not close or open on command;
- closes or opens without command;
- does not make or break the current;
- fails to carry the current;

*Switching in Electrical Transmission and Distribution Systems*, First Edition.
René Smeets, Lou van der Sluis, Mirsad Kapetanović, David F. Peelo, and Anton Janssen.
© 2015 John Wiley & Sons, Ltd. Published 2015 by John Wiley & Sons, Ltd.

- breakdown to earth or between poles;
- breakdown across open pole (internal or external);
- locked in open or closed position; and
- others (requiring intervention within 30 minutes).

Examples of minor failures (mf) are:

- Air or hydraulic-oil leakage in the operating mechanism;
- minor $SF_6$-gas leakage due to corrosion or other causes; and
- change of functional characteristics.

Functional characteristics are, for instance: the closing time, opening time, contact travel characteristics, pressure alarm and lock-out levels, automatic closing or opening at certain pressure levels or at a discrepancy in closing, or opening time between the poles.

Aforementioned definitions have been introduced by CIGRE WG 13.06 (later A3.06). Over the past 40 years, WG 13.06 (A3.06) conducted three international surveys on high-voltage circuit-breaker failures and defects in service. The enquiries covered failures and defects of circuit-breakers with a rated voltage of 63 kV and above.

## 12.1.2   Worldwide Surveys

In the first international enquiry, population and failure data were collected for the years 1974 up to and including 1977. The survey was limited to circuit-breakers that were put into service after 1963, but included various extinguishing technologies, such as bulk oil, minimum oil, air-blast, double pressure $SF_6$-gas, single-pressure $SF_6$-gas, and vacuum. The 102 participating utilities from 22 countries covered in total 77 892 circuit-breaker-years. Population cards and failure cards were used to collect information about population size, service experience, maintenance intervals, maintenance costs, number of switching operations, and failures. Details of failure type, category, cause, origin, consequences, service conditions, and so on were requested.

All information was analyzed per voltage class: 63 to 100, 100 to 200, 300 to 500, and ≥ 500 kV. However, no further distinction was made between the various arc extinguishing technologies. The results from the study were published in 1981 [2]. Summaries and specific investigations were brought before several professional organizations. The study resulted in new mechanical and environmental type tests: an increased number of operating cycles in the mechanical endurance test, a humidity test, and low and high temperature tests.

The same enquiry structure was applied to the second international survey, that was conducted from 1988 up to and including 1991. This was limited to single-pressure $SF_6$ circuit-breakers and to circuit-breakers put into service after 1977. By additional questions on both population and failure cards, distinction could be made between circuit-breakers put into service before and after January 1, 1983. A distinction could also be made between different technologies of the operating mechanism (hydraulic, pneumatic, or spring), between circuit-breaker enclosure (metal-enclosed or non-metal-enclosed), and location (indoors or outdoors). Utilities (altogether 132) from 22 countries contributed to a total population of 70 708 circuit-breaker-years. The final report was published in 1994 [3]. Most important was the question whether the reliability of circuit-breakers had improved, with a specific interest for the mechanical reliability. Another item was the question whether maintenance intervals were larger, compared with the first enquiry, and maintenance costs lower.

In the third survey, there was more focus on the relationship between reliability and age. Therefore, during the observation period from the beginning of 2004 to the end of 2007, it included single-pressure $SF_6$ circuit-breakers of all ages. The total population of 281 900 circuit-breaker-years was overwhelming and 83 utilities of 26 countries participated. The enquiry was one part of a larger reliability survey that also covered earthing switches, disconnectors, instrument transformers, and GIS. With respect to the circuit-breakers, a further distinction was made in the enclosures: single-phase GIS, three-phase GIS, and dead-tank design. In addition, population and failure information was collected on the circuit-breaker application: overhead line, cable, transformer, shunt reactors, shunt-capacitor bank, and bus-coupler. More details of the operational circumstances of dielectric failures were requested (circuit-breaker in open or closed position, or during closing or opening). In 2012, the results of the third worldwide survey were published in a CIGRE Technical Brochure [4], and summarized in [5–7].

## 12.1.3   Population and Failure Statistics

The third enquiry revealed that 54% of the HV circuit-breakers are applied in overhead-line bays and 24% in transformer bays, whereas 10% are used as bus couplers, and 6% to connect cables, mostly to a GIS. Although the circuit-breakers applied to switch shunt reactors and shunt-capacitor banks represent only a small percentage of the total population (1.5 and 3% respectively), they are each responsible for more than 20% of the major failures. These circuit-breakers are operated very frequently. In Figure 12.1 the average number of operating cycles (one close and one open operation) per application is compared with the MF-rate per application, as derived from the results of the third international enquiry. The trend underlines the idea that from the point of view of reliability, a circuit-breaker is mainly a mechanical device.

Concerning the increase in reliability of the operating mechanism, the third enquiry results have to be compared with those of the second enquiry. On average, all technologies for operating mechanisms show great improvement, as the MF-rate attributed to the operating mechanism decreased from 0.29 per 100 circuit-breaker-years to 0.14. Within the whole population of the third enquiry, the spring drive showed the best reliability performance, as can be learnt from Figure 12.2. Looking at the total population, the share of hydraulic mechanisms has decreased from about 50% to less than 20%, while the presence of spring drives increased from 40 to 60%. The reliability of hydraulic and pneumatic systems has increased drastically and has become comparable with or even higher than the reliability of spring drives.

The overall MF-rate improved greatly from the first enquiry (1.58 MF per 100 circuit-breaker-years) to the second (0.67 MF per 100 circuit-breaker-years) and again to the third enquiry (0.30 MF per 100 circuit-breaker-years). The improvement from the first to the second enquiry can be attributed to the change from the old technologies to the single-pressure $SF_6$ technology with its superior interruption performance per arcing chamber. For the same voltage rating and short-circuit current rating, less arcing chambers are required and thus less parts and less operating mechanisms. The improvement from the second to the third survey is assumed to be mainly due to the more efficient design with respect to the mechanical energy consumption, so that the parts of the circuit-breaker are less stressed mechanically.

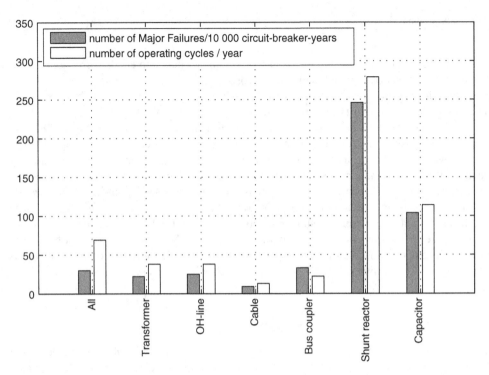

**Figure 12.1** Major-failure rate and number of operating cycles per kind of service. (From the third CIGRE enquiry [4]).

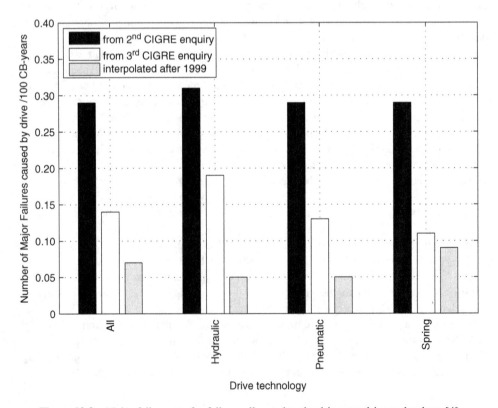

**Figure 12.2** Major-failure rate for failures allocated to the drive-per-drive technology [4].

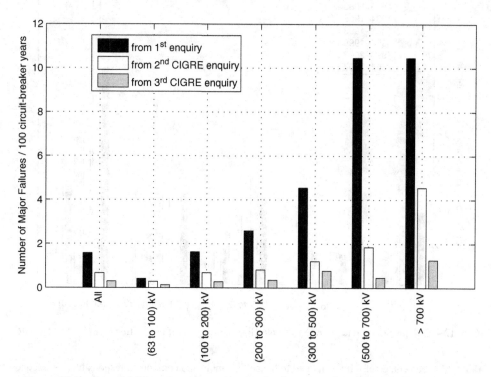

**Figure 12.3**  Overall MF rate by voltage class for the first, second, and third CIGRE enquiries.

Figure 12.3 underlines the idea that more arcing chambers, i.e. more parts, tend to reduce the overall reliability. It shows over the voltage classes a similar tendency for the three surveys. With respect to the enclosure, the third enquiry gives a far lower MF-rate for metal-enclosed circuit-breakers (GIS and dead-tank) than for live-tank circuit-breakers: 0.144 and 0.483 per 100 circuit-breakers, respectively. A similar trend could be seen in the second survey.

The development of the MF-rate by age can be deduced from the data collected in the third enquiry. The tendency of an increasing MF-rate with age, as can be seen in Figure 12.4, is caused by both the wear of older circuit-breakers and the improvement in technology with younger circuit-breakers. Surprisingly, the circuit-breakers of the oldest-age category show a better reliability than those of slightly younger categories. This trend is supposed to be caused by the higher probability that the oldest circuit-breakers have been overhauled or replaced. Therefore, by removing the worst performing circuit-breakers out of the oldest category, the average reliability performance will become slightly better.

The subassemblies and components responsible for MF give the same distribution in the second and in the third enquiry. The operating mechanism is responsible for most major and minor failures, as can be learnt from Table 12.1 that gives the distribution as collected during the second survey. With single-pressure SF$_6$ circuit-breaker technology, the subassemblies and components responsible for the MF give the same distribution regarding the third and the second enquiry.

The failure modes or failure characteristics as collected in the third enquiry are listed in Table 12.2.

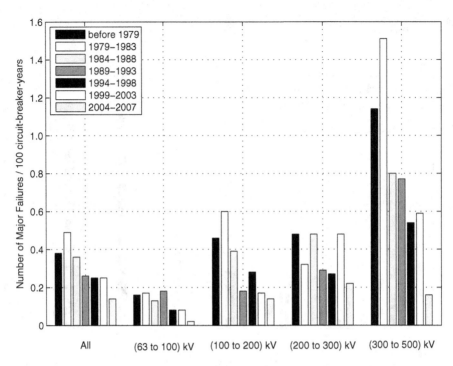

**Figure 12.4** Major-failure rate per rated voltage class and by year of production from the third CIGRE enquiry.

**Table 12.1** Percentage of failures related to the subassembly and component responsible; the second enquiry.

| Subassembly | Components | MF [%] | | mf [%] | |
|---|---|---|---|---|---|
| HV parts | | 21 | | 31 | |
| | Interruption chambers | | 14.0 | | 9.4 |
| | Aux. interrupters, resistors | | 1.3 | | 0.6 |
| | Insulation to earth | | 5.7 | | 20.9 |
| Electrical control and auxiliary | | 29 | | 20 | |
| | Trip/close circuits | | 10.0 | | 1.5 |
| | Auxiliary switches | | 7.4 | | 2.1 |
| | Contactors, heaters | | 7.6 | | 5.4 |
| | Gas-density monitor | | 4.0 | | 10.7 |
| Operating mechanism | | 43 | | 44 | |
| | Compressors, pumps, etc. | | 13.6 | | 18.7 |
| | Energy storage | | 7.6 | | 7.2 |
| | Control elements | | 9.3 | | 11.6 |
| | Actuators, damping devices | | 8.9 | | 5.1 |
| | Mechanical transmission | | 3.8 | | 1.4 |
| Others | | 7 | | 5 | |
| Total | | 100 | 100 | 100 | 100 |

**Table 12.2**  Percentage of MF-rate and mf-rate per failure mode in the third enquiry.

| Major Failure (MF) | MF [%] | Comment |
|---|---|---|
| Does not close on command | 28.2 | Mainly with live-tank circuit-breakers |
| Does not open on command | 16.4 | |
| Closes without command | 0.2 | |
| Opens without command | 5.4 | |
| Fails to carry the current | 1.3 | |
| Dielectric breakdown | 9.9 | Breakdown to earth: 5%;<br>Internal breakdown across open pole during opening operation with failure to interrupt the current: 1.9%;<br>Other breakdowns across open pole: 1.8%;<br>Breakdown between poles: 1.2%. |
| Locked in open or closed position | 25.1 | Alarm has been triggered by the control system |
| Loss of mechanical integrity | 8.1 | Mechanical damage of parts |
| Other | 5.2 | |
| Total | 100 | |

| minor failure (mf) | mf [%] | Comment |
|---|---|---|
| Air or hydraulic-oil leakage | 20.3 | In operating mechanism |
| Small $SF_6$-gas leakage | 35.6 | Large leakage will give the MF-mode "Locked" |
| Leakage of oil in grading capacitors | 1.0 | |
| Change in functional characteristics | 28.4 | 6.8% mechanical; 3.3% electrical;<br>18.3% control and auxiliary systems. |
| Other or no answer | 14.6 | |
| Total | 100 | |

About 10% of MFs were caused by dielectric breakdowns. Four per cent of the MFs (6.5% at the second enquiry) resulted in an explosion or fire; mainly in relation to dielectric breakdowns in live-tank circuit-breakers. The occurrence of explosions of circuit-breakers in various kinds of service is visualized in Figure 12.5. The probability of an explosion is only 0.01 per 100 circuit-breaker-years for the whole population.

Only a small part of the major failures has the characteristic *"Does not break the current"*: 1.9% or 0.006 per 100 circuit-breaker-years. So, the mechanical performance seems to need most attention from the point of view of reliability. The probability that a circuit-breaker has to interrupt, for instance, a short-line fault, is several orders of magnitude lower than the rate of regular operating cycles. Still, reliable fault-current interruption is a very important duty of a circuit-breaker, depending both on its capability to interrupt fault currents and its reliability to *"open and close on command"*.

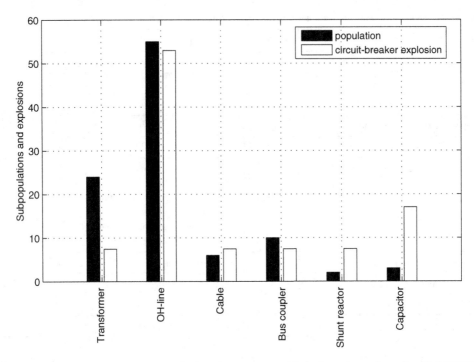

**Figure 12.5** Subpopulations and explosions in percent per kind of service – from the third CIGRE enquiry.

## 12.2 Electrical and Mechanical Endurance

During service life, circuit-breakers have to interrupt fault currents a number of times. Due to the thermal and mechanical stresses on the interruption chamber parts (mainly contacts, nozzle, and gas), the interruption of fault current is associated with a degree of degradation. Electrical endurance is the capability to withstand such repeated stresses during a long service life. Test programs have been defined by IEC separately for MV and HV circuit-breakers to demonstrate with a limited number of tests the electrical endurance during a long service life.

### 12.2.1 Degradation Due to Arcing

The arcing in the circuit-breaker interruption chamber leads to a certain degradation of the internal parts of the interruption chamber.

For SF$_6$ circuit-breakers, manufacturers usually suggest the maximum number of interruptions ($N$) at a given rated short-circuit breaking current ($I_{sc}$) as an equation in the form of:

$$N = AI_{sc}^{B} \tag{12.1}$$

Heavy users of circuit-breakers, such as high-power laboratories, use such equations to determine when their own (laboratory) circuit-breakers should be revised (complete replacement of breaker chamber interior parts).

Typical values for a 170 kV, 63 kA circuit-breaker are: $A = 2 \times 10^4$ and $B = -1.8$ ($I_{sc}$ in kA), suggesting that this circuit-breaker could interrupt 11 times its rated short-circuit breaking current.

Using this method, it follows that a circuit-breaker should undergo revision after a weighted summation of current has reached the value of 1 after $N$ high-power tests, each with a current $I_j$:

$$\sum_{j=1}^{N} \frac{1}{A} I_j^{-B} = 1 \tag{12.2}$$

In standard practice at the KEMA high-power laboratory of DNV GL, one auxiliary circuit-breaker chamber is revised 5 to 6 times per year.

Extended research into 420 kV, $SF_6$-puffer circuit-breaker degradation due to arcing has led to an empirically-derived current dependence of arcing stresses, with $F_{60}(I)$ being the *equivalent fraction of arcing stress* due to test-duty T60 (60% of $I_{sc}$) [8].

$$F_{60}(I) = 9.35 \left( \frac{I}{I_{sc}} \right)^3 \text{ for } I < 0.35 I_{sc} \qquad F_{60}(I) = 2.38 \left( \frac{I}{I_{sc}} \right)^{1.7} \text{ for } I \geq 0.35 I_{sc} \tag{12.3}$$

From this can be derived $F_{60}(0.1) = 0.01$ and $F_{60}(1) = 2.4$, so that a small-current interruption (test-duty T10) causes 0.01 of the wear of a 60% $I_{sc}$ stress. On the other hand, one interruption at full $I_{sc}$ causes 2.4 times the rate of degradation at 60% of $I_{sc}$.

There are various components that degrade due to electrical stress. It is DNV GLs' experience that the results of degradation affect the capability of each switching duty differently. For switching duties in which very-fast-rising transient recovery voltages are involved, material loss from the nozzle (*ablation*) leading to widening of the nozzle throat, and thus loss of $SF_6$ gas blast pressure, is the dominant reason for failure.

Detailed measurements of near-current-zero arc conductivity in the course of a large number of short-line-fault tests show that there is often a trend of increasing arc conductivity as tests proceed. This suggests a gradual widening of the nozzle and reduction of blast-pressure, which ultimately limits the ability to clear short-line faults [9].

For duties with the highest currents and long arc duration (asymmetrical-current interruption), the effect of arcing on the contacts, that is, contact material losses and the interaction of the metallic vapour with the direct environment, may be the dominant deterioration factor.

It is very difficult to give general rules, since some laboratory circuit-breakers end their recommended service life in spite of virtually undamaged nozzles, whereas other designs show severe nozzle damage after the same accumulated arcing stress and they are still functioning. Some end their life with virtually unaffected contacts.

Experiments suggest that the main reason for service-life termination is the wear of contacts and the associated pollution [10].

Apparently, each of the various degradation processes (loss of contact material, nozzle ablation, gas pollution, etc.) has its own impact on the service life of circuit-breakers depending on:

- the technology (puffer, self-blast);
- the current (many times a smaller current or few times a large current);
- the arcing time [11], and so on.

## 12.2.2   Electrical-Endurance Verification

Service experience indicates that the HV components of modern circuit-breakers seldom show unacceptable deterioration or wear. In order to avoid excessive maintenance of the main contacts, nozzles, or other HV components inside the interruption chambers, the electrical-endurance type-test gives an indication of the allowable number of short-circuit currents to be interrupted.

IEC has defined test programmes [12] to verify the endurance of the circuit-breaker chamber parts against repeated arcing. Two circuit-breaker classes as a function of electrical endurance are defined:

- E1 (*basic electrical endurance*): circuit-breaker has proved basic electrical endurance by having passed the standardized fault-breaking test-duties. No additional test programme has been defined.
- E2 (*extended electrical endurance*): circuit-breaker designed so as not to require maintenance of the interrupting parts of the main circuit during its expected operating life, and only minimal maintenance (lubrication, replenishment of gas, and cleaning of external surfaces) of its other parts.

There are therefore two choices: either to use a circuit-breaker having maintainable internal parts and perform maintenance as needed during its expected operating life, or to install a circuit-breaker rated class E2 having passed a more onerous test programme to verify its capability.

The E2 test-programme for HV circuit-breakers is based on a maintenance-free period of 25 years [13]. The test programme consists of a number of breaking tests for simulation of the wear, followed by a series of acceptance tests to prove the basic interruption capability. Table 12.3 shows the number of switching operations necessary to verify the extended electrical-endurance qualification.

For circuit-breakers with rated voltage up to and including 52 kV, E1 and E2 test requirements are included in [12]; for voltages above and including 72.5 kV, these requirements are formulated in [13].

The regular short-circuit current-interrupting test duties as defined in [12] give already a considerable cumulative number of kiloamperes interrupted. As long as no maintenance has been carried out between the test duties T10 to T100, no additional test is required: class E1 can then be assigned.

Essentially, the electrical-endurance type test consists of a large number of short-circuit current interruptions (see Table 12.3), followed by dielectric tests to assess the condition of the sample. The condition assessment and its evaluation are similar to that applied for the type test duties T10 to T100.

When the first edition of IEC TR 62271-310 on HV circuit-breaker E2-endurance-test requirements was issued in 2004, experts published their doubts about the statistical interpretation of the above results [14]. In the same year, CIGRE started to investigate the calculation background of fault-current probabilities. For that purpose, a Monte Carlo model was set up for line faults whereas field data of Working Group 13.08 served as input in combination with information on impedances of overhead lines and substations [15]. Statistical distribution functions (Weibull, Gaussian, and uniform) were fitted to the data.

**Table 12.3** Extended electrical-endurance (E2) test program.

| Test current/duty | | $U_r \leq 52$ kV IEC 62271-100 (2012) [12] Number of operating sequences | | | $U_r \geq 72$ kV IEC 62271-310 (2008) [13] Number of operating sequences | | |
|---|---|---|---|---|---|---|---|
| Percentage of $I_{sc}$ | Operating sequence | (list 1) | (list 2) | (list 3) | (25 kA) | (40 kA) | (63 kA) |
| 10% | O | 84 | 12 | – | 9 | 9 | 9 |
| | O-CO | 14 | 6 | – | | | |
| | O-CO-CO | 6 | 4 | 1 | | | |
| 30% | O | 84 | 12 | – | | | |
| | O-CO | 14 | 6 | – | | | |
| | O-CO-CO | 6 | 4 | 1 | | | |
| 60% | O | 2 | 8 | 15 | 15 | 10 | 7 |
| | O-CO-CO | 2 | 8 | 15 | | | |
| | O-CO-CO | 2 | 4 | 2 | | | |
| 100% (symmetrical) | | | | | | | |
| No. of equivalent 100% $I_{sc}$ stresses[a] | | 24.0 | 29.4 | 31.5 | 6.3 | 4.2 | 3.0 |

[a] This is the number of calculated switching (single-opening) operations with 100% current giving the equivalent wear, using Equation 12.3 of the summed tests listed above.

Several hundred thousands of simulations constructed a statistical distribution of fault currents. The results were verified with Japanese utility data on fault-current distribution [16]. To translate the distribution to an equivalent number of rated short-circuit-current interruptions, the equivalent electrical stress functions in Equation 12.3 were used.

Table 12.3 lists the electrical-endurance test requirements resulting from the outcome of the Monte Carlo simulations for several voltage classes and short-circuit current ratings for rated voltages $\geq 72.5$ kV.

The electrical-endurance type test essentially consists of accelerated wear tests and acceptance tests [13]. The wear tests are designed for fault-current interruptions at 60 and 10% of the rated short-circuit current without TRV stresses. These tests are immediately followed by acceptance tests composed of a slightly modified T10 test, an L75 test with a fault current reduced to 60%, capacitive-current switching tests with a reduced capacitive voltage factor, and a dielectric condition check. The fault-current clearing tests at the acceptance stage are added to those of the wear stage to achieve a number of 60% interruptions that is equivalent to the number of 100% interruptions.

From a user's perspective, regarding the majority of applications, the regular type test T100 with $\geq 5$ times a 100% fault-current interruptions gives already enough confidence that the electrical endurance during service conditions is covered. Only in special cases, where one could expect a high number or large fault currents to be interrupted, could an extended electrical-endurance test be recommended. Alternatively, the intervals for overhaul could be reduced.

This leads to other considerations on mechanical and electrical endurance. A period of 25 years has been taken into account in the standards, and is on the basis of Table 12.3. Nowadays, however, circuit-breakers are expected to have a service life beyond 40 years, maybe "maintenance free". Users have to be aware that the latter service periods (over 40 years) result in about 50% higher mechanical and electrical stresses than those indicated as "number of equivalent 100% $I_{sc}$ stresses" in Table 12.3. From a statistical point of view, only in cases where the number of operating cycles or fault clearings is exceptional, users have to be careful.

From Equation 12.3, the number of equivalent 100% $I_{sc}$ stresses can be calculated (see the last row in Table 12.3). It can then easily be established that the extended electrical-endurance requirements are far more severe for distribution circuit-breakers than for transmission circuit-breakers. This reflects the higher fault incidence in distribution systems, compared with transmission systems.

## 12.2.3  Mechanical Endurance

On the population cards used for the first and the second enquiry, specific questions were asked about the average number of operating cycles per year. In the third enquiry this question appeared on the failure cards. The statistical evaluation of the answers was different for each survey and resulted in an average of 27, 42, and 69 operation cycles per year for the first, second, and third surveys, respectively. Presumably, an average of about 30 operating cycles per year is a reasonable estimate [7]. The 90 percentile shows to be rather constant: about 80 operating cycles per year. This is a relevant figure to determine mechanical-endurance requirements.

Indeed, assuming a maintenance-free period of 25 years, 90% of the circuit-breakers will have to face 2000 operating cycles at most. For circuit-breakers applied for regular duties, the type test of mechanical-endurance class M1 defined in IEC standard 62271-100 [12], requires

2000 operating cycles without maintenance (or only maintenance to the instructions in the manufacturer's manual). After the test, the condition of the test object has to be as new (or within tolerances as given in the manual). Circuit-breaker applications that require far more operating cycles per year have to be specified and tested to class M2, extended electrical endurance. The extended mechanical-endurance type test requires 10 000 operating cycles.

For reliability assessment, and for identification of the weak spots and endurance limits, manufacturers are testing prototypes to a much larger number of cycles: tens of thousands of CO-cycles. These development tests aimed at increase in reliability lead to adequate information on necessary maintenance intervals and maintenance actions [17].

## 12.3   CIGRE Studies on Life Management of Circuit-Breakers

Circuit-breaker *life management* covers all periods of the life of a group of circuit-breakers in the following terms: specification, development, testing, acceptance, erection, on-site commissioning, inspection, maintenance, diagnostics and monitoring, refurbishment, dismantling and disposal, and all necessary administrative procedures [18]. More confined, the term life management is applied to the decision-making process with respect to the *residual life* of a circuit-breaker. Residual life refers to the circuit-breaker's remaining technical life or better, to the remaining life of a population of circuit-breakers. Apart from the residual life, other aspects may play a decisive role in the choice between refurbishment, replacement, or continuing service operation without any special action. Such aspects are, for instance, of an economic, legal, organizational, or environmental nature. Changes in network conditions are amongst the most important parameters that influence end-of-life decisions.

In general, because of the relatively low capital cost and operating-cost levels of new circuit-breakers, life extension beyond 40 years is exceptional. Obsolete technology and the lack of spare parts, tools, and know-how contribute to the choice for replacement rather than refurbishment – though other factors, such as the location and enclosure (for example GIS), or the combination with other equipment, may hamper replacement.

Nevertheless, practice shows that in existing applications, certain designs of MV and HV circuit-breakers have proved to be very robust and reliable. For those designs, even after several decades of operation, a significant extension of lifetime can be assumed through a programme of refurbishment and re-testing, see for an example Section 7.2, saving substantial costs without compromising on reliability.

Residual-life assessment requires knowledge about the condition of individual circuit-breakers and of their sub-populations. Maintenance, monitoring, diagnostic tests, inspections, and registration of stresses in service – all of these supply the information needed. The exchange of service experience between users, manufacturers, and other experts will support a more mature condition assessment. Statistical methods to analyse the service experience and determine the *bath-tub curves* are available [19, 20].

### 12.3.1   Maintenance

Maintenance is defined as the combination of all technical, supervision, and administrative actions intended to retain an item in or restore it to a state in which it can perform a required function.

Supervision actions, including visual inspections and diagnostic tests, are to be seen as maintenance activities. The international enquiries conducted by CIGRE Working Groups 13.06, 13.08, and A3.06 show that most utilities apply time-based maintenance to their circuit-breakers. However, over time, they tend to increase the intervals between maintenance actions, such as inspections, diagnostic tests, and major overhaul. In fact, utilities apply condition assessment in order to adapt the time interval between maintenance actions. Often they use risk management to determine the intervals per sub-population of circuit-breakers [21].

Compared with the old technologies, there is a huge increase in maintenance intervals with single-pressure $SF_6$-gas technology. Also, the maintenance costs have decreased considerably. Major overhaul nowadays is foreseen after 20 years or more. In fact, it is recommended not to dismantle circuit-breakers, unless there is evidence that interruption chambers need internal inspection and/or overhaul. To assess the internal condition without dismantling, diagnostic techniques come into play.

## 12.3.2  Monitoring and Diagnostics

Table 12.4 gives an overview of common monitoring and diagnostic-test techniques. The most important monitoring functions are $SF_6$-gas pressure, lockouts, the presence of auxiliary power, the condition of trip and closing coils/circuits, pole discrepancy, the number of operating cycles,

**Table 12.4**  Common monitoring, diagnostic tests, and inspection items for HV circuit-breakers.

| Monitoring | Permanent observation, mostly with some indication or alarm |
|---|---|
| $SF_6$-gas density | $SF_6$-gas pressure and temperature, electronic in future |
| Number of operating cycles | Counter in circuit-breaker cubicle (usually not remotely observed) |
| Sum of short-circuit currents | Alternatively, $\Sigma(I^2 t)$, for instance, by means of modern protection relays |
| Number of motor starts | Hydraulic and pneumatic drives |
| Accumulated running time | Hydraulic oil pumps and pneumatic air compressors |
| Lockout | Alarm that circuit-breaker is locked in open or closed position, maybe including cause |
| Auxiliary power | Power/voltage available |
| Trip and closing circuits | Condition of trip and closing coils and circuits |
| Position of main contacts | Through the position of auxiliary contacts |
| Stroke (velocity of contacts) | Sometimes, the travel characteristics of kinematic chain are monitored |
| Charging energy | Often in some way, the accumulator pressure or spring position is monitored |
| Recharging time | Often the time required to charge the accumulator is monitored |
| Heating | Heating or other environmental conditions may be monitored |

**Table 12.4** (*Continued*)

| Diagnostic tests | Activity with special equipment to assess the condition |
|---|---|
| $SF_6$-gas quality | Humidity check, impurities, decomposition products |
| $SF_6$-gas leakage | Consumption of $SF_6$-gas |
| Bushings | Power factor, partial discharges, capacity |
| Contact position | Dynamic contact resistance, contact timing (operating time) |
| Contact travel, damping | Kinematic chain stroke and velocity, in combination with main contact touch |
| Contact resistances | Resistance measurement, thermography |
| Dynamic pressure drop | Hydraulic or pneumatic drive |
| Recharging time | Accumulator, spring position |
| Release valves | Hydraulic or pneumatic drive |
| Starting, running current | Electric motors in hydraulic, pneumatic, and spring drives |
| Hot spot temperature | Thermography |
| Vibrations patterns | Vibration measurements |
| Continuity of coils | Trip and closing coils and circuits |
| | |
| Inspection items | Essentially a visual inspection, that may include partial dismantling |
| Surface of insulation | Voids, tracking, cracks, condensation, decomposition, impurities |
| Oil, grease | Leaks, levels, lubrication |
| Contacts | Burning, arcing marks, wear, ageing |
| Kinematic chain | Fractures, play, corrosion, rigidity |
| Auxiliary circuits | Power supply, overheating, auxiliary contacts and circuits, alarms, displays |
| Structure | Corrosion, pollution, painting, signs of overheating and arcing, leakages, etc. |

the number of motor operations (hydraulic and pneumatic drives), and the total running time of pumps or compressors. Quite often, only a general alarm is transmitted to the control centre while more detailed information is available at the substation or in the control cubicle [18, 22].

The increased reliability of circuit-breakers relieves somewhat the urgency for monitoring techniques for modern circuit-breakers. The emphasis should be more on applying monitoring techniques to older designs. But, in general, for economic reasons, it is better to consider the use of diagnostic test techniques instead of the modification of old circuit-breakers for the implementation of a monitor.

### 12.3.3   Life Management of Circuit-Breakers for Frequent Load-Switching

Apart from mechanical and electrical endurance – for regular or extended use, as discussed in Section 12.2 – a third stress factor has to be mentioned: restrikes and re-ignitions. Especially, circuit-breakers applied to switch shunt reactors and capacitor banks may face restrikes and re-ignitions, as described in Sections 4.2 and 4.3.3.2. Because such circuit-breakers have to switch very often, the occurrence of a restrike or re-ignition is rather realistic and consequential wear of nozzles, main contacts, and other vital parts is probable. Adequate testing (mechanical, capacitive, and/or small inductive currents [23]) and prudent consideration of the application of controlled switching and diagnostic tools to monitor restrikes and re-ignitions are recommended. Shorter inspection and maintenance intervals, compared with those of general applications, are to be considered.

## 12.4   Substation and System Reliability Studies

The results of the circuit-breaker reliability studies are also valuable for substation- and system-reliability investigations. The information of the first and second enquiry has been used to calculate for each voltage class the number of MF per command to open/close and the number of other MFs per year [24]. For substation- and system-reliability studies, such a distinction is necessary. The same approach can be followed to calculate the MF-rates with the information from the third worldwide enquiry [7].

Comparison of the number of MFs per operating cycle and per year (for all voltages) between the first, second, and third enquiry can be gained from Table 12.5. The last column gives the MFs per year for failures without a command to close or open. The last but one column gives the overall MF-rate per year. All the other columns are related to MFs with a command to close or open. The failure mode *"Locked in open or closed position"* is divided amongst the failures *"without command"* (50%) and *"does not open on command"* (13%) or *"does not close on command"* (37%). This procedure is the same as applied in the second enquiry [24]. The failure modes *"unknown"*, *"other"*, and *"loss of mechanical integrity"* are proportionally divided over the other modes. When comparing the failure modes, a much better performance can be noticed from the first to the second and from the second to the third survey. Only the number of the failure mode *"does not open on command"* has not significantly improved.

**Table 12.5**   Circuit-breaker reliability data for system studies; all voltages $\geq$ 63 kV.

| CIGRE enquiry | Does not open (a) | Does not break (b) | Does not close (c) | Does not make (d) | MF with command (a + b + c + d) | Operation cycles per year | MF with command | Overall MF | MF without command |
|---|---|---|---|---|---|---|---|---|---|
| | MFs per 10 000 operating cycles | | | | | | MF per year | | |
| 1st [2] | 0.84 | 0.11 | 2.01 | 0.10 | 3.06 | 27 | 0.0081 | 0.0158 | 0.0077 |
| 2nd [3] | 0.30 | 0.08 | 0.89 | 0.04 | 1.32 | 27 | 0.0035 | 0.0067 | 0.0032 |
| 3rd [4] | 0.26 | 0.02 | 0.49 | 0.01 | 0.78 | 27 | 0.0021 | 0.0030 | 0.0009 |

Summarizing, in approximate numbers, the occurrence of circuit-breaker failure-mode is as follows:

- once per 50 000 commands to open, a circuit-breaker will fail to open;
- once per 500 000 commands to open, it will not break the current;
- once per 25 000 commands to close it will not close or make the current;
- once per 1000 years, a circuit-breaker will show another MF;
- once per 3000 years, a circuit-breaker will show a dielectric failure other than "does not break the current";
- once per 10 000 years, a circuit-breaker may show a fire or explosion.

Such figures prompt experts to look for new substation configurations. The old-technology circuit-breakers faced problems with their reliability as well as availability due to the high consumption of maintenance time. Important substations have been designed as double-busbar schemes with an auxiliary busbar, or three busbar schemes, or even double-breaker schemes, in order to by-pass a circuit-breaker when overhauling it.

Nowadays circuit-breakers require far less maintenance and the necessity to by-pass a circuit-breaker has disappeared. Moreover, the overall reliability of three disconnectors per circuit-breaker has become lower than that of the circuit-breaker itself, although the performance of disconnectors has also been greatly improved. Following reference [25], the MF-rate is 0.2 per 100 disconnector-years (0.29 for air-insulated disconnectors and 0.05 for GIS-disconnectors). Similar values are applicable to earthing switches.

As a result, a general trend to realize simpler substation lay-out can be seen, though the number of rigorous examples is limited. A typical case in point is the design and application of disconnector-circuit-breakers with the aim of creating substations without disconnectors [26].

# References

[1] IEC 62271-1 (2007) High-voltage switchgear and controlgear – Part 1: Common Specifications.
[2] Mazza, G. and Michaca, R. (1981) The first international enquiry on circuit-breaker failures and defects in service. *Electra*, **79**, 21–91.
[3] CIGRE Working Group 13.06 (1994) Final Report of the second international enquiry on high voltage circuit-breaker failures and defects in service. CIGRE Technical Brochure 83.
[4] CIGRE Working Group A3.06 (2012) Final Report of 2004-2007 International Enquiry on Reliability of High Voltage Equipment, Part 2 – Reliability of High Voltage SF6 Circuit-Breakers. CIGRE Technical Brochure 510.
[5] CIGRE Working Group A3.06 (2012) Final Report of the 2004-2007 International Enquiry on Reliability of High Voltage Equipment. Electra No. 264, pp. 49–51.
[6] Runde, M. (2013) Failure frequencies for high-voltage circuit breakers, disconnectors, earthing switches, instrument transformers, and gas-insulated switchgear. *IEEE Trans. PowerDeliver.*, **28**(1), 529–530.
[7] Janssen, A.L.J., Makareinis, D. and Sölver, C.-E. (2014) International surveys on circuit-breaker reliability data for substation and system studies. *IEEE Trans. Power Deliver.*, **29**(2), 808–814.
[8] Pons, A., Sabot, A. and Babusci, G. (1993) Electrical endurance and reliability of circuit-breakers, common experience and practice of two utilities. *IEEE Trans. Power Deliver.*, **8** (1), 168–174.
[9] Smeets, R.P.P., Kertész, V., Nishiwaki, S. and Suzuki, K. (2002) Performance Evaluation of High-Voltage Circuit-Breakers by Means of Current Zero Analysis. IEEE/PES T&D Conference Asia Pacific, Yokohama, Japan.
[10] Jeanjean, R., Salzard, C. and Migaud, P. (2002) Electrical Endurance Tests for HV Circuit-Breakers. IEEE Winter Meeting.

[11]  Osawa, N. and Yoshioka, Y. (2003) Analysis of Nozzle Ablation Characteristics of Gas Circuit-Breakers. IEEE T&D Conference Asia Pacific.

[12]  IEC 62271-100 (2012) High-voltage switchgear and controlgear – Part 100: Alternating-current circuit-breakers. Ed. 2.1.

[13]  IEC TR 62271-310 (2008) High-voltage switchgear and controlgear – Part 310: Electrical endurance testing for circuit-breakers above a rated voltage of 52 kV. 2nd edition.

[14]  Janssen, A.L.J. and Sölver, C.-E. (2004) The statistics behind the electrical endurance type test for HV circuit-breakers. CIGRE Conference, Paper A3-304.

[15]  CIGRE Task Force A3.01 (2005) Statistical analysis of electrical stresses on high-voltage circuit-breakers in service. *Electra*, **220**, 24–26.

[16]  Smeets, R.P.P. and Ito, H. (2005) Electrical Endurance of Circuit-Breakers in Service. CIGRE A3/B3 Colloquium, Paper 102, Tokyo.

[17]  Janssen, A.L.J., Heising, C.R., Sanchis, G. and Lanz, W. (1996) Mechanical endurance, reliability compliance and environmental testing of high voltage circuit-breakers. CIGRE Conference, Paper 13-101.

[18]  CIGRE Working Group 13.08 (2000) Life management of circuit-breakers. CIGRE Technical Brochure 165.

[19]  Jongen, R.A. (2012) Statistical Lifetime Management for Energy Network Components. PhD. Thesis, Technical University Delft.

[20]  Janssen, A. and Jongen, R. (2009) Residual Life Assessment and Asset Management Decision Support by Hazard Rate Functions. 6th Southern Africa Regional Conference and CIGRE SC A2, A3 & B3 Joint Colloquium, Capetown, Paper C 206.

[21]  Balzer, G., Strnad, A., Neumann, C. *et al.* (2000) Life cycle management of circuit-breakers by application of reliability centered maintenance. CIGRE Conference, Paper 13-103.

[22]  CIGRE Working Group 13.09 (2000) User Guide for the Application of Monitoring and Diagnostic Techniques for Switching Equipment for Rated Voltages of 72.5 kV and above. CIGRE Technical Brochure 167.

[23]  Kapetanović, M. (2011) *High Voltage Circuit-Breakers*, ETF – Faculty of Electrotechnical Engineering, Sarajevo, ISBN 978-9958-629-39-6.

[24]  Heising, C.R., Colombo, E., Janssen, A.L.J. *et al.* (1994) Final report on high-voltage circuit-breaker reliability data for use in substation and system studies. CIGRE Conference, Paper 13-201.

[25]  CIGRE Working Group A3.06 (2012) Final Report of 2004-2007 International Enquiry on Reliability of High Voltage Equipment, Part 3 – Disconnectors and Earthing Switches. CIGRE Technical Brochure 511.

[26]  Lundquist, J., Andersson, P.-O., Olovsson, H.-E. *et al.* (2004) Applications of disconnecting circuit-breakers. CIGRE Conference, Paper A3-201.

# 13

# Standards, Specification, and Commissioning

## 13.1 Standards for Fault-Current Breaking Tests

In general, a standard is a set of rules that establishes uniform behaviour, criteria, methods, processes and/or practice with the aim to ensure the quality of specific equipment. For almost every switching situation described in this book, international standards have been developed.

The following functions can be attributed to standards related to switchgear:

- Definition of the target for designers and developers of switchgear. Defining clear performance criteria for all relevant functions, standards greatly enhance product reliability and effectiveness and enable easy exchange of apparatus in various locations in the network.
- Reference to network operators. The standards are designed to cover approximately 90% of the possible conditions including faults. Operators should compare the fault parameters that can arise in their particular system with the standardized parameters in order to decide whether the equipment can handle the fault situation or whether mitigating measures are needed. In some cases, even custom-made equipment can be ordered to meet specific requirements, or testing can be required with non-standardized, specific duties.
- Guidelines for testing institutions. Test circuits are developed that can provide adequate stresses. This aspect is covered in Chapter 14.

The set of standards that is by far most used in the world is the series issued by the *International Electrotechnical Commission* (IEC). Regarding fault-current making and breaking, as well as capacitive-current switching, which are common switching duties for circuit-breakers, the relevant standard is IEC 62271-100 [1] (before 2002 known as IEC Publication 56, first issued in 1937, second edition in 1954, third in 1971 and fourth in 1987). General, common, characteristics of switchgear, not related to switching performance, are standardized by IEC 62271-1 [2].

*Switching in Electrical Transmission and Distribution Systems*, First Edition.
René Smeets, Lou van der Sluis, Mirsad Kapetanović, David F. Peelo, and Anton Janssen.
© 2015 John Wiley & Sons, Ltd. Published 2015 by John Wiley & Sons, Ltd.

Other major standardizing bodies harmonize with IEC; currently, that is, in 2014, harmonization of IEC and the United States standards from IEEE/ANSI (Institute of Electrical and Electronics Engineers/American National Standards Institute) is well on its way [3].

The most important American standards are the following:

- IEEE Std. C37.04-1999: "IEEE Standard Rating Structure for AC High-Voltage Circuit-Breakers";
- ANSI C37.06-1997: "AC High-Voltage Circuit-Breakers Rated on a Symmetrical Current Basis – Preferred Ratings and Related Required Capabilities";
- IEEE Std. C37.09-1999: "IEEE Standard Test Procedure for AC High-Voltage Circuit-Breakers Rated on a Symmetrical Current Basis".

A large number of national standards all over the world are based on or derived from the IEC standards.

### 13.1.1  Background and History of the Standardized IEC TRV Description

During normal operation of a system, the energy transferred by the electromagnetic field to the load is equally divided over the electric field and the magnetic field. Interruption of the current eliminates the magnetic energy and electrical energy remains (see Section 1.6). The entire system has to adjust itself to its new operating state while both subsystems on either side of a switching device generate voltage transients, usually oscillating. The actual waveform of the transients is determined by the configuration of the system, that is, its distributed and lumped circuit elements.

When a switching device (whether it is a HV circuit-breaker, switch, or fuse) interrupts a current in a power system, the device experiences initially the *transient recovery voltage* (TRV), followed by the *recovery voltage* (RV) across its terminals.

The TRV is the voltage that appears across the contacts of a switching device immediately after the current interruption and during the subsequent transient state. The TRV duration is of the order of milliseconds, and it typically contains high-frequency oscillatory, exponential, or combined waveforms that are damped out due to dissipation of the intrinsic energy in the resistive parts of the circuit. The transient components can be single frequency, double frequency, or multi-frequency oscillations or triangular shapes. The rate-of-rise of the TRV and its peak value are of vital importance for a successful breaking operation of the interrupting device.

In the period following the transient recovery voltage, the interrupting device must be able to withstand also the power-frequency RV that stresses the gap between the contacts.

In the early years of switchgear and fuse design, the TRV was an unknown phenomenon, and the recovery voltage was regarded to consist of the power-frequency RV only. The duty on circuit-breakers was commonly expressed in terms of the circuit voltage prior to short circuit and the magnitude of the current in the arc. It was, however, experienced that, in practice, other circuit characteristics would affect the duty to an important extent. Improved measurement equipment, such as the cathode-ray oscillograph and later the cathode-ray oscilloscope, made measurements with a higher time resolution possible and revealed the existence of high-frequency oscillations immediately after current interruption: the TRV was discovered [4].

This resulted in system studies of transmission and distribution networks. Many investigations were carried out in different countries to determine the parameters of the TRV across circuit-breakers clearing short circuits in networks. Theoretical studies of power-systems, using analogue computer simulations, called *transient network analysers* (TNA), were verified by on-site tests in different networks. This brought insight into the TRV waveforms, frequencies of oscillations and peak values. Better understanding of the transient phenomena in the network has led to improved testing in high-power laboratories, more accurate measurements of the current-zero phenomena, and consequently resulted in more-reliable switchgear with a higher interrupting capability.

The use of $SF_6$ as extinguishing medium allowed a leap forward in the short-circuit performance of HV circuit-breakers. However, it was further discovered that the single- and multiple-frequency representations of TRVs are justified only for a short time until voltage reflections or travelling waves return from a remote open circuit, a fault location, or other discontinuity. It was no longer sufficient to simulate the networks with lumped elements and the distributed parameters of the networks had to be taken into account. The severity of short-line-fault TRVs appeared well understood and described.

During that period, the end of the 1950s, effort was made to represent the TRV oscillations by standardized waveforms to be able to create TRVs by lumped elements within the walls of high-power laboratories. The four-parameter waveform was proposed by Hochrainer in 1957 [5], and the possibilities to create the waveform in the high-power laboratory were investigated by Baltensperger [6]. Relevant literature can be found in [7–9]. In the early 1960s, several network studies were undertaken in Japan and Europe and attempts were made to better define the transient phenomena for analogue modelling. The parameters of TRVs were then specified, however, in different ways in various national standards. Therefore, Subcommittee 17 on High-Voltage Switchgear of the International Electrotechnical Commission (IEC) asked CIGRE ("Conférence International des Grands Réseaux Électriques" or "International Council on Large Electric Systems") Study Committee on circuit-breakers (named Study Committee 3 in those days) to promote new extensive investigations on an international base. The Working Group (WG) 3.1 of CIGRE, set up for this purpose in 1959, decided to start a complete investigation of TRVs associated with the opening of the first pole of a circuit-breaker clearing a three-phase unearthed fault in some large 245 kV networks.

Two of these networks were fully investigated: the Italian network in its 1962 situation and the existing French network of 1965 [10, 11]. Some 2000 simulated TRV wave shapes associated with short-circuit currents up to 45 kA were collected. Based on this collection, a classification of current ratings was proposed. In publication 56-2 (1971) on HV alternating circuit-breakers, IEC recommended characteristic values for simulation of the TRVs by the four-parameter method and by the two-parameter method (see Section 13.1.2). The values in the relevant tables were mainly based on studies of the 245 kV systems. For the higher operating voltages up to 765 kV, the values were extrapolated because actual data were not available at that time.

In the meantime, studies were made of TRVs in systems operating at a maximum voltage of 420 kV and above [12]. In some cases, these studies revealed deviations from the TRV parameters stipulated by IEC 56-2 (1971). In the light of this, CIGRE Study Committee 13 (Switching Equipment) commissioned WG 13.01 to study the problems of TRVs in EHV systems. It was concluded that the rate-of-rise of 1 kV $\mu s^{-1}$, as a basis for earthed three-phase terminal faults, was somewhat low, and the real dielectric stresses could be better characterized by 2 kV $\mu s^{-1}$ and a first-pole-to-clear factor $k_{pp} = 1.3$.

It was decided in IEC-SC-17A to revise the TRV tables of the IEC 56-2 and IEC 56-4. New values based on the above-mentioned CIGRE WG 13.01 studies were proposed during the years 1976–1979. Because these studies dealt mainly with rated voltages of 300 kV and above, and questionnaires received from utilities gave no indication about serious failures of MV circuit-breakers, IEC-SC-17A decided in 1979 only on the standard TRV values for rated voltages of 100 kV and above. The TRV values for rated voltages below 100 kV remained unchanged. CIGRE SC-13 decided in 1979 to set up a task force for collecting data and establishment of TRV parameters for MV circuit-breakers. The report of the task force was published in 1983 [13].

In 1981, WG 13.05 was initiated by CIGRE SC13 to study the TRV conditions caused by clearing transformer faults and series-reactor limited faults. WG 13.05 collected data on transformer natural frequencies and ratings. The report of the WG was published in 1985 [14]. A combined CIGRE/CIRED (CIRED: "Congrès International des Réseaux Électriques de Distribution" or "International Conference on Electricity Distribution") WG CC-03 undertook from 1994 until 1998 the work of investigating the TRVs in networks up to 100 kV. A summary of the report of WG CC-03 is published in 1998 [15] and extensive results are available [16].

Other relevant results of research into the characteristics of TRVs have been published in reference [17, 18].

At present (2014), standards for switching devices are available up to the rated voltage of 1200 kV, mainly because of new systems in China (1100 kV) and India (1200 kV) that have already been put into operation or are planned (see Section 10.8). The basis for the UHV TRV parameters was laid by CIGRE WG A3.22 and reported in 2008 and 2010 [19, 20].

## 13.1.2   IEC TRV Description

IEC uses the approach of defining limiting line segments or envelopes of TRVs in the voltage–time plane. The TRV parameters are defined for each short-circuit test-duty as a function of the rated voltage $U_r$ and the first-pole-to-clear factor $k_{pp}$ (see Section 3.3.2). The envelopes to be observed in testing are described as follows (IEC 62271-100, 4.102.2):

- If the TRV is expected to approximate a damped single-frequency oscillation, experienced particularly at rated voltages below 100 kV, a *two-parameter* TRV is defined. This waveform is adequately represented by an envelope consisting of two line segments determined by:
  - the *reference voltage* $u_c$, that is, the TRV peak value; and
  - the time $t_3$ to reach $u_c$.
    Geometrically it implies a line segment from the point (0 ; 0) to ($t_3$ ; $u_c$), and a horizontal line starting from ($t_3$ ; $u_c$).
    In addition, there is a *delay line* of TRV, starting after the *time delay* $t_d$ at ($t_d$ ; 0) and ending at the the point ($t'$ ; $u'$) of the reference voltage $u'$ and the time $t'$ to reach $u'$. This is outlined in Figure 13.1a, showing the TRV envelope for a 24 kV circuit-breaker, interrupting its full rated short-circuit breaking current (test-duty T100, see Section 13.1.3).
- If the TRV has a multi-frequency nature, generally at rated voltages of 100 kV and above, a *four-parameter* TRV is defined. It contains first a period of high rate-of-rise, followed

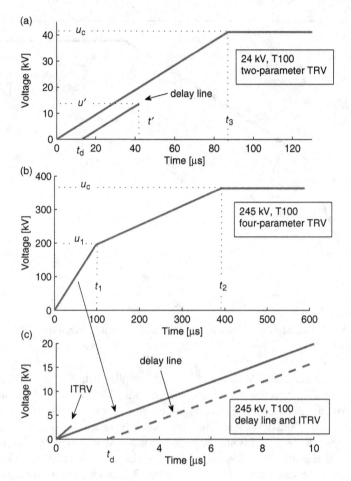

**Figure 13.1** IEC envelopes constructed from the parameters $t_1$, $u_1$, $t_2$, $u_c$, $t_3$ for rated voltages of 24 and 245 kV and 100% short-circuit current (test-duty T100).

by a period of lower rate-of-rise. This waveform is adequately represented by an envelope consisting of three line segments defined by four parameters:
- the first reference voltage $u_1$;
- the time $t_1$ to reach $u_1$;
- the second reference voltage $u_c$, that is, TRV peak value; and
- the time $t_2$ to reach $u_c$.

Geometrically it implies a line segment from the point $(0 ; 0)$ to $(t_1 ; u_1)$, a subsequent line segment from $(t_1 ; u_1)$ to $(t_2 ; u_c)$, and a horizontal line starting from $(t_2 ; u_c)$. This is outlined in Figure 13.1b, showing the TRV envelope for a 245 kV circuit-breaker interrupting its full rated short-circuit-breaking current.

In addition, there is a delay line of TRV, starting after the time delay $t_d$ at $(t_d ; 0)$ and ending at the the point $(t' ; u')$ of the *reference voltage* $u'$ and the time $t'$ to reach $u'$. The *Initial* TRV (ITRV, see Section 3.6.1.2) is shown in the enlarged Figure 13.1c.

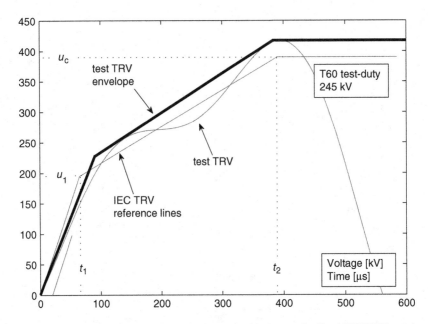

**Figure 13.2**    Comparison of an inherent test-TRV with the standardized IEC TRV envelope.

In order to guarantee that the TRV recorded in a test complies with the relevant IEC reference envelope, it must be verified that the envelope of the test-TRV shows up completely to the left of and above the reference lines; in addition, the initial part of the test-TRV must appear completely to the left of the delay line. The function of the delay line is to guarantee that the initial rate-of-rise of the test-TRV is not lower than in practice. It is the duty of test-institutions to produce TRVs that approach as close as possible the IEC reference lines, in order to perform tests that are neither too severe nor too light. In Figure 13.2, an inherent (prospective) test-TRV is compared with the IEC reference lines. The envelope of the inherent TRV shows its inability to reach sufficient RRRV in the initial part from 0 to $t_1$.

## 13.1.3   IEC Test-Duties

In the standards, a number of test-duties are prescribed, each of them covering a particular fault situation. The test duties define currents and TRVs, as well as arcing times and operation sequences to be applied in testing for verification of circuit-breaker capability to cope with the specific conditions and stresses.

The following technical terms and acronyms are in common use:

*Terminal Faults* (TFs)
This implies a fault directly at the circuit-breaker terminals. In the IEC standard, for voltages below 100 kV, a distinction is made between cable systems (commonly termed S1) and line systems (S2). Due to the relatively small stray capacitance

of line systems, more severe conditions regarding the RRRV are applicable here. The following test-duties are defined:

- T10: Covering faults, where current is supplied by a single transformer with relatively high short-circuit impedance. The fault current is standardized to 10% of the rated short-circuit breaking current $I_{sc}$ and is applied only symmetrically. The specified TRV is considered a single-frequency TRV created by the oscillation circuit of the transformer stray capacitance and inductance. This two-parameter TRV is characterized by a high value of the RRRV. The envelope of the single-frequency TRV is described by $t_3$ and $u_c$.
- T30: Equivalent to T10, but with 30% of the rated short-circuit breaking current $I_{sc}$.
- T60: Covering faults, where fault current is supplied by transformer(s) and overhead line(s). The TRV has basically two components, a fast one originating from the substation side, described by $t_1$ and $u_1$, and a slower one originating from the line side, described by $t_2$, $u_c$. The fault current is standardized to 60% of the rated short-circuit breaking current $I_{sc}$, and is applied only symmetrically. The RRRV is relatively modest because the equivalent capacitance of the main current-source circuit is in parallel to the capacitances of the various sources.
- T100: Covering faults where fault current is supplied by multiple sources, as in heavily meshed grids. The maximum (100%) of the rated short-circuit breaking current $I_{sc}$ is associated with this test-duty and has to be applied both symmetrically (T100s) and asymmetrically (T100a). The RRRV is modest, because the equivalent capacitance of the multiple circuits that supply the fault current in parallel is relatively large. Due to multiple sources, the fault current is very high.

*Transformer-Limited Faults* (TLFs)

This implies faults located downstream from the circuit-breaker in radial networks with the current supplied by a transformer (see Section 3.4). The TLF duty (IEC 62271-100, Annex M) is defined for systems of rated voltages below 100 and above 800 kV. The fault current is standardized as 30% of the rated short-circuit breaking current $I_{sc}$ for voltages below 100 kV and as 10 and 12.5 kA for voltages above 800 kV.

- *Transformer-Secondary Fault* (TSF): In this case, the circuit-breaker is at the primary (HV or EHV) side of the transformer and the fault at the secondary (MV or HV) side.
- *Transformer-Fed Fault* (TFF): In this case, the circuit-breaker is at the secondary (MV or HV) side of the transformer and the fault is located behind the circuit-breaker.

The inherent RRRV is very high. This is because only the natural frequency of the transformer, as an isolated network element, creates a single-frequency oscillation (at the load side in the transformer-secondary fault and at the source side in the transformer-fed fault). In the case where a long cable connects the transformer to the circuit-breaker (longer than a few tens of meters), the cable capacitance reduces the transformer natural frequency; as a result, the RRRV may be considerably reduced.

*Single-Phase and Double-Earth Faults*
The arcing time in this condition is longer than in three-phase conditions, and the combination of current and recovery voltages are non-standard. The following test duties are defined:

- *Single-Phase-Fault Test* (SEF): Covering applications in effectively earthed systems where circuit-breakers are expected to clear single-phase faults. The tests imply a breaking current of $I_{sc}$ and recovery voltage of $U_r/\sqrt{3}$.
- *Double-Earth-Fault Test* (DEF): Covering the cases of earth faults in non-effectively earthed neutral systems. The faults are on two different phases, one of which occurs on one side of the circuit-breaker and the other on the opposite side. This situation of two simultaneous faults is more likely to occur than it might seem at first sight. A single-earth fault in a non-effectively earthed neutral systems leads to a voltage rise in the other phases, enhancing the probability of the second fault.
Current to be interrupted is $\frac{1}{2}\sqrt{3}\,I_{sc}$ and recovery voltage $U_r$.

*Overhead-Line Faults*
*Short-line fault* (SLF): Covering faults on overhead lines at a short distance (a few kilometres) from the substation (see Section 3.4). Fault current is standardized to 90, 75, and 60% of the rated short-circuit breaking current $I_{sc}$. The TRV is standardized separately for source and load (line) side because tests are carried out (unlike the terminal fault tests) with separate source-side and load-side circuits. The source circuit is specified in the same way as for the terminal fault test-duties, that is, with a relatively low RRRV. This is adequate because the SLF duties are intended to verify the thermal breaking capacity. Failure to pass this test is normally a (thermal) re-ignition immediately following current zero (see Section 3.6.1.4).

The load-side circuit should represent the transient behaviour of a short overhead line, having a surge impedance $Z_0 = 450\ \Omega$ and *amplitude factor* $k_{af} = 1.6$ (see Section 3.4).

*Out-of-Phase Faults* (OP)
This covers out-of-phase faults. The rated out-of-phase breaking current $I_d$ is standardized to 25% of the rated short-circuit breaking current $I_{sc}$. Test-duty OP1, with test current $0.3I_d$ and OP2, with test-current $I_d$ are defined.

A very important verification of a circuit-breaker function is to ensure its capability to interrupt in the complete arcing window. The arcing window is the range of possible arcing times from minimum arcing time (determined by the circuit-breaker) and the maximum arcing time (determined by circuit and circuit-breaker). During each test-duty, it is necessary to demonstrate this basic arcing window. The breaking capability at an arcing time just above minimum, at maximum and at the intermediate arcing time must be verified.

The tests must be performed at the rated operating sequence.

### 13.1.4 IEC TRV Parameters Selection and Application

The selection of the correct TRV characteristics (RRRV, $k_{af}$, and $k_{pp}$) is relatively complicated because it depends on the test-duty, rated voltage of the system, its lay-out (cable- or overhead-line system) and earthing. With the TRV parameters, the TRV envelope parameters can then be calculated for each test-duty.

From Table 13.1, the TRV characteristics (RRRV, $k_{af}$, and $k_{pp}$) can be taken as a function of these factors. For system voltages < 100 kV, a distinction is made between cable systems (i.e. without lines) and line systems (may include cables); the line systems have the most severe TRVs because of a smaller capacitance compared with cable systems. In this voltage range, RRRV is given here as second-order functions of the rated system voltage $U_r$. These functions are interpolated from the tabulated IEC values and deviate less than 5% from the values in IEC 62271-100.

For system voltages 100 kV $\leq U_r \leq$ 170 kV, two options for system earthing are given: effectively earthed ($k_{pp} = 1.3$) and non-effectively earthed ($k_{pp} = 1.5$). Below this range, all the systems are assumed non-effectively earthed, and above this range, the systems are assumed effectively earthed. The (three-phase) faults are assumed to be earthed.

With the values of the TRV characteristics (RRRV, $k_{af}$, and $k_{pp}$), the TRV envelope parameters ($t_1$, $t_2$, $t_3$, $u_1$, $u_c$, $t_d$, $t'$, and $u'$) can be calculated for each situation using the equations in Table 13.2. The voltages are given in per unit value as:

$$1 \text{ p.u.} = U_r \sqrt{\frac{2}{3}} \tag{13.1}$$

Next, TRV envelopes can be constructed according to the method outlined in Section 13.1.2.

As an example, Figure 13.3 shows a collection of the most important TRV envelopes regarding a 170 kV circuit-breaker to be applied in a system with the first-pole-to-clear factor $k_{pp} = 1.3$. The terminal-fault duties T10 and T30 have a two-parameter and steep TRV envelope, whereas the others (T60 and T100) have a four-parameter and a slower TRV envelope. The OP test-duty has the highest peak value but the slowest rate-of-rise. The single-phase SLF duty has a very low peak of TRV.

## 13.2 IEC Standardized Tests for Capacitive-Current Switching

The aim of capacitive-current switching tests is to provide confidence that the circuit-breaker will restrike only in exceptional cases. The level of confidence in IEC is standardized in two classes: C1: low probability of restrike and C2: very low probability of restrike.

This implies that no circuit-breaker can be claimed to be restrike free. Testing is aimed at creating conditions that enable restrikes, such as a small voltage jump $\Delta U$ (see Section 4.2.2 Equation 4.1) and short arcing times.

Table 13.3 gives an overview of the requirements in single-phase tests.

**Table 13.1** IEC standardized TRV characteristics RRRV, $k_{af}$ and $k_{pp}$.

| Rated voltage $U_r$ [kV] | 3.6 to 72.5 | | 100 to 170 | | 245 to 800 | > 800 |
|---|---|---|---|---|---|---|
| Neutral earthing | Unearthed | | | | Earthed | Earthed |
| Cable/Line system | Cable | Line (≥ 17.5 kV) | Unearthed line | Earthed line | Earthed line | Earthed line |
| **RRRV [kV/μs]** — T100 | $-0.00012\,U_r^2 + 0.0174\,U_r + 0.1223$ | $-0.000071\,U_r^2 + 0.0154\,U_r + 0.7216$ | 2 | Not applicable | 2 | 2 |
| T60 | $-0.00030\,U_r^2 + 0.0421\,U_r + 0.3055$ | $-0.00011\,U_r^2 + 0.0249\,U_r + 1.1378$ | 3 | | 3 | 3 |
| T30 | $-0.00060\,U_r^2 + 0.0875\,U_r + 0.6744$ | $-0.00016\,U_r^2 + 0.0411\,U_r + 2.0453$ | 5 | | 5 | 5 |
| T10 | $-0.00051\,U_r^2 + 0.0823\,U_r + 0.8074$ | $-0.00018\,U_r^2 + 0.0432\,U_r + 2.1111$ | 7 | | 7 | 7 |
| TLF | $-0.00139\,U_r^2 + 0.1911\,U_r + 1.2525$ | | Not applicable | | | 15.4 to 18 |
| SLF source | Not applicable | $-0.000037\,U_r^2 + 0.0093\,U_r + 0.4983$ | 2 | | 2 | 2 |
| SLF line 40 kA @ 50 Hz | Not applicable | Test-duty $L_{90}$ | 7.2 | | | 5.3 |
| SLF line 40 kA @ 50 Hz | | Test-duty $L_{75}$ | 6.0 | | | 4.4 |
| OP | $-0.000090\,U_r^2 + 0.0129\,U_r + 0.0900$ | $-0.000060\,U_r^2 + 0.0115\,U_r + 0.4692$ | 1.67 | 1.54 | 1.54 | 1.54 |
| **$k_{af}$** — T100 | 1.4 | 1.54 | 1.4 | 1.54 | 1.4 | 1.5 |
| T60 | 1.5 | 1.65 | 1.5 | | 1.5 | 1.5 |
| T30 | 1.6 | 1.74 | 1.54 | | 1.54 | 1.54 |
| T10 | 1.7 | 1.8 | 1.53 | | 1.53 | 1.76 |
| TLF | 1.6 | | Not applicable | | | 1.53 |
| SLF source | 1.54 | | 1.4 | | 1.4 | 1.5 |
| OP | 1.25 | | 1.25 | | 1.25 | 1.25 |
| **$k_{pp}$** — T100 to T30 | 1.5 | | 1.5 | 1.3 | 1.3 | 1.2 |
| T10 | 1.5 | | 1.5 | | 1.5 | 1.2 |
| TLF | 1.5 | | Not applicable | | | 1.2 |
| SLF source | 1 | | 1 | | 1 | 1 |
| OP | 2.5 | | 2.5 | | 2 | 2 |

**Table 13.2**  IEC TRV-envelope parameter calculation for fault-current breaking duties.

| Test-duty | TRV parameter | | | | | | |
| --- | --- | --- | --- | --- | --- | --- | --- |
| | $u_1$ [p.u.] | $t_1$ [μs] | $u_c$ [p.u.] | $(t_3; t_2)$ [μs] | $t_d$ [μs] | $u'$ [kV] | $t'$ [μs] |
| Rated voltage $U_r < 100$ kV (two-parameter representation) | | | | | | | |
| T100 Cable syst. | Not applicable | | $k_{pp} \cdot k_{af}$ | $u_c/\text{RRRV}$ | $0.15\, t_3$ | $u_c/3$ | $t_d + t_3/3$ |
| T100 Line syst. | | | | | $0.05\, t_3$ | | |
| T60 to T10, TLF | | | | | $0.15\, t_3$ | | |
| SLF Source | | | | | $0.05\, t_3$ | | |
| OP | | | | | $0.15\, t_3$ | | |
| Rated voltage $(100 \le U_r \le 800)$ kV (four-parameter representation, except T30 and T10) | | | | | | | |
| T100 | $0.75\, k_{pp}$ | $u_1/\text{RRRV}$ | | $4\, t_1$ | $2$ μs | $u_1/2$ | $t_d + u'/\text{RRRV}$ |
| T60 | | | | $6\, t_1$ | $2$ μs to $0.30\, t_1$ | | |
| T30, T10 | Not applicable | | $k_{pp} \cdot k_{af}$ | $u_c/\text{RRRV}$ | $0.15\, t_3$ | $u_c/3$ | $t_d + t_3/3$ |
| SLF source | $0.75\, k_{pp}$ | $u_1/\text{RRRV}$ | | $4\, t_1$ | $2$ μs | $u_1/2$ | $t_d + u'/\text{RRRV}$ |
| OP | | | | $2\, t_1$ to $4\, t_1$ | $2$ μs to $0.10\, t_1$ | | |
| Rated voltage $U_r > 800$ kV (four-parameter representation, except T30 and T10, out-of-phase) | | | | | | | |
| T100 | $0.75\, k_{pp}$ | $u_1/\text{RRRV}$ | | $3\, t_1$ | $0.28\, t_1$ | $u_c/4$ | $t_d + u'/\text{RRRV}$ |
| T60 | | | | $4.5\, t_1$ | $0.30\, t_1$ | | |
| T30, T10, TLF | Not applicable | | $k_{pp} \cdot k_{af}$ | $u_c/\text{RRRV}$ | $0.15\, t_3$ | $u_c/3$ | $t_d + t_3/3$ |
| SLF source | $0.75\, k_{pp}$ | $u_1/\text{RRRV}$ | | $3\, t_1$ | $2$ μs | $u_c/4$ | $t_d + u'/\text{RRRV}$ |
| OP | Not applicable | | | $u_c/\text{RRRV}$ | $2$ μs to $0.05\, t_3$ | $u_c/3$ | |

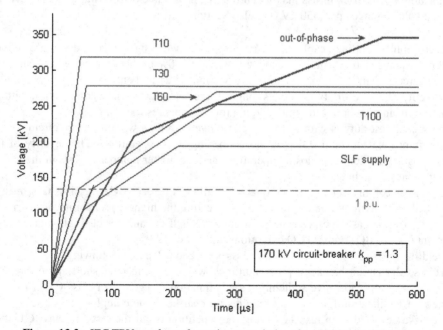

**Figure 13.3**  IEC TRV envelopes for various test-duties of a 170 kV circuit-breaker.

**Table 13.3** Number of single-phase test operations for classes of restrike probability.

| Test duty | Current [%] of $I_{cap}$ | $\Delta U$ [%] | Class C1 No pre-conditioning | | Class C2 Pre-conditioning with 60% of $I_{sc}$ | |
|---|---|---|---|---|---|---|
| | | | Operations | $t_{arc\ min}$ | Operations | $t_{arc\ min}$ |
| Line duty 1 (LC1) | 10 to 40 | < 2 | 24 O | 6 O | 48 O | 12 O |
| Line duty 2 (LC2) | 100 | < 5 | 24 CO | 6 CO | 24 O + 24 CO | 6 O + 6 CO |
| Cable duty 1 (CC1) | 10 to 40 | < 2 | 24 O | 6 O | 48 O | 12 O |
| Cable duty 2 (CC2) | 100 | < 5 | 24 CO | 6 CO | 24 O + 24 CO | 6 O +6 CO |
| Bank duty 1 (BC1) | 10 to 40 | < 2 | 24 O | 6 O | 48 O | 12 O |
| Bank duty 2 (BC2) | 100 | < 5 | 24 CO | 6 CO | 120 CO | 84 CO |
| Maximum restrikes allowed (incl. repetition of test series) | | | 3 | | 1 | |

In this table, $I_{cap}$ is the rated capacitive-switching current, which is different for each of the test-duties listed:

- line charging (LC) current $I_l$ from 10 A (< 100 kV) to 900 A (at 800 kV), 1200 A (at 1100 kV), and 1300 A (at 1200 kV);
- cable charging (CC) current $I_c$ from 10 A to 500 A (at 550 kV);
- capacitor bank charging (BC) single-bank current $I_{sb}$ and back-to-back current $I_{bb}$ of 400 A in the entire voltage range up to 550 kV;
- capacitor back-to-back inrush making current $I_{bi}$, peak value of 20 kA at 4.25 kHz over the entire voltage range up to 550 kV (see also Section 4.2.6).

For each application (line-, cable-, and capacitor-bank-switching), two test-duties are defined. The first requires testing with a reduced current, 10 to 40% of the rated capacitive current, a strong source circuit (< 2% voltage jump, see Section 4.2.2), and only opening (O) operations. The severity of this test-duty is in the stronger source circuit, which in combination with the small current and its shorter minimum arcing time favours restrike(s).

The second test-duty is with full rated capacitive current, a weaker source circuit (< 5% voltage jump) and opening (O) plus close–opening (CO) operations. The severity of this test-duty is in the additional closing operations and the higher number of tests in the case of capacitor-bank-switching.

As can be seen in Table 13.3, one quarter of all tests (except BC2 for class C2) must be performed with minimum arcing time $t_{arc\ min}$, creating the highest probability of restrike. If tests are performed as three-phase tests for class C2, half the number of single-phase tests is sufficient (except BC2 with 80 CO operations, instead of 120).

The differences in requirements of the classes C1 and C2 are the following:

In class C2, a circuit-breaker is pre-conditioned with 60% of its rated short-circuit-breaking current $I_{sc}$ prior to capacitive switching as a simulation of ageing; in class C1, it is tested "as new". Classification for class C2 requires no restrike in the complete test-series, or – if one restrike occurred – no restrike during the repetition of all the tests. In class C1, these requirements are less stringent: one restrike is allowed during the complete series or two restrikes in the complete series and then no more than one additional restrike in the repetition of all series.

Note that the higher value of the voltage jump $\Delta U$ enhances the probability of a re-ignition and reduces the probability of a restrike (see Section 4.2.2). Since restrike may have serious implications, higher values of voltage jump are generally beneficial. In order to realize conditions in which the majority of the cases are covered, such as in type tests, the voltage jump in test circuits must be kept below a certain value. In agreement with Equation 4.1, this implies a low ratio $I_{cap}/I_{sc}$. This is why in the testing of capacitive switching capability, the source circuits with short-circuit power corresponding to the real service situation should be applied. High-power source circuits are needed in spite of the far lower current applied in the test. IEC prescribes capacitive test-duties to be carried out with $I_{sc} > 50\,I_{cap}$ and $I_{sc} > 20\,I_{cap}$.

In many cases, three-phase testing has to be replaced by single-phase testing, mostly because of test-facility limitations. In that case, a careful selection of the single-phase source voltage must be made, depending on the system earthing. A multiplication factor (*capacitive voltage factor $k_c$*) is introduced, by which the single-phase-to-earth voltage $U_r/\sqrt{3}$ of the source is to be multiplied. The test voltage measured at the circuit-breaker location immediately before the opening must be not less than $k_c U_r/\sqrt{3}$. The following values for $k_c$ are standardized:

- $k_c = 1.0$ for solidly earthed systems without significant phase-to-phase interaction (i.e. in practice, capacitor banks with earthed neutral and screened cables);
- $k_c = 1.2$ for effectively earthed systems with phase-to-phase coupling (typically systems > 52 kV and belted cables;
- $k_c = 1.4$ for non-effectively earthed neutral systems or capacitor banks with isolated neutral. This is the value applied most often in testing.

  Since in this case the testing is single phase, the maximum recovery voltage corresponds to $2k_c = 2.8$. This is higher than the 2.5 maximum value for three-phase interruption; refer to Section 4.2.3. On the other hand, the rate-of-rise of the three-phase recovery voltage is higher than the single phase (1 – cosine) voltage.
- $k_c = 1.7$ for non-effectively earthed neutral systems in the presence of a phase-to-earth fault (rarely applied).

## 13.3  IEC Standardized Tests for Inductive-Load Switching

The main aim of inductive-load switching tests is to obtain data for prediction of overvolt-ages that can be expected in a particular application. Inductive-load switching is an example of strong interaction between the circuit-breaker and circuit and, therefore, test results and overvoltages observed should be strictly associated with the real test circuit only. Generally, because of the compactness of test circuits, compared with the real grid, overvoltages in inductive-load tests tend to be higher than in service. Testing requirements for inductive-load switching are laid down in IEC 62271-110 [21].

There is no standardized transformer magnetizing-current switching test. Due to the non-linearity of the transformer core, it is not possible to model correctly the switching of trans-former magnetizing current by linear components in a test laboratory. Testing with an available transformer, such as a test-laboratory transformer, would only be valid for that particular transformer and could not be representative for other transformers. In addition, the switching of transformer magnetizing current is usually less severe than any other inductive-current switching duty.

## 13.3.1 Shunt-Reactor Switching

Shunt-reactor switching tests can be performed in test-laboratories; test-duties and test circuits are prescribed [23].

Two test circuits are defined: load circuit 1 (LC1) and load circuit 2 (LC2). The higher-current circuit LC1 creates the highest-frequency TRV causing the higher probability of re-ignition, whereas the lower-current circuit LC1 has the highest inductance with a higher energy $\frac{1}{2}Li_{ch}^2$ stored at current chopping and causing a higher TRV suppression peak (see Section 4.3.3.2).

The switching of lower inductive-load currents is more severe than switching of higher inductive-load currents because, at a given voltage level, the lower current implies a higher inductance. Therefore, the current values of test circuit LC2 do not cover lower currents in practice. If lower currents than those in LC2 are involved, it is recommended to test at the real currents to be interrupted in the given application.

Shunt-reactor switching tests need a single-frequency TRV, the frequency of which can be directly derived from the reactor inductance (determined by the rated voltage $U_r$ and load current) and the standardized value for the stray capacitance $C_L$ of the load, that is, of the reactor. From this, the time to peak of the (1 − cosine) function $t_p$ can be calculated as the half-period of this oscillation. The associated value of $t_3$ (the time to peak of the IEC two-parameter TRV envelope) is related to $t_p$, depending on the amplitude factor $k_{af}$ (the damping of the oscillation). In Figure 13.4 $t_p$ and $t_3$ are visualized.

In the graph of Figure 13.5, the ratio $t_3/t_p$ is shown as a function of the amplitude factor $k_{af}$.

The factor 870 for the $t_3$ calculation in Table 13.4 stems from this graph – in fact the factor 0.87 is used, multiplied by 1000 to account for entering $C_L$ in nF and $U_r$ in kV.

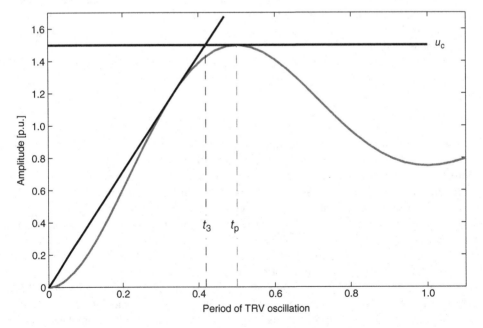

**Figure 13.4**   Visualization of $t_3$ and $t_p$.

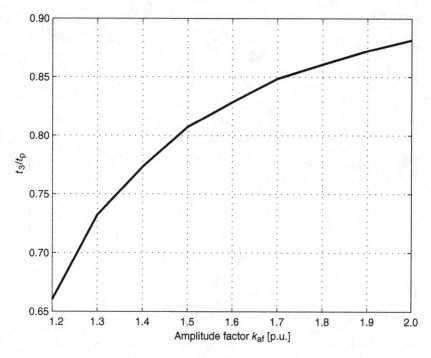

**Figure 13.5**   Relationship $t_3/t_p$ as a function of amplitude factor $k_{af}$.

An important observation is that the value for $u_c$ is a prospective (inherent) parameter only. In real tests, the peak TRV value is always higher because current chopping adds the suppression peak to the TRV peak.

Since the observed overvoltages in shunt-reactor tests are strongly inherent to the test circuit, some characteristics of the chopping and re-ignition behaviour are required to be entered in test reports. These characteristics can then be used to predict overvoltages in a given application according to methods outlined by CIGRE [24] and summarized in Section 4.3.3.3.

**Table 13.4**   IEC Shunt-reactor test parameters.

| Load circuit | | | | LC1 | LC2 | TRV | |
|---|---|---|---|---|---|---|---|
| $U_r$ [kV] | $k_{pp}$ [p.u.] | $k_{af}$ [p.u.] | $C_L$ [nF] | $I$ [A] | | $u_c$ [p.u.] | $t_3$ [μs] |
| 12 to 36 | | | 0.8 | 1 600 | 500 | | |
| 52 to 72.5 | 1.5 | | 1.75 | 630 | 200 | | |
| 100 to 170 | | 1.9 | 1.75 | 315 | | $k_{pp} \cdot k_{af}$ | $870\pi \times \sqrt{\dfrac{U_r k_{pp} C_L}{\sqrt{3}\,\omega I}}$ |
| 245 to 800 | 1.0 | | 2.6 | | 100 | | |
| > 800 | | | 9.0 | | | | |

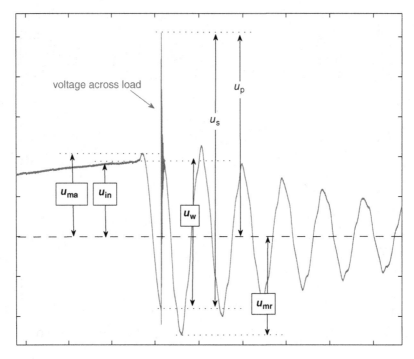

**Figure 13.6** Oscillogram of voltage across inductive load from an IEC shunt-reactor switching-test showing parameters necessary for overvoltage calculation.

The following four parameters have to be recorded for this calculation (indicated as boxed parameters in Figure 13.6), the values refer to the example oscillogram in Figure 13.6:

- $u_{in}$   initial voltage (to neutral) at the moment of current chopping (185 kV);
- $u_{ma}$ suppression peak voltage to earth (208 kV);
- $u_{mr}$ recovery peak voltage of the load side to earth (247 kV), if higher than $u_{ma}$; and
- $u_w$  voltage across circuit-breaker at re-ignition (367 kV).

Other interruption parameters defined in IEC 62271-110 [21] are:

- $u_s$   maximum peak-to-peak overvoltage excursion at re-ignition (693 kV, oscillating at 1.2 MHz); and
- $u_p$   maximum overvoltage to earth (510 kV).

The level of overvoltages produced is not relevant for the decision whether or not the shunt-reactor test requirements of IEC have been met.
The criteria to pass a test are:

- the circuit-breaker shall consistently interrupt the current with (one or more) re-ignitions at one current zero crossing only (this is a very severe requirement to meet); and
- re-ignitions shall always occur between arcing contacts only.

## 13.3.2  Medium-Voltage Motor Switching

Motor switching tests are applicable to all three-pole circuit-breakers having rated voltages $\leq 17.5$ kV, which may be used for the switching of three-phase asynchronous squirrel-cage or slip-ring motors.

The test circuit for motor-switching tests is described in detail in IEC 62271-110 [21]. This description differs from all other standards regarding switching, where normally envelopes for TRVs are prescribed without any guidance of how to accomplish them. The reason for the detailed description of the test circuit consists in the very high frequencies that are involved and the way the phases interact, for example, to produce virtual current chopping (see Section 4.3.3.4). Therefore, this test represents a real motor switching application, with a real cable, real busbar, with only the motor represented by a linear *LR* circuit. This is considered the only way to verify the switching behaviour under this condition.

Also in this test, IEC sets no limits to the overvoltages resulting from re-ignition or (virtual) chopping in a test circuit, since these are only relevant to the specific application. Overvoltages between phases may be as significant as phase-to-earth overvoltages.

The test circuit consists of a three-phase source, a 6 m busbar connection to the circuit-breaker and 100 m of belted cable to the motor substitute circuit. Two motor-substitute circuits are defined, one for 100 A and one for 300 A, as well as two source circuits with capacitance to earth of 40 nF and 2 µF respectively.

The following four parameters have to be recorded: (see [21]):

- $u_{ma}$ suppression peak voltage to earth;
- $u_{mr}$ recovery peak voltage to earth (if $> u_{ma}$);
- $u_s$  maximum peak-to-peak overvoltage excursion at re-ignition; and
- $u_p$  maximum overvoltage to earth.

The criterion of re-ignitions being allowed at one current zero only, as specified for shunt-reactor switching, is not applicable to motor switching.

## 13.4  Specification and Commissioning

### 13.4.1  General Specifications

Utility circuit-breaker specification can take a number of forms with varying degrees of manufacturer involvement. The two most common forms are the functional or performance type and the detailed specification type.

The functional specification type is used mainly in turnkey projects where the contractor delivers whole substations as the deliverable product. Such a specification might simply read as:

> 245 kV $SF_6$ gas circuit-breaker, 40 kA, 2 000 A, 900 kV LIWL, live-tank, suitable for use at ambient temperatures in the range from –50 °C to +40 °C and moderate seismic qualification level, all in accordance with IEC 62271-100.

This statement calls for the manufacturer's standard IEC compliant circuit-breaker with inherent fittings but with exceptional low ambient temperature and seismic requirements. In general, circuit-breaker specifications should use the requirements of standards to the greatest degree possible by reference rather than by transcription.

The more comprehensive detailed specification is used by most utilities. This type can be as follows:

- Supply only: circuit-breakers are supplied by the manufacturer, and installed and commissioned by the user.
- Supply and installation: circuit-breakers supplied and installed by the manufacturer and commissioned by the user.
- Supply, installation, and commission: circuit-breakers supplied, installed, and commissioned by the manufacturer and then turned over to the user.

The detailed specification differs from the functional specification in that it incorporates requirements unique to the user. These requirements can, for example, relate to ambient conditions, clearances, interfaces and so on. A recommended approach to specification writing is to use a hierarchical format starting at a high level and working down into exact detail, as shown in Figure 13.7.

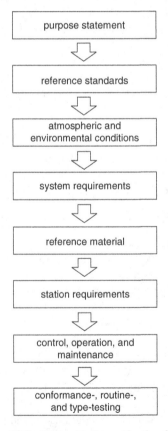

**Figure 13.7**   Hierarchical specification format.

For ease of consideration by the manufacturer, a convenient approach is to provide a listing of specific requirements in a tabular format rather than in a textual format.

## 13.4.2   Circuit-Breaker Specification

The sequence of requirements in IEC 62271-100 can be used as a basis for a hierarchical specification format for each particular circuit-breaker. The proposed specification parts are:

- General: Statement as to the purpose and intended use of the circuit-breakers.
- Standards: Listing of the applicable international or national standards.
- Atmospheric and environmental conditions: Listing by either reference to standards or specific in the event of particular conditions, such as extreme low or high ambient temperatures, seismic zone, pollution, and so on.
- Service conditions: System level requirements.
- Drawings and reference material: Applicable user drawings, internal station design, protection and control standards and related material.
- Ratings: Rating details for each of the circuit-breakers covered by the specification.
- Special ratings or features: Ratings beyond standard values, such as extra-long transmission-line switching; provision of features related to operation or monitoring.
- Design and construction: This part covers a wide range of user specific requirements:
  Gases and liquids: Gas quality, tightness and allowable leakage rate, and provision for gas filling, removal, and sampling; liquid filling and drainage facilities.
  - Earthing: Earthing pads are required at all points intended to operate at earth potential.
  - Bushings and insulators: Applies to dead tank and live tank circuit-breakers, respectively; user states the required colour and creepage distance.
  - Support structures: Steel support structures are supplied with the circuit-breaker and users typically specify a minimum clearance between the top of the footing on which the structure rests and the base of the insulators or bushings.
  - Operating mechanisms: It can be spring, pneumatic, or hydraulic type. The user states requirements, or alternatively, has a preference for one type of stored energy and control for safe operation, that is, block a close operation if there is insufficient stored energy to complete a close–open operation.
  - Control cabinet: User requirements include lighting, power receptacles, spare terminal blocks and auxiliary switches, access and degree of protection.
  - Operational features: User requirements include AC and DC control voltages, duty cycles, phase-disagreement, anti-pump interlocking, gas density monitoring, and remedial action, as applicable.
  - Maintenance facilities: Usually limited to the provision of means to attach travel-curve measuring devices and for $SF_6$ gas sampling.
- Conformity testing: This testing applies to particular requirements specified by the user. For example, most users require some form of trip-circuit continuity verification. Typically, a low-level trickle current is used for this purpose and a conformity test is required to demonstrate that the trip coil can reset against the continuous trickle current.
- Type and routine testing: Reference should be made to IEC 62271-100 [1] noting additions or exceptions to the stated testing.

Controlled switching is discussed in Section 11.4.1 and circuit-breakers applied with controllers for this purpose will have extra specification provisions. For shunt capacitor-bank switching and transmission-line high-speed auto-reclosure, reference can be made to IEC 62271-302 [22]; for shunt reactor switching, reference can be made to both IEC 62271-302 and IEC 62271-110 [21].

### 13.4.3    Information to be given with Requests for Offers

Section 9 of IEC 62271-100 can be followed in this regard [1]. Additional information may be required based on the particular application, for example, description of the load circuit for shunt-reactor switching and for the method of offer evaluation. The evaluation can be done on a device capital-cost only, on the installed cost, or on a life-cycle cost basis. For the last-named case, detailed maintenance requirements and schedules are needed.

### 13.4.4    Information to be provided with Submitted Offers

Chapter 9 of IEC 62271-100 and the information of Section 13.4.3 above can be followed in this regard.

### 13.4.5    Circuit-Breaker Selection

Chapter 8 of IEC 62271-100 and also the application guide IEC 62271-306 [23] should be followed in this regard.

The short-circuit current rating should be selected on the basis of actual fault currents likely to be achieved within the lifetime of the circuit-breaker. A one-rating-fits-all approach is not recommended because circuit-breakers are type tested to their ratings and not to the application requirements. Thus, a 40 kA rated circuit-breaker applied on a system with a 20 kA short-circuit prospective level does not have a type test to support its use at the lower current. This is particularly an issue for self-blast circuit-breaker technologies, and users who practise a "blanket order" procurement strategy should consider additional type testing to confirm performance at lower fault currents.

### 13.4.6    Circuit-Breaker Commissioning

Chapter 10 of IEC 62271-100 should be followed in this regard. Commissioning is important for a number of reasons because it:

- initiates the warranty;
- ensures compliance and may identify specification oversights;
- may identify manufacturing defects and shipping damage;
- provides data for benchmarking future maintenance; and
- confirms that plant is safe to operate.

# References

[1] IEC 62271-100 (2012) High-voltage switchgear and controlgear – Part 100: Alternating-current circuit-breakers. Ed. 2.1.

[2] IEC 62271-1 (2007) High-Voltage Switchgear and Controlgear – Part 1: Common Specifications. Ed. 1.0.

[3] Das, J.C. and Mohla, C.D. (2013) Harmonizing with the IEC: ANSI/IEEE Standards for High-Voltage Circuit Breakers. *IEEE Ind. Appl. Mag.*, **19**(1), 16–26.

[4] Park, R.H. and Skeats, W.F. (1931) Circuit-breaker recovery voltages, magnitudes and rates of rise. *Trans. A.I.E.E.*, **50**(1), 204–239.

[5] von Hochrainer, A. (1957) Das Vier-Parameter-Verfahren zur Kennzeichnung der Einschwingspannung in Netzen. *ETZ*, **78**(19), 689–693.

[6] Baltensperger, P. (1960) Definition de la tension transitoire de rétablissement aux bornes d'un disjoncteur par quatre paramètres, possibilités des stations d'essais de court-circuit. Bulletin de l'Association Suisse des Électriciens 3.

[7] Catenacci, G. (1956) Le frequenze proprie della rete Edison ad A.T. L'Elettrotrechnica XLIII(3).

[8] Baatz, H. (1956) *Ueberspannungen in Energieversorgungsnetzen*, Part C, Chapters 1–5, Springer-Verlag, Berlin.

[9] Pouard, M. (1958) Nouvelles notions sur les vitesses de rétablissement de la tension aux bornes de disjoncteurs à haute tension. Bulletin de la Société Francaise des Électriciens, 7th series VIII(95), pp. 748–764.

[10] Catennacci, G., Paris, L., Couvreux, J.-P. and Pouard, M. (1967) Transient recovery voltages in French and Italian high-voltage networks. *IEEE Trans. Power Ap. Syst.*, **PAS-86**(11), 1420–1431.

[11] Barret, J.P. (1965) Dévelopements récents des méthodes d'étude des tensions transistoires de manoeuvre sur les réseaux a haute tension. *Rev. Gen. Electr.*, **74**, 441–470.

[12] Braun, A. *et al.* (1976) Characteristic values of the transient recovery voltage for different types of short circuits in an extensive 420 kV system. *ETZ-A*, **97**, 489–493.

[13] Slamecka, E. on behalf of WG CC03 of CIGRE SC 13 (1983) Transient recovery voltages in medium-voltage networks. *Electra*, **88**, 49–88.

[14] Parrot, P.G. (1985) A review of transformer TRV conditions. *Electra*, **102**, 87–118.

[15] Sluis, van der, L. (1998) Transient recovery voltages in medium-voltage networks. *Electra*, **181**, 139–151.

[16] CIGRE Working Group CC 03 (1998) Transient Recovery Voltages in Medium Voltage Networks. CIGRE Technical Brochure 134.

[17] Wagner, C.L. and Smith, H.M. (1984) Analysis of transient recovery voltage (TRV) rating concepts. *IEEE Trans. Power Ap. Syst.*, **PAS-103**, 3354–3362.

[18] Ozaki, Y. (1994) Switching surges on high-voltage systems. Central Research Institute of Electric Power Industry, Tokyo, Japan.

[19] CIGRE Working Group A3.22 (2008) Field Experience and Technical Specifications of Substation Equipment up to 1 200 kV. CIGRE Technical Brochure 362.

[20] CIGRE Working Group A3.22 (2010) Background of Technical Specifications for Substation Equipment Exceeding 800 kV AC. CIGRE Technical Brochure 456.

[21] IEC 62271-110 (2012) High-voltage switchgear and controlgear – Part 110: Inductive load switching.

[22] IEC 62271-302 (2010) High-voltage switchgear and controlgear – Part 302: Alternating current circuit-breaker with intentionally non-simultaneous pole operation.

[23] IEC 62271-306 (2012) High-voltage switchgear and controlgear – Part 306: Guide to IEC 62271-100, IEC 62271-1, and IEC standards related to alternating current circuit-breakers.

[24] CIGRE Working Group 13.02, (1995) Interruption of Small Inductive Currents, CIGRE Technical Brochure 50, ed. S. Berneryd.

# 14

# Testing

## 14.1 Introduction

Testing of switching equipment is usually performed with one of the following objectives:

*Research and development.* These tests are normally carried out in test-laboratories of manufacturers or, when the necessary power and/or voltage is not available, in third-party laboratories. Apart from custom-designed special equipment, the aim is the development of products that will be ultimately submitted to type-tests. Research and development test requirements can vary depending on the stage of development, but the final aim is usually a design capable of withstanding the standardized stresses of a type-test.

*Acceptance.* In these tests, verification is requested of withstanding non-standardized stresses that may occur under special conditions of the electrical network, environment, operation, and so on. Regarding switching, examples are TRVs beyond the standardized limits (for example in the application of series current-limiting reactors or filter banks in converter stations) or other special conditions (for example exceptional short-line fault conditions, missing current zeros, high DC time constants, etc.). Usually, equipment has already been type-tested before being subjected to the additional acceptance tests. The user of switchgear proposes the test-requirement based upon his knowledge of abnormal conditions in his system. Manufacturers and independent laboratories perform such tests.

*Type-test certification.* Type tests are aimed to demonstrate the capability of a single sample of a batch of identical products to conform to a certain standard. Once this has been demonstrated, a type-test certificate is issued by a certification authority. A certificate contains a record of a series of type tests carried out strictly in accordance with a recognized standard. It is a proof that the component tested has fulfilled all the requirements of a recognized standard. If the equipment tested has fulfilled the requirements of this standard, the relevant ratings assigned by the manufacturer are endorsed by the certifying authority. The certificate is applicable

*Switching in Electrical Transmission and Distribution Systems*, First Edition.
René Smeets, Lou van der Sluis, Mirsad Kapetanović, David F. Peelo, and Anton Janssen.
© 2015 John Wiley & Sons, Ltd. Published 2015 by John Wiley & Sons, Ltd.

only to equipment of a design identical to the tested one. The certifying authority is responsible for the validity and the contents of the certificate.

The responsibility for conformity of any apparatus having the same designation as the one tested rests with the manufacturer. The certificate contains the essential drawings and a description of the equipment tested.

Type-tests are carried out in independent, duly accredited test-laboratories. Several of these in the world apply the rules of the Short-circuit Testing Liaison (STL), an organization of test laboratories and authorities looking after a uniform interpretation of standards throughout the world by providing practical guidelines [1]. STL, a purely technical institute, makes use of the competence and expertise of its members to provide these guidelines. This voluntary society also defines the test-report templates to ensure that equipment users can easily compare the results of the different member laboratories. The society also defines rules and procedures to assure the quality of the test results and certified products.

Five categories of tests are distinguished by STL to verify [2]:

- short-circuit making and breaking performance;
- switching performance, normally the capacitive-current-switching performance;
- dielectric performance;
- temperature-rise performance and measurement of the main-circuit resistance; and
- mechanical performance.

STL members issue certificates on these five items related to the specified rated values, except for the mechanical performance, for which there is no rating defined. By the repetition of the type-test duties after a period of time, usually after five years, assurance can be obtained that the manufacturing quality of the circuit-breaker or switchgear, quality of material and workmanship is maintained.

In contrast to type-tests, carried out on one sample of a batch, *routine tests* are tests to which each individual piece of equipment is subjected. They are for the purpose of revealing faults in material and construction. They do not impair the properties and reliability of the test object.

The following subsections highlight only those test methods intended to verify the *breaking capacity* of circuit-breakers, that is, the capability to interrupt short-circuit currents. A detailed overview with many practical examples of all making, breaking, and switching test-methods can be found in reference [3].

## 14.2 High-Power Tests

### 14.2.1 Introduction

High-power-test technology has been a main driver behind the development of HV circuit-breakers. A large number of test circuits and test methods have been developed since 1911 when

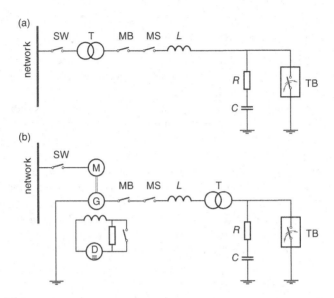

**Figure 14.1** Simplified single-line diagrams of high-power laboratory layout: (a) network-supplied; (b) generator supplied. SW – source circuit-breaker; MB – master breaker; $L$ – current-limiting reactor; M – motor; G – generator; T – short-circuit transformer; M – make switch; $R$, $C$ – TRV shaping resistor, capacitor; D – excitation circuit; TB – circuit-breaker under test.

AEG built the world's first high-power laboratory in Kassel, Germany, with a short-circuit power of 150 MVA.

Based on the way the power for testing is supplied, three types of high-power laboratories can be distinguished:

- Laboratories directly supplied from the network, as outlined in Figure 14.1a. The advantage is that a large amount of short-circuit power can be available. In power systems, there are a relatively small number of locations suitable for high-power testing. To avoid instability of the supplying network during the short-circuit test, the available short-circuit power at the location of the test laboratory should be approximately ten times the maximum power used during an actual test. The network must not be disturbed by frequent short-circuits and high overvoltages and, often, the test parameters cannot be adjusted as prescribed. Especially, the match of test voltage is not always easy. In addition, the available testing power of even the most powerful network laboratories falls far short of that required for present-day HV circuit-breakers, the interrupting capability of which corresponds to tens of gigavoltamperes.
- Laboratories having their own power source(s), that is, generators separate from the network, shown in Figure 14.1b. Conditions very close to actual electric power systems can be simulated in generator-supplied test laboratories. Testing in these laboratories has numerous advantages in comparison with network-supplied laboratories. The main advantage is flexibility. The main disadvantage is that huge investments are necessary.
- Laboratories that have current and voltage supplied from pre-charged capacitor banks. These have limited power and are exclusively used for research and development.

**Figure 14.2**   Short-circuit generator (in the background) and frequency converter (foreground) in order to enable high-power testing with short-circuit current of variable power frequency (reproduced with permission of KEMA laboratories).

The power of even the largest high-power laboratories is not sufficient for testing the majority of HV circuit-breakers. Therefore, alternative test methods, such as *synthetic testing* (see Section 14.2.3) are in use to impose adequate current and voltage stresses on circuit-breakers.

A short-circuit generator, Figure 14.2, is a specially designed three-phase generator that supplies the short-circuit power for the test in a generator test station. The characteristics of such generators differ markedly from those of their conventional counterparts used to generate electric power commercially.

Short-circuit generators have very low reactance, high mechanical strength of windings against electrodynamic and centrifugal forces, and they are suited for impulse excitation. A motor is used to speed up the rotor to its synchronous rated mode of operation before the test. During the short-circuit test, basically short-circuiting the stator, the excitation and the kinetic energy of the rotor mass supply the power. Because it takes approximately 20 minutes to bring the generator up to speed, the power supplied by the network is considerably lower than the (apparent) power used for testing.

The largest high-power laboratory using generators, and at the same time the largest overall high-power test facility in the world is KEMA laboratories, operated by DNV GL, Arnhem, the Netherlands, with 8400 MVA (at 50 Hz) and 10 000 MVA (at 60 Hz) three-phase power available in the test-bays (see Figure 14.3). An expansion of the installed power of 50% is planned to be available in 2015. The power of 8400 MVA is sufficient to test a circuit-breaker in a three-phase circuit up to 145 kV and a maximum rated short-circuit breaking current of 31.5 kA. Above these levels, circuit-breakers can be tested in a single-phase circuit; alternatively, single-unit tests, two-part tests, and/or synthetic (indirect) test methods are standardized as described in more detail below.

Large network-supplied laboratories are at les Renardières, France, operated by EdF, and Rondissone, Italy, operated by CESI. In India, a large network-supplied laboratory is under construction at Bina.

Other large, well-known manufacturer-independent high-power laboratories are [4]: CESI located in Milano, Italy; IPH located in Berlin, Germany, operated by CESI; Zkušebnictví

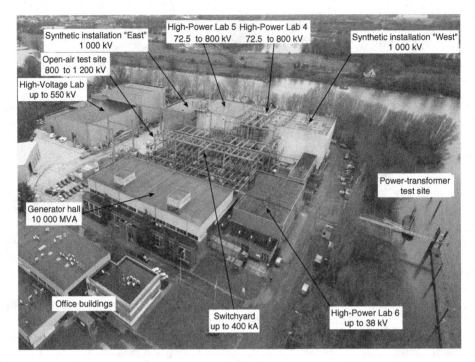

**Figure 14.3** KEMA High-Power Laboratory (HPL) in Arnhem, the Netherlands, 2013 (reproduced with permission of KEMA laboratories).

located in Běchovice near Prague, the Czech Republic, operated by DNV GL; VEIKI located in Budapest, Hungary; ICMET located in Craiova, Romania; the laboratory of the Federal Grid Company of the United Energy System of Russia (FGC of UES), located in Moscow, Russia; XIHARI in Xi'an and EETI in Suzhou, China; CPRI located in Bangalore, India; KERI located in Changwon, South Korea; KEMA laboratory US located in Chalfont, Pennsylvania, USA; PowerTech located in Vancouver, Canada; LAPEM located in Irapuato, Mexico, and CEPEL located near Rio de Janeiro, Brazil.

In addition, most large switchgear manufacturers have their own high-power laboratories: Siemens located in Berlin, Germany; ABB located in Ludvika, Sweden and Baden, Switzerland; Alstom located in Villeurbanne, France; Schneider Electric located in Grenoble, France and Ormazabal Group in Bilbao, Spain; Toshiba Co., Mitsubishi Electric and Hitachi located in Japan.

Methods of testing the breaking capacity of circuit-breakers can be divided into direct tests and indirect tests, the latter being known as *synthetic tests* [5].

*Direct tests* are tests where the applied current and voltage are obtained from a single power source – which can be either one or more short-circuit generators in parallel – the power system, or a combination of these. A direct test is the preferred test method, described in Section 14.2.2.1. The single power source must be capable of delivering the short-circuit current as well as transient and power-frequency recovery voltage. Basically, the direct test allows a test of a three-pole circuit-breaker with full current and full voltage, preferably in a three-phase circuit. Direct tests can be performed as field tests or as laboratory tests.

*Synthetic (indirect) tests* are recognized equivalent tests for verification of the breaking capacity of HV-, EHV- and UHV circuit-breakers by other means than with full short-circuit power available from a single power source. Instead, several sources are employed, as described in Section 14.2.3.

If, due to limitations of the testing facility, the short-circuit performance of the circuit-breaker cannot be proved in a three-phase circuit, several methods employing either direct or synthetic test methods may be used – singly or in combination:

- *single-phase tests* – tests on only one pole of a three-pole circuit-breaker with appropriate stresses as specified for each of the poles of a three-pole circuit-breaker (see Section 14.2.2.2);
- one (or more) *unit(s) tests* – tests on a unit (interrupting chamber) or on a group of units at the current specified for the test on the complete pole of a circuit-breaker and at the appropriate fraction of the applied voltage specified for the test on the complete pole of the circuit-breaker (see Section 14.2.2.3); and
- two- (or multi-) *part tests* – tests carried out in two or more successive parts when all requirements for a given test duty cannot be met simultaneously (see Section 14.2.2.4).

## 14.2.2  Direct Tests

### 14.2.2.1  Direct Three-Phase Tests

In the three-phase test circuit depicted in Figure 14.4, the master circuit-breaker MB is placed directly behind the three-phase short-circuit generator G. The role of this master breaker is to interrupt the short-circuit current in the case of a failure of the tested circuit-breaker TB and in the case of a short-circuit in the test installation. Master breakers and their control shall operate extremely fast (typically within 10 ms) in order to allow study of the root cause of a failure of the TB to pass a certain test. Otherwise, the test-object will be destroyed during an unsuccessful test, leaving no possibility for analysis.

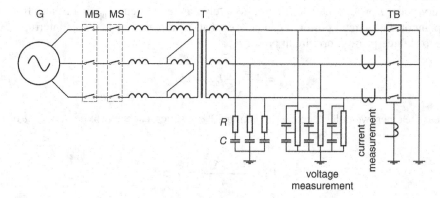

**Figure 14.4**  Diagram of a direct three-phase test circuit. G – generator; MB – master breaker; MS – make switch; *L* – current-limiting reactor; T – short-circuit transformer; *R, C* – TRV shaping resistor, capacitor; TB – circuit-breaker under test.

When the tested circuit-breaker has to perform a break operation, it is initially in a closed position, the master breaker is closed, but the make switch MS is open. The adjustable reactor $L$ is used to add extra reactance in the circuit to realize the required test current at the chosen test voltage. Because the terminal voltage of the short-circuit generators is relatively low, between 10 and 17 kV, specially designed short-circuit transformers T are necessary to transform the generator voltage to the higher testing voltage. The TRV-adjusting elements $R$ and $C$ are connected at the HV side of the transformer. After the generator is driven to the nominal power frequency and the rotor is excited, the make switch is closed, initiating a short-circuit current flowing through the tested circuit-breaker. Upon an opening command, the tested circuit-breaker will try to interrupt the current. After successful interruption, the tested circuit-breaker is stressed by the transient recovery voltage coming from TRV-adjusting elements together with the inductance formed by the total circuit reactance consisting of the synchronous reactance of the generator, the reactor, and the leakage reactance of the transformer.

When the tested circuit-breaker has to perform a make-break operation, it must make a short circuit. Before a make-break test, the tested circuit-breaker is in the open position and it closes, after the make switch has been closed to apply the open-circuit voltage to the tested circuit-breaker terminals.

For any test in the high-power laboratory, a number of recordings of currents and voltages are required at different time scales. Other time-varying parameters, such as the position of the tested circuit-breaker contacts or pressure inside the compression volume can also be recorded. For the recordings, multi-channel digital transient recorders are available, which have sampling ranges up to 25 Msamples/s and a vertical resolution of 12 bits. The great advantage is that once the data are stored in digital form, they can be analysed by a computer and stored permanently for future study.

### 14.2.2.2   Direct Single-Phase Tests

A single-phase test as a direct-test method may be performed in place of three-phase conditions encountered in service. However, care must be taken to ensure that relevant stresses are applied, since the circuit-breaker poles are generally not equally stressed during interruption.

As explained in Section 3.3.2, the first pole-to-clear has to recover against a power-frequency recovery voltage of $k_{pp}$ [p.u.] In a non-effectively earthed system, $k_{pp} = 1.5$, and from the viewpoint of short-circuit power, it can be seen that in this case, the first pole interrupts 50% of the total three-phase short-circuit power $P_{sc}$:

$$P_1 = 1.5 \frac{U_r}{\sqrt{3}} I_{sc} = 0.5 P_{sc} \tag{14.1}$$

whereas the other two poles face the remaining 50% in series, each approximately 25% of $P_{sc}$:

$$P_2 \approx P_3 \approx \frac{U_r I_{sc} \sqrt{3}}{4} = 0.25 P_{sc} \tag{14.2}$$

The conditions for the first pole-to-clear represent the most severe stress regarding TRV, but the arcing times of the last two poles-to-clear are longer. As it is important that both the arcing window and associated TRV are applied during the direct single-phase test, the use of the arcing time of the last pole-to-clear in combination with the TRV of the first pole-to-clear is a possible alternative. Using this alternative makes it possible to carry out only three breaking operations per test duty, as both IEC and ANSI/IEEE standards require, but prolongation of arcing times may be necessary. However, it is recognized that such single-phase tests are more severe than three-phase tests [6] since, in the three-phase condition, the highest TRV is applied only on the first pole-to-clear, which has shorter arcing times.

Another alternative test procedure, which reproduces exactly the stresses of a three-phase interruption, is also possible, but a greater number of interruptions is required to provide a full demonstration. For the first pole-to-clear with $k_{pp} = 1.3$, nine instead of three breaking operations per test-duty are required, and for $k_{pp} = 1.5$, six operations are required for complete verification of the performance. This is because three valid operations are required for demonstration of satisfactory performance for each of the separate phase conditions of arcing time and TRV.

Moreover, during the direct single-phase test, the procedures may give rise to an invalid test that can overstress and damage the circuit-breaker. Therefore, particularly for non-solidly earthed systems, it is recommended to use synthetic test methods instead of direct single-phase tests [7].

The description above is relevant for tests with symmetrical short-circuit currents. The test procedure for asymmetrical current is even more complicated, as interruption can occur either after a minor loop or after a major loop of current.

For single-phase testing, it is necessary to determine whether it is justified to subject only one pole to a test. The interrupting process is affected by electrodynamic forces, the gas exhaust from the adjacent poles, the contact speed, the pressure of the extinguishing medium, and so on. The energy output of the operating mechanism should be properly considered since the single-phase operation requires only a reduced operating force.

In other words, the design of the circuit-breaker and the interaction of poles are particularly important when the three poles are driven by a common mechanism and are in a common enclosure. In this case, short-circuit performance can only be verified by three-phase testing. This is explained in Figure 14.5. In this figure, the course of the contact travel is given for a no-load operation, single-phase, and three-phase breaking test on the same circuit-breaker. The increasing stress on the mechanism, comparing no-load operation (no arcing interrupter) with single-phase breaking (only one arcing pole) and three-phase breaking (three arcing poles) is evident, showing appreciable slowing down of the contact travel in the three-phase arcing condition. These interactions do not apply to circuit-breakers having independently operated poles.

### 14.2.2.3   Unit Testing

When a modular concept is used in the design of HV circuit-breakers with two or more identical interrupter units in series, performance of a complete pole, and possibly of the whole circuit-breaker, can be evaluated from the performance of one or more units. For unit testing to be valid, all units must be identical in construction. They must operate simultaneously and must not influence each other.

The test current must be the same as in service and the test voltage should be the appropriate proportion of the rated voltage $U_r$. This proportion depends on the number ($n$) of

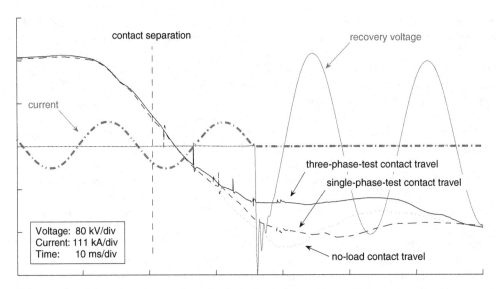

**Figure 14.5** Contact travel curves in no-load, single-phase breaking, and three-phase breaking tests.

series units and on the voltage distribution across the units. The test voltage for a unit should correspond to the voltage share that would appear across the highest stressed unit. In order that natural frequencies remain the same as in the full circuit, the circuit inductance $L$ should be proportionally lower ($L/n$) and the capacitance $C$ should be proportionally increased ($nC$), as illustrated in Figure 14.6 for a circuit-breaker with three identical interrupters in series per pole. In order to guarantee the correct voltage across a unit, evaluation has to be carried out on the distribution of voltages across the various units within a frequency range from power frequency to TRV frequency. This involves calculation of the voltage transients, taking into account all capacitors and stray capacitances, such as those to earth.

The tests are then carried out with the voltage of the highest stressed unit, which is normally the unit connected to the source circuit; this voltage is always higher than $U_r/n$.

Especially in designs without grading capacitors, where the voltage distribution is determined only by stray elements, a significant influence of the voltage distribution is observed

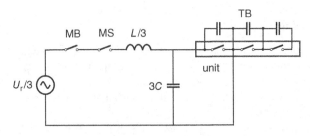

**Figure 14.6** Basic principle of unit tests.

for different locations of the circuit-breaker in the test laboratory because of different stray capacitances to the earthed environment.

#### 14.2.2.4   Two-Part or Multi-Part Testing

If all TRV requirements for a given test duty cannot be met simultaneously, the test may be carried out in two or more successive parts regarding the initial portion of the TRV and the relevant parameters specified for the TRV envelope.

The number of tests for each part is the same as the number required for the test duty. The arcing times in separate tests, forming part of a multi-part test, are required to be the same with a margin of ±1 ms.

### 14.2.3   Synthetic Tests

#### 14.2.3.1   Introduction

Circuit-breaker testing requires high and increasing short-circuit power. The short-circuit ratings of many circuit-breakers are so high that verification of their breaking capacity is no longer possible with direct test circuits. Figure 14.7 shows the increase over years of the short-circuit power per interrupter unit [8]. The most powerful interruption chambers can interrupt

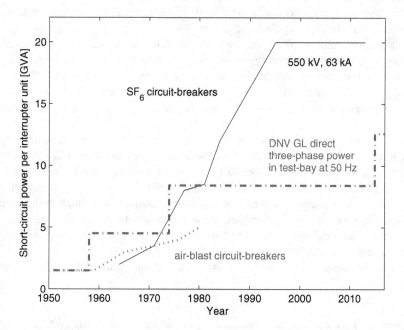

**Figure 14.7**   Increase of short-circuit power per interrupter unit and test power of DNV GL high-power laboratory.

63 kA at 550 kV, which would require a single-phase direct power source of 26 GVA[1]. This is approximately three times the maximum output power of the largest test laboratory available in the world in 2013. Therefore, synthetic (indirect) testing of circuit-breakers [9] has been developed as an economical, technically correct, and equivalent alternative to direct testing.

In synthetic tests, all or a major portion of the current is obtained from one source (current circuit) whereas the applied voltage and/or the recovery voltages (transient and power-frequency) are delivered wholly or in part by one or more separate sources (voltage circuits).

Synthetic testing methods are based on the fact that during the breaking operation, when high current flows through the circuit-breaker, there is only a relatively small arc voltage between the open contacts and when high (T)RV appears across the contacts, there is basically only voltage across the circuit-breaker without current. These two phenomena occur in two consecutive periods.

Therefore, synthetic testing methods use two energy sources for simulating the electrical stresses on a circuit-breaker:

- *high-current source*, one (or more short-circuit generators), which supplies the power-frequency current at reduced voltage of 15 to 60 kV during the high-current interval; and
- *high-voltage source*, a capacitor bank with a tuned TRV circuit, which supplies the transient recovery voltage and the recovery voltage at relatively low power during the high-voltage interval.

It is essential during synthetic tests that circuit-breakers are subjected to conditions as severe as those of full direct-test methods. It has been internationally recognized, that synthetic test methods are suitable for type testing of HV circuit-breakers because of the proven high degree of equivalence with direct short-circuit tests.

The synthetic methods need not be seen only as valid substitutes for direct tests. When properly applied, synthetic tests bring significant advantages to testing. One of the advantages is that synthetic tests are non-destructive because, at failure to interrupt, only the capacitor-bank energy is supplied to the failed part instead of the generator energy. This makes synthetic testing an ideal development tool – also because current and voltage stresses can be controlled separately.

However, there are operating conditions, such as capacitor-bank current switching, where synthetic tests show a considerable drawback because in the case of restrikes in a capacitive synthetic circuit, the internal circuit-breaker parts are exposed to much lower stresses than in reality. In service, the full load capacitor discharges across the gap, often already rather wide open. Due to the high energy released, the possible consequence is much higher compared with the restrike in a capacitive synthetic circuit where the discharge energy is quite low. Thus, after a restrike, the probability to pass capacitive tests in a successive test-series is larger in a synthetic capacitive circuit than in a direct capacitive circuit.

Because of the importance of synthetic testing, a relevant standard has been published by IEC [10].

---

[1] The power necessary to test a circuit-breaker pole in a single-phase test is given by $k_{pp}(U_r/\sqrt{3})I_{sc}$.
The system short-circuit power per phase is given by $(U_r/\sqrt{3})I_{sc}$.
The short-circuit power needed to test a three-pole circuit-breaker is given by $\sqrt{3}U_r I_{sc}$.

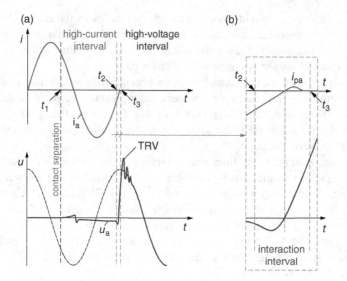

**Figure 14.8** Intervals during the interruption process. (a) high-current and high-voltage interval; (b) interaction interval.

### 14.2.3.2 Intervals During the Interruption Process

During the interruption process, there are in general three significant time instants recognized in Figure 14.8:

$t_1$ instant of contact separation;
$t_2$ start of a significant change in arc voltage; and
$t_3$ cessation of the current flow, including the post-arc current, if any.

Considering the current and voltage stresses during the interruption process, there are three intervals after the instant of contact separation:

- *high-current interval* from $t_1$ to $t_2$;
- *interaction interval* from $t_2$ to $t_3$; and
- *high-voltage interval* or recovery interval after $t_3$.

The *high-current interval* is the time from the contact separation to the start of the significant change in arc voltage, that is, the sharp increase to the arc-voltage extinction peak. During the high-current interval, the tested circuit-breaker is stressed by the synthetic test circuit in such a way that the initial conditions for the interaction interval are the same as under service conditions.

The *interaction interval* is the time from the start of the significant change in arc voltage, prior to current zero, to the time when the current, including the post-arc current, if any, ceases to flow. During the interaction interval, the short-circuit-current stress changes into dielectric stress, and the circuit-breaker behaviour can significantly influence the current and voltages in the circuit. As the current decreases to zero, the arc voltage may rise to charge the parallel capacitance and distort the current passing through the arc, as explained in Section 3.6.1.6. After current zero, the post-arc conductivity may result in additional damping of the transient recovery voltage, which may influence the voltage across the circuit-breaker and the energy supplied to the residual post-arc plasma in the contact gap.

The interaction between the circuit and the circuit-breaker during the interaction interval is of extreme importance to the interrupting process and it presents a critical period in terms of possible thermal failure of the circuit-breaker. It is, therefore, important that the interrupter is stressed during this time interval in the same way in the synthetic circuit as under operating conditions in the power system.

The *high-voltage interval* is the time after cessation of the current flow. During the high-voltage interval, the gap of the tested circuit-breaker is stressed by the recovery voltage. In synthetic testing, the recovery voltage is, in principle, a DC voltage, which exponentially decays due to the discharge of the capacitor bank. Such a decaying DC recovery voltage imposes an unrealistic stress on the circuit-breaker compared with the AC recovery voltage in the power system. DC recovery voltage must be avoided in synthetic testing as much as possible, since space charges can accumulate and stress the interrupter's internal parts differently from the AC voltage case.

A further complication in synthetic testing is the requirement that recovery voltage must not decay too fast and shall not be less than the equivalent instantaneous value of the power-frequency recovery voltage during a period equal to 1/8 of a cycle of the rated power frequency. Whether an exponentially decaying DC, an AC, or a combined DC and AC recovery voltage is used, it should be kept as close as possible to $U_r\sqrt{2}/\sqrt{3}$. In any case, the recovery voltage must not fall below $0.5U_r\sqrt{2}/\sqrt{3}$ within 100 ms after interruption [10].

Especially in vacuum circuit-breakers at higher voltage ratings, having sometimes a tendency to late restrikes (see Section 4.2.4), it is important to keep the recovery voltage as close as possible to $U_r\sqrt{2}/\sqrt{3}$ during a period of 300 ms, as required in IEC 62271-100. Hybrid test circuits that combine the synthetic test method (during the TRV period) with a direct test method (during the recovery voltage) have been demonstrated [11, 12]. Such a solution should be considered for synthetic vacuum- and generator-circuit-breaker testing.

### 14.2.3.3  Intervals during the Making Process

Prior to making, a circuit-breaker shall withstand the voltage applied across its terminals. During the making process, there are, in general, three significant time instants recognized in Figure 14.9:

$t_0$ instant of pre-strike;
$t_1$ instant of contact touch; and
$t_2$ instant of reaching fully closed (latched) position.

Considering the current and voltage stresses during the making process, there are three intervals after the instant of pre-strike between contacts:

- high-voltage interval prior to $t_0$;
- pre-arcing interval from $t_0$ to $t_1$;
- latching interval from $t_1$ to $t_2$.

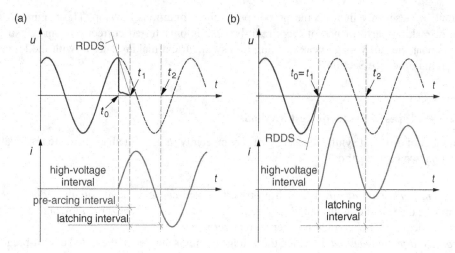

**Figure 14.9**  Intervals during the making process. Left: making at voltage peak leading to symmetrical current; right: making at voltage zero followed by asymmetrical current. RDDS – rate-of-rise of dielectric strength.

The *high-voltage interval* is the time from the application of voltage, with the circuit-breaker in the open position, to the moment of breakdown, that is, *pre-strike* of the contact gap. Therefore, during the high-voltage interval, the tested circuit-breaker shall be stressed by the voltage circuit in such a way that the initial conditions for the pre-strike are the same as under service conditions.

The *pre-arcing interval* is the time, during the closing of the circuit-breaker, from the moment of pre-strike across the gap to the touching of the contacts. During pre-arcing, the circuit-breaker is subjected to electrodynamic forces due to the current and to deteriorating effects due to the arc energy.

Two typical cases may occur depending on the moment of closing, Figure 14.9a and b:

(a) Breakdown occurs near the crest of the applied voltage establishing an almost symmetrical current with maximum pre-arc duration and energy.
(b) Breakdown occurs near the zero crossing of the applied voltage establishing an asymmetrical current with a negligible pre-arc energy, but with maximum electromagnetic force.

The *latching interval* is the time, during the closing of the circuit-breaker, from the touching of the contacts to the moment when the contacts reach the fully closed (*latched*) position. During this interval, the circuit-breaker has to close in the presence of the electrodynamic and gas-dynamic forces due to the current and contact friction forces.

Taking the fundamental principle of synthetic testing methods into consideration, its initial purpose was to verify interruption capability. Unrealistic conditions arise during making tests if the circuit-breaker makes the high-current circuit with full asymmetrical current but at reduced voltage. Considering the severe stresses due to the pre-arc in service, it is also necessary to test the rated short-circuit-making capability at full voltage and symmetrical current because only then do the proper pre-arcing conditions develop. The symmetrical-current-making performance of the circuit-breaker is only tested correctly if the pre-strike occurs near the full voltage peak, leading to a symmetrical making current with the longest pre-arcing time (see Section 3.8.1).

### 14.2.3.4   Types of Synthetic Test Methods

A number of different synthetic test circuits are currently in use, but all of them are practically based on two basic methods:

- *current-injection method*, in which a properly tuned capacitor-bank circuit is discharged through the tested circuit-breaker before power-frequency current zero, allowing automatically adequate TRV conditions after current zero; and
- *voltage-injection method*, in which the voltage circuit is applied to the tested circuit-breaker immediately after power-frequency current zero.

Synthetic test circuits are commonly used for single-phase testing, but they are also available for three-phase testing. Three-phase testing uses a three-phase current source and two or even three voltage sources, one of which provides the transient recovery voltage for the first-pole-to-clear and the other supplies the transient recovery voltage for the second- and third-pole-to-clear, in the case of a non-earthed neutral application.

### 14.2.3.5   Current-Injection Methods

The current-injection methods rely upon the superposition of currents shortly before current zero of the power-frequency short-circuit current. A current of smaller amplitude but higher frequency, supplied from the voltage circuit, is superimposed to the power-frequency current through the tested circuit-breaker. In this way, the current zero in the tested circuit-breaker (TB) occurs later than in the *auxiliary circuit-breaker* (AB) placed in series with the tested circuit-breaker outside of the injection current loop (see Figure 14.10). The auxiliary circuit-breaker thus separates the high-voltage source, which provides the injection current, from the

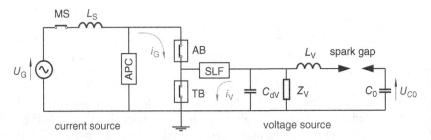

**Figure 14.10**   Typical current-injection circuit with voltage circuit in parallel to the circuit-breaker under test. For symbols, see the text.

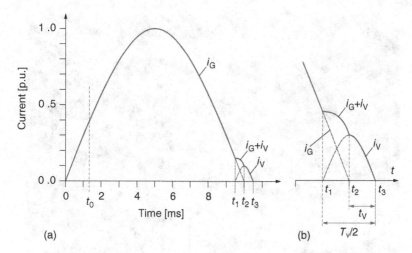

**Figure 14.11** (a) Current through the circuit-breaker under test in the parallel current-injection circuit; (b) current injection timing.

medium-voltage current-source circuit. At the interruption of the current in the tested circuit-breaker, the TB is already connected to the voltage circuit that delivers the TRV. Thus, there will be a natural change-over of stresses from high current to high voltage without delay.

For the current-injection method, there are two principal possibilities: parallel and series current injection. Figure 14.10 shows the simplified circuit diagram of a current-injection circuit with the voltage circuit connected in parallel to the tested circuit-breaker TB. The synthetic test circuit used for the *parallel current injection* is often referred to as a Weil–Dobke circuit [13]. Figure 14.11 illustrates the current through the tested circuit-breaker in the parallel current-injection test circuit.

Before the test, both the auxiliary circuit-breaker (AB) and the tested circuit-breaker (TB) are in a closed position. Closing of the make switch (MS) initiates the flow of short-circuit current $i_G$ supplied by the current source. Approximately at the same time, the time instant $t_0$, the auxiliary and tested circuit-breakers separate their contacts.

An *arc-prolongation circuit* (APC) (see Section 14.2.3.10), also called the *re-ignition circuit*, is necessary to produce realistic arc duration in synthetic tests. Without this circuit, the current would be interrupted already at an earlier current zero, because at that current zero the TRV is far too low, being supplied by the MV high-current circuit only.

At time instant $t_1$, the spark gap is fired and the main capacitor bank $C_0$, charged before the test to a certain voltage, discharges through the impedance $L_V$ and injects a high-frequency current $i_V$, which adds up to the current $i_G$. The time instant of injection is selected by means of a current-dependent control circuit (*injection timer*). During the time span from $t_1$ to $t_2$, the current circuit and the injection circuit are connected in parallel to the tested circuit-breaker, and the current through the tested circuit-breaker is the sum ($i_V + i_G$).

At time instant $t_2$, the power-frequency short-circuit current $i_G$ reaches zero. Because the driving voltage of the current source is rather low, it is relatively easy for the auxiliary circuit-breaker to interrupt current $i_G$. Thus, it separates the high-current circuit from the high-voltage circuit. When the tested circuit-breaker interrupts the injected current at time instant $t_3$, it is

**Figure 14.12**    Photo of (part of) a synthetic voltage source for producing (T)RV in HV circuit-breaker tests (reproduced with permission of KEMA laboratories).

stressed by the TRV which is provided by the voltage circuit and controlled by the parameters $C_{dV}$, $Z_V$, and $L_V$. The spark gap is still conducting during this period.

In Figure 14.12, the voltage source of a synthetic installation is shown providing TRV up to a 1000 kV peak.

An *artificial line* for the testing of HV circuit-breakers under short-line-fault (SLF) conditions (see Section 3.6.1) can be added in the HV circuit. Such a device with a limited number of discrete components creates a voltage wave shape very similar to that coming from a faulted short overhead-line section [14].

During the period of current injection, the voltage on the main capacitor bank reverses in polarity. By the time the TRV oscillations have damped out, the remaining recovery voltage has a DC character, and this puts a higher stress on the tested circuit-breaker than the power-frequency AC-recovery voltage in the direct test circuit. To overcome this, $Z_V$ should contain a reactor with a high quality-factor, tuned with the main capacitor bank $C_0$ to the power frequency. Preferably, a power-frequency AC source should be added to provide constant AC recovery voltage. The latter is called a "Skeats" circuit [15].

Should the tested circuit-breaker fail to interrupt at time instant $t_3$, then the subsequent current flow consists only of the injected current of the voltage source, thereby limiting possible damage to the circuit-breaker. This is a significant advantage when performing development tests.

In order to provide maximum equivalence with the conditions in reality, it is essential that the rate-of-change of current is the same as in the service condition:

$$\left.\frac{di_v}{dt}\right|_{i=0} = \left.\frac{di_g}{dt}\right|_{i=0} \tag{14.3}$$

Apart from the correct rate-of-change of the injected current, the frequency of the injected current shall preferably be of the order of 500 Hz, that is, between 250 and 1000 Hz [13].

In order to prevent undue influence on the wave shape of the power-frequency current, the lower limit of the frequency of the injected current is 250 Hz.

The maximum allowed frequency of the injected current, 1000 Hz, is determined by the interval of significant change of arc voltage. This interval shall be smaller than the time during which the arc is supplied only by the injected current. To achieve this, the period of the injected-current frequency should be at least four times the interval of significant change of arc voltage.

The moment of current injection shall be such that the time during which the tested circuit-breaker is supplied only by the injected current is no longer than a quarter of the period of the injected-current frequency $(T_v)$, with a maximum of 500 μs. Attention should be paid to the possibility of overstressing the circuit-breaker if the time during which the tested circuit-breaker is supplied only by the injected current $(t_v)$ is shorter than 200 μs, because up to $t_2$, the $di/dt$ is too high due to the summation of currents delivered by the current and voltage source.

Irrespective of fulfilling all requirements mentioned above regarding the injected current, the prolongation of the arcing time influences the thermal energy content of the arc. The influence depends on the interrupting principle of the tested circuit-breaker. Therefore, the equivalence of the parallel current-injection method should be considered in relation to a specific circuit-breaker design. Computer calculations for a self-blast $SF_6$ interrupter show noticeable (> 5%) reduction of both the blast pressure inside the compression volume and the gas-mass flow through the nozzle in the current zero caused by artificial prolongation of arcing time of approximately 1 ms. Puffer interrupters are not so sensitive to the prolongation of arcing time [16].

Comparability tests and testing experience over the years have proved that the parallel current-injection method gives the best equivalence with the in-service situation. It is used by many high-power laboratories in the world and preferred above another method of current injection *series current injection* [3].

### 14.2.3.6 Voltage-Injection Methods

Voltage-injection methods also use a current source and a separate voltage source like parallel current-injection methods. The difference is that the current source in the voltage-injection method delivers not only the short-circuit current but also provides the initial part of the TRV.

The high-voltage part of the TRV is injected from the voltage source after the final current zero near the peak of the recovery voltage that is produced by the current source. A coupling capacitor $C_V$, connected across the auxiliary circuit-breaker (AB), is large enough to effectively apply the transient recovery voltage of the current source to the tested circuit-breaker.

Figure 14.13 shows the series voltage-injection synthetic test circuit with the voltage circuit connected in parallel to the auxiliary circuit-breaker.

Before the test, both the auxiliary circuit-breaker (AB) and the tested circuit-breaker (TB) are in the closed position. After the make switch (MS) closing, the current source supplies the short-circuit current for the circuit-breaker (TB) under test. During the whole interaction interval, the arcs in the auxiliary circuit-breaker and in the tested circuit-breaker are connected

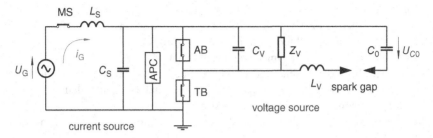

**Figure 14.13** Typical voltage-injection circuit with voltage injection in series with the circuit-breaker under test. For symbols see text.

in series. Both circuit-breakers interrupt the current simultaneously at the instant $t_1$ in Figure 14.14. When they have interrupted the short-circuit current, essentially the entire transient recovery voltage of the current circuit is applied to the tested circuit-breaker via the parallel capacitor $C_V$. In this way, the capacitor transmits the necessary energy for the post-arc current. At instant $t_2$, just before the peak of this voltage, the spark gap is fired, and the voltage oscillation of the voltage circuit is added to the recovery voltage of the current source across the tested circuit-breaker. The resulting TRV across the auxiliary circuit-breaker is then practically supplied by the voltage circuit only.

Because the injected current is small, the energy in the voltage circuit is relatively low, and the main-bank capacitance $C_0$ can be much smaller than that of the current-injection circuit. Voltage-injection methods are, therefore, more economical. During the interaction interval, however, the circuit parameters of the voltage-injection circuit differ from the circuit parameters of the current-source circuit. This is not the case for the current-injection circuit, and it is for this reason that the voltage-injection methods do not have general applicability. Therefore, according to IEC 62271-101 [13], voltage-injection methods are only permitted if there are no ITRV requirements or if these requirements are covered by SLF testing without time delay. Voltage injection also requires very accurate timing, which is rather difficult to achieve.

The voltage circuit can be also connected in parallel with the tested circuit-breaker instead of the auxiliary circuit-breaker. However, this method is not in common use.

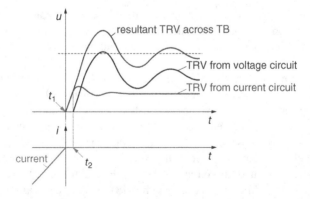

**Figure 14.14** Current and TRV of the series voltage-injection test circuit.

### 14.2.3.7 Three-Phase Synthetic Test Methods

In principle, the standards allow tests to be performed on a single-phase basis only in the case of circuit-breakers consisting of three independent single-pole devices. In any case, the type tests must verify the capability of a circuit-breaker to interrupt three-phase faults. Therefore, three-phase tests are desired and preferable whenever possible. This is of particular importance for circuit-breakers having their three poles in a common enclosure and also for those with individually enclosed poles but operated by a common operating mechanism. The use of the three-phase test procedure is necessary to reproduce the requested stresses in terms of arcing window, asymmetry, and duration of minor, major, and extended loops on the three poles. Preferably, all the above stresses should be applied in the same test. If this is impossible, a multi-part test procedure may be the only available method.

While single-phase synthetic test methods are well established, three-phase synthetic testing is rather new and technically challenging. The principle that the current is obtained from a current source and the recovery voltage after a chosen current zero from a voltage source, as it is in single-phase synthetic testing, is also applied in three-phase synthetic testing [17, 18]. These circuits use a three-phase current source and, generally, two voltage sources. One voltage source provides the TRV for the first pole-to-clear and the other gives the recovery voltage for the second- and third poles-to-clear, in the case where these poles clear simultaneously as in non-effectively earthed systems.

A three-phase synthetic circuit with current injection in all phases, shown in Figure 14.15, consists of the following components:

- a three-phase current source G;
- voltage source 1 with current injection parallel circuit connected across the first pole-to-clear;

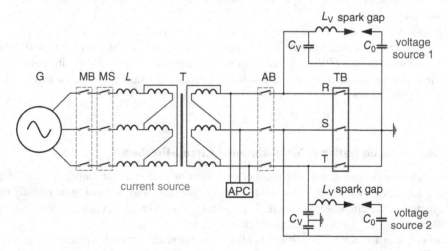

**Figure 14.15** Three-phase synthetic circuit with current injection in all phases for $k_{pp} = 1.5$. For symbols see text.

(a)                                                      (b)

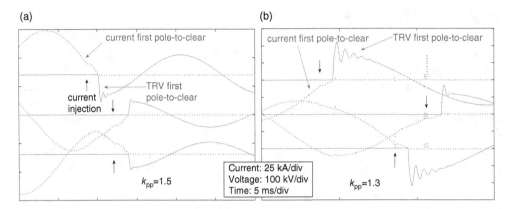

**Figure 14.16** Result of a three-phase synthetic test with $k_{pp} = 1.5$ (a) and $k_{pp} = 1.3$ (b).

- voltage source 2, as above, connected across the other two poles in series that interrupt the current at the same moment because of the absence of earthing at the source side;
- a three-pole auxiliary circuit-breaker (AB);
- a three-pole tested circuit-breaker (TB); and
- arc-prolongation circuits (APC) connected to each phase of the current circuit to prevent an early interruption by the tested circuit-breaker and to assure the longest possible arcing time, see Section 14.2.3.10.

The recovery voltage can be adequately distributed between the two last clearing poles with grading capacitors. The operation of the circuit is rather complicated and, therefore, it is only available in a limited number of high-power laboratories [19]. An example of such a test result is shown in Figure 14.16a.

It should be noted that a first-pole-to-clear factor lower than 1.5, as is the case of earth faults in effectively-earthed systems, cannot be covered by this circuit since each phase must be provided with its own TRV after each current zero. Synthetic testing of the circuit-breakers for such systems requires three separate voltage sources. An example of such a test result is shown in Figure 14.16b.

### 14.2.3.8 Synthetic Testing of Metal-Enclosed Circuit-Breakers

In the cases where circuit-breakers consist of more than one interruption chamber, a part of the tests that has to verify the fault-current interruption capability may be performed on only half (or a quarter) of the circuit-breaker pole. For live-tank circuit-breakers equipped with grading capacitors, this is generally a valid approach.

However, in order to meet the requirements for metal-enclosed switchgear according to IEC 62271-203 [20], the correct dielectric stresses between live parts and enclosure must also be guaranteed during the interruption process. Consequently, any test other than on the complete circuit-breaker is considered technically incorrect.

In unit testing of metal-enclosed EHV and UHV circuit-breakers (or half-pole testing in the case of two units), it is very difficult to correctly represent all the mechanical, gas-dynamic, electrodynamic, and dielectric stresses with reference to the full-pole service conditions. The stresses to be considered when single-phase testing replaces three-phase testing are:

- dielectric stresses during current interruption;
- gas-dynamic stresses;
- electrodynamic stresses.

These stresses are described in more detail below:

(a) Dielectric stresses during current interruption.
    Half-pole tests on metal-enclosed circuit-breakers do not represent the correct (full) dielectric stresses between live parts and enclosure, at least not in the short-circuit current tests (Figure 14.17a), since there is only half the test voltage between live internal parts and the enclosure.

    In testing metal-enclosed circuit-breakers with grading capacitors (including live-tank type), unit tests may not represent the transient stresses that occur in service due to unequal dielectric behaviour of the arcing chambers. In unit tests, stresses on grading capacitors (such as occur in pre-strikes) and on the breaker chambers are not represented correctly. This is essential since recent work of CIGRE identified grading capacitors as a major contributor to circuit-breaker failures [21].

    A dielectric-stress safety margin of some percent is usually applied in unit testing. This is to allow for an unequal voltage distribution due to stray capacitance across the circuit-breaker units. Because the capacitance of grading capacitors is normally much larger than the stray capacitances, the unequal voltage distribution is covered by a safety margin of a few percent above 50% (for a two-chamber circuit-breaker). In the case of designs without grading capacitors, this small safety margin is not adequate.

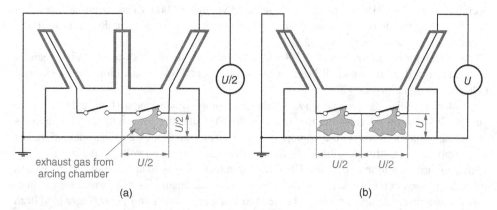

(a)                                                                    (b)

**Figure 14.17**  (a) Half-pole test of metal-enclosed circuit-breaker: voltage across interrupter is OK, voltage between live parts and enclosure is not OK. (b) Full-pole test of metal-enclosed circuit-breaker: voltage across interrupters is OK, voltage between live parts and enclosure is OK.

It is obvious that tests performed on a half-pole of a circuit-breaker do not imply all the fore-mentioned stresses, and consequently, they cannot provide adequate evidence for a satisfactory performance of the test object in service. Such evidence can be provided only by tests on a full-pole assembly.

(b) Gas-dynamic stresses

Depending on the design, which may or may not use a combined gas compartment for the arc extinction medium, the dynamic gas pressure and gas flow of the interruption chambers may mutually affect the gas-dynamic phenomena that interfere with the extinction process.

The hot exhaust gases may also deteriorate the dielectric withstand capability of the space surrounding the interruption chambers, that is, the space between poles, across the chambers, and to the enclosure, as illustrated in Figure 14.17. With GIS and dead-tank circuit-breakers, the gas-dynamic phenomena and the influence of hot, ionized, contaminated exhaust gas have to be taken into account for a decision to perform tests on a unit- or full-pole assembly and to which side of the circuit-breaker the largest dielectric stress has to be applied.

(c) Electrodynamic stresses

Tests performed on a half-pole assembly, without using the other unit as an auxiliary circuit-breaker, require an equivalent conducting path for the short-circuit current, simulating correctly its influence on the arc in the unit under test. For the same reason, three-phase tests are necessary on three-phase metal-enclosed circuit-breakers due to the compact design. The high-current connections in the direct vicinity of the test object also have to be designed with great care, taking into consideration realistic electrodynamic stresses on the arc and the circuit-breaker structure.

Vacuum circuit-breaker arcs, that need the complete contact surfaces, may be sensitive to phase-to-phase forces on the arcs that constrain their activity to certain contact areas only. Therefore, testing of such devices in a three-phase circuit, especially regarding compact arrangements such as in dead-tank designs, is highly recommended.

### 14.2.3.9  Synthetic Testing of UHV Circuit-Breakers

Circuit-breakers for UHV (ultra-high voltage, rated voltage > 800 kV) are required especially for various large-scale transmission projects. The testing of these is a challenge in itself and several solutions have been proposed [22].

The considerations in Section 14.2.3.8 resulted in new synthetic UHV test circuits specially designed to test dead-tank and GIS circuit-breakers under full-pole conditions with the metal enclosure at earth potential [23].

In order to realize this, a two-stage synthetic solution was adopted (Figure 14.18). One synthetic installation provides the first stage of the TRV and the second installation delivers the top part superimposed on the voltage waveform from the first stage. This method makes UHV short-circuit testing possible even for circuit-breakers at rated voltage of 1200 kV.

An alternative solution to perform UHV short-circuit testing is to apply suitable voltages to both sides of the circuit-breaker that has to be placed on an insulating platform. This method, however, has disadvantages, such as the need to bring all control and power signals to high potential, mechanical instability, and long assembly time.

In Figure 14.19, an oscillogram is shown of IEC test-duty T10 with the voltage peak value of 2060 kV for the first pole-to-clear factor $k_{pp} = 1.5$. In the absence of standardized values at

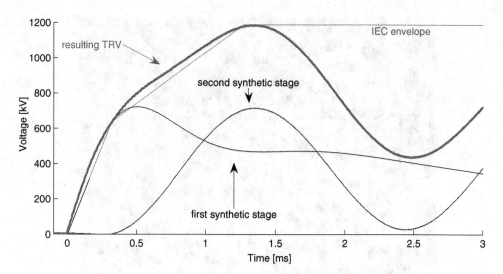

**Figure 14.18** Oscillogram of two-stage synthetic four-parameter TRV for 800 kV circuit-breakers and IEC TRV envelope.

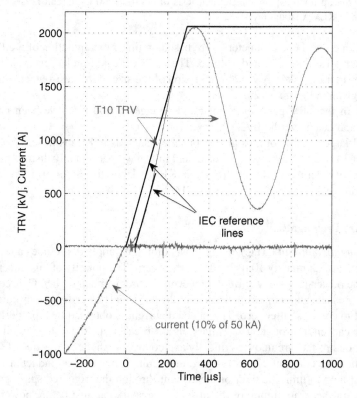

**Figure 14.19** Realized TRV of T10 test-duty ($k_{pp}$ = 1.5) for a 1100 kV circuit-breaker and IEC TRV envelopes based on 2 × 550 kV.

**Figure 14.20**  Set-up for full-pole short-circuit tests of a 1100/1200 kV circuit-breaker (reproduced with permission of KEMA laboratories).

the time of testing, the TRV parameters were based on linear extrapolation of the IEC 62271-100 requirements for 550 kV rated voltage. The circuit-breaker under test was a dead-tank four-chamber circuit-breaker. A visual impression of the size of this test-object situated in the laboratory is given in Figure 14.20.

The values of the TRV parameters for rated voltages > 800 kV have been proposed by CIGRE [24] and taken over by IEC.

An unusual large number of arcs placed in series, such as in 8 gaps in series, results in at least a total of 14 to 18 kV of arcing voltage, and this may unacceptably distort the current in the test circuit. In order to avoid such distortion, KEMA laboratories performs these tests with a generator voltage of 60 kV, see also Section 14.2.3.11.

### 14.2.3.10  Arc Prolongation

The *arc-prolongation circuit* (APC), being most commonly applied, provides a rapidly rising pulse of current approximately 10 μs before current zero. The polarity of this injected pulse is the same as the power-frequency current, however, its $di/dt$ is much higher. Consequently, the tested circuit-breaker cannot interrupt the sum of the pulse and power-frequency current, which is thus forced to continue after zero. The injected current is obtained by a properly triggered discharge of a capacitor through the auxiliary circuit-breaker and the tested circuit-breaker. A series resistor damps the circuit over-critically and controls both the peak value of the injected current and the time constant of this re-ignition circuit. Sometimes an inductance is built in the re-ignition loop to limit the $di/dt$ of the current through the triggered spark-gap. A basic requirement is the accurate timing of the firing of the spark gap just before power-frequency current zero. For synthetic testing of vacuum circuit-breakers, $di/dt > 3$ kA μs$^{-1}$ is required because vacuum interrupters have the capability to interrupt $di/dt$ values below this value [25].

Several such circuits may be used for prolonging the arcing time through several loops of current. In principle, the same re-ignition circuit can be applied to re-ignite both the tested and the auxiliary circuit-breaker.

### 14.2.3.11  Voltage of the Current-Source Circuit

A critical aspect of synthetic testing is the voltage level at which the short-circuit current is supplied. In testing of $SF_6$ circuit-breakers, there are at least two breakers in series, the test circuit-breaker and the auxiliary circuit-breaker, each with a number of arcs in series, one arc for each extinction chamber.

This implies that the total arc voltage of all these series arcs reduces the available voltage that drives the test current [26]. Thus, especially for tests at EHV/UHV levels, where up to four arcs in series are active in each circuit-breaker (that is, at least eight in total) the test current is reduced after contact separation of the circuit-breakers. In order to keep the reduction of the test current by the arc voltage within an acceptable limit, the voltage of the current-supply source must be chosen as high as possible – for EHV/UHV testing in the order of 40 to 60 kV.

For testing circuit-breakers with very low arc voltage, such as vacuum breakers, this argument does not hold, and rather low-voltage sources, for example, 10 kV, can be satisfactory.

# References

[1] STL (2011) Guide to the interpretation of IEC 62271-100: Edition 2008, issue 9.
[2] STL (2011) General Guide, issue 6.
[3] Kapetanović, M. (2011) *High Voltage Circuit-Breakers*, ETF – Faculty of Electrical Engineering, Sarajevo, ISBN 978-9958-629-39-6.
[4] (2013) Worldwide Directory of HV/HP Laboratories. INMR, issue 99, quarter 1, vol. 21, pp. 107–140.
[5] Slamecka, E. (1966) *Prüfung von Hochspannungs-Leistungsschaltern*, Springer Verlag, Berlin.
[6] IEC 62271-306 (2012) High-voltage switchgear and controlgear – Part 306: Guide to IEC 62271-100, IEC 62271-1, and IEC standards related to alternating current circuit-breakers.
[7] STL (2005) Guide to the Interpretation of IEC 62271-100: Second Edition: 2003 High-Voltage Switchgear and Controlgear – Part 100: High-Voltage Alternating Current Circuit-Breakers, Issue 3, 11.11.
[8] Fröhlich, K. (2002) Medium- and High Voltage circuit breakers – State of the Art. Proc. 21st Int. Conf. on Elec. Contacts, pp. 492–503.
[9] Thoren, B. (1950) Synthetic Short-Circuit Testing of Circuit-Breakers. CIGRE Report.
[10] IEC 62271-101 (2012) High-Voltage Switchgear and Controlgear – Part 101: Synthetic testing. Ed. 2.0.
[11] Smeets, R.P.P., te Paske, L.H., Kuivenhoven, S. *et al.* (2009) The Testing of Vacuum Generator Circuit-Breakers. CIRED Conference, Paper No. 393.
[12] Smeets, R.P.P., te Paske, L.H., Kuivenhoven, S. *et al.* (2014) Interruption Phenomena and Testing of Very Large SF6 Generator Circuit-Breakers. CIGRE Conference, paper A3-307.
[13] Slamecka, E. (1953) The Weil circuit. *AEG Mitteilungen*, **43**, 211–216.
[14] van der Linden, W.A. and van der Sluis, L. (1983) A new artificial line for testing high-voltage circuit breakers. *IEEE Trans. Power Ap. Syst.*, **PAS-102**(4), 797–804.
[15] Skeats, W.F. (1936) Special tests on impulse circuit breakers. *Elec. Eng.*, **55**, 710–717.
[16] Kapetanović, M. and Ahmethodžić, A. (2000) Behavior of Interrupters Based on Principles Using Arc-Energy in Direct and Synthetic Test Circuits. CIGRE Conference, paper 13-202, Paris.
[17] Dufournet, D. (2000) Three-phase short circuit testing of high-voltage circuit breakers using synthetic circuits. *IEEE Trans. Power Deliver.*, **15**(1), 142–147.
[18] van der Sluis, L. and van der Linden, W.A. (1987) A three phase synthetic test circuit for metal-enclosed circuit-breakers. *IEEE Trans. Power Deliver.*, **2**(3), 765–771.

[19] de Lange, A.J.P. (2000) *High Voltage Circuit Breaker Testing with a Focus on Three Phases in One Enclosure Gas Insulated Type Breakers*. Ph.D. Thesis, Delft University of Technology, ISBN 90-9014004-2.

[20] IEC 62271-203 (2011) High-voltage switchgear and controlgear – Part 203: Gas-insulated metal-enclosed switchgear for rated voltages above 52 kV. Ed. 2.0.

[21] CIGRE Working Group A3.18 (2007) Electrical Stresses on Circuit Breaker Grading Capacitors During Shunt Reactor Switching. CIGRE A3 Colloquium, paper PS3-09.

[22] Sheng, B.L. and van der Sluis, L. (1996) Comparison of synthetic test circuits for ultra-high-voltage circuit breakers. *IEEE Trans. Power Deliver.*, **11**(5), 1810–1815.

[23] Smeets, R.P.P., Kuivenhoven, S., Hofstee, A.B. *et al.* (2008) Testing of UHV (> 800 kV) Circuit Breakers. CIGRE Conference, paper A3-206.

[24] CIGRE Working Group A3.22 (2008) Technical Requirements for Substation Equipment exceeding 800 kV. CIGRE Technical Brochure 362.

[25] Smeets, R.P.P., Kuivenhoven, S. and te Paske, L.H. (2010) Testing of Vacuum Circuit Breakers for Transmission Voltage and Generator Current Ratings. CIGRE Conference, Paris, paper A3-309.

[26] van der Sluis, L. and Sheng, B.L. (1995) The influence of the arc voltage in synthetic test circuits. *IEEE Trans. Power Deliver.*, **10**(1), 274–279.

# List of Abbreviations

| | |
|---|---|
| AC | alternating current |
| ACSR | aluminium conductor steel reinforced |
| AIS | air-insulated substation |
| AMF | axial magnetic field |
| APC | arc-prolongation circuit |
| ATP | alternative transient program |
| CB | circuit-breaker |
| CFC | chlorinated fluorocarbon compounds |
| CIGRE | Conférence International des Grands Réseaux Électriques |
| CO | close–opening (operating sequence) |
| COV | continuous operating voltage (of an arrester) |
| CSC | current source conversion |
| CT | current transformer |
| DC | direct current |
| DEF | double-earth fault |
| EHV | extra-high voltage |
| EMC | electromagnetic compatibility |
| EMI | electromagnetic interference |
| EMTDC | ElectroMagnetic Transients for DC |
| EMTP | ElectroMagnetic Transients Program |
| EPA | Environmental Protection Agency |
| EPRI | Electric-Power Research Institute |
| FACTS | flexible AC transmission systems |
| FFC | fully fluorinated compounds |
| FRA | frequency-response analysis |
| FTP | frequency-domain transient program |
| GIL | gas-insulated transmission lines |
| GIS | gas-insulated switchgear |
| HSES | high-speed earthing switch |
| GWP | global warming potential |
| HV | high voltage |

*Switching in Electrical Transmission and Distribution Systems*, First Edition.
René Smeets, Lou van der Sluis, Mirsad Kapetanović, David F. Peelo, and Anton Janssen.
© 2015 John Wiley & Sons, Ltd. Published 2015 by John Wiley & Sons, Ltd.

| | |
|---|---|
| IEC | International Electrotechnical Commission |
| IEEE | Institution of Electrical and Electronics Engineers |
| IGBT | insulated-gate bipolar transistor |
| ITRV | initial transient recovery voltage |
| LCC | line commutation conversion |
| LIPL | lightning-impulse protection level |
| LLF | long-line fault |
| LV | low voltage |
| MOLTL | maximum operable length at thermal limit (of a HV cable) |
| MMCB | miniature moulded-case circuit-breaker |
| MOSA | metal oxide surge arrester |
| MOV | metal oxide varistor |
| MR | multiple re-ignitions |
| MRTB | metallic return transfer breaker |
| MTTF | mean time-to-failure |
| MTS | mixed-technology substations |
| MV | medium voltage |
| NGO | non-governmental organization |
| NSDD | non-sustained disruptive discharge |
| O | opening operation |
| ODE | ordinary differential equation |
| OP1 | out-of-phase (fault), test-duty 1 |
| OP2 | out-of-phase (fault), test-duty 2 |
| p.u. | per-unit value |
| PT | potential transformer |
| PTFE | polytetrafluoroethylene (teflon) |
| PWM | pulse-width modulation |
| r.m.s. | root-mean-square value |
| RDDS | rate of decrease of dielectric strength |
| RLF | reactor-limited fault |
| RMF | radial magnetic field |
| RRRV | rate-of-rise of recovery voltage |
| RV | recovery voltage |
| SCOF | self-contained oil-filled |
| SEF | single-earth fault |
| SFO | slow-front overvoltage |
| $SF_6$ | sulfur hexafluoride |
| SIL | surge-impedance loading |
| SIPL | switching-impulse protection level |
| SLF | short-line fault |
| SPAR | single-pole auto-reclosure |
| STL | short-circuit testing liaison |
| TF | terminal fault |
| TFF | transformer-fed fault |
| TLA | transmission line arrester |
| TLF | transformer-limited fault |
| TLV | threshold limit value |
| TMF | transversal magnetic field |
| TNA | transient network analyser |

| TPAR | three-phase auto-reclosure |
| TRV | transient recovery voltage |
| TSF | transformer-secondary fault |
| UHV | ultra-high voltage |
| VCC | virtual current chopping |
| VFTO | very-fast transient overvoltage |
| VSC | voltage source conversion |
| WCI | Western Climate Initiative |
| WG | Working Group |

# Index

Ablation, 72, 169, 212, 355
Acceptance tests. *See* Testing, 386
Advanced opening pole. *See* Circuit-breaker, 138
Air-blast circuit-breaker. *See* Circuit-breaker, 129
Air-break disconnector. *See* Disconnector
Ampacity of a HV cable. *See* Power system
Amplitude factor. *See* Transient Recovery Voltage (TRV)
Angle of impedance, 145
Anode spots. *See* Vacuum arc, 244
Anode-fall voltage. *See* Arc, 174
Arc
    anode-fall voltage, 174
    arc voltage, 9, 70, 75–79, 127, 164, 168–170, 172, 174, 232, 234, 235, 237, 238, 244, 246, 263, 264, 278, 396, 397, 403, 411
    arcing time, 15, 35, 100, 106, 124, 129, 131, 136, 137, 177, 181, 211, 238, 264, 277, 287, 321, 323, 333, 343, 355, 393, 403, 406, 411
    cathode-fall voltage, 174
    electrode voltage drops, 174
    free-burning arc, 175
    plasma, 11, 13, 21, 71, 80, 164, 165, 168, 177, 180, 183, 226–228, 230, 232–235, 244, 246, 253, 271
    secondary arc, 118, 139
    $SF_6$ arc, 184
    vacuum arc, 124
Arc chutes. *See* Circuit-breaker, 169, 171
Arcing contacts. *See* Circuit-breaker
Arcing stress. *See* Circuit-breaker, 355
Arcing time. *See* Making operation, 87

Arc-Prolongation Circuit (APC). *See* Testing, 401
Arc-voltage extinction peak. *See* Breaking operation, 77
Arrester. *See* Surge arrester, 316
Artificial line. *See* Testing, 402
Asymmetrical-current peak factor. *See* Short-circuit current, 31
Auto-reclosure. *See* Circuit-breaker
Auxiliary circuit-Breaker (AB). *See* Testing, 355, 400
Axial Magnetic Field (AMF) principle. *See* Vacuum circuit-breaker, 235

Back-to-back configuration. *See* Capacitor banks, 88
Bipolar AMF contact system. *See* Vacuum circuit-breaker, 239
Break time. *See* Circuit-breaker
Breaking operation, 2, 3, 96, 393
    arcing time, 323, 339
    arc-voltage extinction peak, 77, 397
    delayed (missing) current zero, 173, 262, 263
    fault-current breaking, 365
    fault-current breaking (interruption), 48
    high-current interval, 396, 397
    high-voltage interval, 396–398
    instant of contact separation, 168, 321, 323, 397
    interaction interval, 397, 398, 403
    interruption of short-circuit current, 48
    mechanical opening time, 323
    multiple re-ignitions, 128, 130, 131, 136, 251
    multiple restrikes, 102, 251, 342

*Switching in Electrical Transmission and Distribution Systems*, First Edition.
René Smeets, Lou van der Sluis, Mirsad Kapetanović, David F. Peelo, and Anton Janssen.
© 2015 John Wiley & Sons, Ltd. Published 2015 by John Wiley & Sons, Ltd.

Printed in the United States
By Bookmasters